THE GEOLOGICAL ASSOCIATION OF CANADA SPECIAL PAPER 19

HISTORY OF CONCEPTS IN PRECAMBRIAN GEOLOGY

Papers presented at a symposium organized
by the International Committee on the History
of Geological Sciences held in Montreal, August 23-24, 1972,
and additional papers.

Edited by
W.O. Kupsch and **W.A.S. Sarjeant**

1979

W.O. Kupsch and W.A.S. Sarjeant
Department of Geological Sciences
University of Saskatchewan,
Saskatoon, Sask. S7N 0W0

International Standard Book Number 0-919216-13-7

Printed by: The Hunter-Rose Company, Toronto, Ontario

CONTENTS

Dr. James Buckland Mawdsley, retired Head of the Department of Geological Sciences and Dean of Engineering, University of Saskatchewan, died in Edmonton, Alberta on December 3, 1964 while attending a Canadian Regional Conference on Permafrost. At the time of his death he was the Director of the Institute for Northern Studies, University of Saskatchewan.

Jim, as he was known to his colleagues (he was "J.B." to his students), was born on July 22, 1894, near Siena, Italy, the son of British-American parents. At the age of 10, his family decided to move and settled at Gainsborough in southeastern Saskatchewan. After his public and high school training in Saskatchewan he entered McGill University in 1913. His career, like that of many of his contemporaries, was interrupted by the *First World War.* Twice wounded in France, first with the *Princess Patricia Canadian Light Infantry* and then as a pilot with the *Royal Flying Corps,* he was awarded the *M.B.E.* at the close of the war. In 1919 he returned to

McGill and two years later graduated in Mining Engineering (Geology Option). His summers were spent working in mines at Butte, Montana; Sudbury, Ontario; and Leadville, Colorado. He continued his geological studies at Princeton University and, under the influence of Dr. A.F. Buddington, he became interested in Precambrian geology. Under Buddington's direction and working for the Geological Survey of Canada during the summers Jim Mawdsley undertook a study of the deposits of ilmenite for his Ph.D. thesis entitled "St. Urbain area, Charlevoix District, Québec". His work was later published as Memoir 152 by the Geological Survey of Canada.

On graduation from Princeton in 1924, he became a member of the permanent staff of the Geological Survey of Canada and, together with Drs. H.C. Cooke and W.F. James, carried out extensive field work in the Noranda-Malartic and Chibougamou Precambrian areas of northwestern Quebec. Although his ideas were never published, he was probably the first person to recognize the relationship of sulphide mineral deposits to near-surface volcanism in the Noranda-Rouyn area of Quebec.

In 1929, after five years of service, Jim resigned from the Geological Survey of Canada to become Professor and Head of the Department of Geology at the University of Saskatchewan. During his 32 years as Head of the Department he endeared himself to hundreds of students as an outstanding teacher and wise counsellor and instilled in many of them a love of the North and a keen interest in the Precambrian geology of the Canadian Shield.

In addition to his academic duties, his professional activities included field work in northern Saskatchewan for the Geological Survey of Canada and the Saskatchewan Department of Mineral Resources. Private consulting assignments took him to other parts of the Canadian Shield, to the United States, and Great Britain. For several years he worked on a report dealing with the Precambrian mineral deposits of Québec and saw this published in 1949 as part of Geological Report Number 20, volume III, "Geology of Québec-Economic Geology" by the Québec Department of Mines.

During and for several years after the *Second World War* Dr. Mawdsley was actively connected with military training on the campus of the University of Saskatchewan. He started as *Captain* and later became *Major* in the *Canadian Officers Training Corps.* He then took over command of the *Canadian Airforce Unit* with the rank of *Squadron Leader,* and, finally, in 1943 he was assigned the command of the *University Naval Training Division* of *H.M.C.S. Unicorn,* a post he held until his retirement with the rank of *Captain* in 1958.

In 1933, he was elected a Fellow of the Royal Society of Canada and was Chairman of Section IV for the year 1954-55. He was President of the Geological Association of Canada during 1955-56 and of the Canadian Institute of Mining and Metallurgy for 1961-62. In 1953, he was awarded the Institute's Barlow Memorial Medal for his paper "Uraninite-bearing deposits, Charlebois Lake area, northern Saskatchewan." He was a Fellow and Director of the Royal Canadian Geographical Society; a Charter Associate of the Arctic Institute of North America; a Fellow of the Geological Society of America; and a member of the Society of Economic Geologists, the Engineering Institute of Canada, and the Association of Professional Engineers of Saskatchewan.

As Director of the Institute for Northern Studies from 1961 until his untimely death in 1964, he played a major role in its organization and development and exerted a very great influence on research in northern Saskatchewan and other parts of the Canadian North.

Recognized as an able administrator, scientist and teacher, perhaps his greatest service will prove to be the influence he had on all those who worked or studied under him. In them he not only instilled a feeling of scientific curosity regarding Precambrian geology but also a keen interest and love of the North.

A.R. Byers

JAMES BUCKLAND MAWDSLEY

Le docteur James Buckland Mawdsley, retiré comme Chef du Département des Sciences géologiques et comme Doyen de Génie à l'Université de Saskatchewan, fut décédé à Edmonton en Alberta, le 3 décembre, 1964, lors de la Conférence canadienne régionale sur le Pergélisol. A son décès, il était le Directeur de l'Institut des Etudes du Nord, à l'Université de Saskatchewan.

Jim, comme ses collègues le nommaient (ses étudiants le nommaient J.B.) naquit le 22 juillet 1894, près de Sienne, en Italie, le fils de parents britanno-américains. Quand il eut 10 ans, sa famille décida de déménager et s'installa à Gainsborough dans le sud-est de Saskatchewan. Il fit ses études primaires et secondaires en Saskatchewan et s'enrolla à l'Université McGill en 1913. Sa carrière fut interrompue par la *Première Guerre Mondiale,* comme le fut celle de plusieurs de ses contemporains. Blessé deux fois en France, premièrement avec *l'Infanterie Légère canadienne du Princess Patricia* et ensuite comme pilote avec le *Royal Flying Corps,* on lui conféra le *M.B.E.* à la fin de la guerre. En 1919, il retourna à McGill et deux ans plus tard, fut gradué en Génie minier (Option de Géologie). Il passa ses étés à travailler dans les mines à Butte, Montana; Sudbury, Ontario; et Leadville, Colorado. Il continua ses études géologiques à l'Université Princeton et sous l'influence du Dr. A.S. Buddington, sous sa direction et travaillant pour la Commission géologique du Canada durant les étés, il devint intéressé à la géologie du Précambrien. Jim Mawdsley entreprit l'étude des dépôts d'ilmenite pour une thèse doctorale intitulée "St. Urbain area, Charlevoix District, Quebec" qui fuit ensuite publiée par la Commission géologique du Canada, étant le Mémoire 152.

En graduant de Princeton en 1924, il devint membre du personnel permanent de la Commission géologique du Canada, et avec les docteurs A.C. Cooke et W.F. James, réalisa les travaux de terrain détaillés dans les régions précambriennes de Noranda-Malartic et de Chibougamau, dans le nord-ouest de Québec. Quoique ses idées ne furent jamais publiées, il fut probablement le premier à reconnaître les relations entre les dépôts de minérais de sulfides et le volcanisme près de la surface dans la région de Noranda-Rouyn au Québec.

En 1929, après cinq années de service, Jim laissa la Commission géologique du Canada pour devenir professeur et Chef du Département de Geologie à l'Université de Saskatchewan. Lors de ses 32 années comme chef du départment, il devint cher à des centaines d'étudiants comme professeur extraordinaire et sage conseiller, et inspira chez plusieurs d'eux un amour du Nord et un intérêt vif dans la géologie précambrienne du Bouclier canadien.

En plus de ses devoirs académiques, ses activités professionnelles inclurent du travail de terrain dans le nord de Saskatchewan, pour la Commission géologique du Canada et le Départment des Ressources minérales de Saskatchewan; ses travaux consultatifs privés le conduirent à d'autres parts du Bouclier canadien, aux Etats-Unis et en Grande Bretagne. Durant plusieurs années il prépara un rapport traitant de dépôts de minérais précambriens du Québec, et en 1949 publia "Geology of Quebec – Economic Geology" comme partie du Rapport géologique Numéro 20, volume III part le Département des Mines du Québec.

Durant plusieurs années après la *Deuxième Guerre Mondiale,* le docteur Mawdsley se trouva actif dans l'entraînement militaire sur le campus de l'Université de Saskatchewan. Il commenca comme *capitaine* et ensuite devint *major* dans le *Corps-Ecole d'Officiers canadien.* Ensuite il devint *chef d'escadrille* dans *l'Aviation Canadienne* et enfin, en 1943 se trouva assigné le commandement de la *Division universitaire d'Entraînement naval* de *H.M.C.S. Unicorn,* un poste qu'il garda jusqu'à sa retraite avec le rang de *capitaine,* en 1958.

En 1933, il fut élu comme membre associé de la Société royale du Canada et fut président de la Section IV pour l'année 1954-1955. Il fut président de l'Association géologique du Canada durant 1955-1956 et de l'Institut canadien des Mines et Métallurgie en 1961-1962. En 1953 on lui décerna le Barlow Memorial Medal de l'Institut pour son oeuvre "Uraninite-bearing deposits, Charlebois Lake area, northern Saskatchewan". Il fut membre fondateur de l'Institut arctique de l'Amérique du Nord; membre associé de la Société géologique d'Amérique; et membre de la Société de Géologues économiques, de l'Institut du Génie du Canada et de l'Association des Ingénieurs professionnels de Saskatchewan.

Comme Directeur de l'Institut des Etudes du Nord depuis 1961 jusqu'à sa mort en 1964, il joua un role majeur dans son organisation et dans son développement, et il exerca une grande influence sur la recherche dans le nord de la Saskatchewan et dans d'autres parties du Nord canadien.

Reconnu comme étant administrateur scientifique et professeur capable, il se peut que son plus grand service soit l'influence qu'il exerca sur tous ceux qui ont étudié ou travaillé avec lui. C'est chez ceux-là qu'il inspira non seulement un sentiment de curosité scientifique envers la géologie du Précambrien, mais aussi un intérêt vif et un amour pour le Nord.

A.R. Byers

PREFACE

The roots of the present *Special Paper* can be traced back to the Twenty-fourth International Geological Congress held in Montreal, Canada, August, 1972. In conjunction with the Congress several scientific organizations held symposia. The International Committee on the History of Geological Sciences (INHIGEO) organized Symposium 106: *History of the Concepts of Precambrian Geology*. The program follows.

Organizer: J.B. Waterhouse
Wednesday, August 23, 1972

Chairman: W.O. Kupsch

1. V.V. TIKHOMIROV: Development of concepts of physico-geographical environments in the Precambrian (The ideas of J. Walther, A.P. Pavlov, L.S. Berg).
2. J.H.V. ANDERSON: The concept of Pre-cambrian geology and the recognition of Pre-cambrian rocks in Scotland and Ireland.
3. D.R. BOWES and A.M. HOPGOOD: Significance of structural trend in Precambrian terrain.
4. D.F. BRANAGAN: Understanding the Australian Precambrian: An historical review.
5. T.D. FORD: The history of study of the Precambrian rocks of Charnwood Forest, England.

Thursday, August 24, 1972

Chairman: V.V. Tikhomirov

6. W.O. KUPSCH: Boundary of the Canadian Shield.
7. J.S. STEVENSON: Geological concepts developed in the Precambrian of Sudbury, Ontario.
8. I.V. BATJUSHKOVA: Evolution of ideas on the Precambrian history of continents and oceans.
9. G.W. WHITE: William Maclure's concept of Primary Rocks ("Basement Complex").
10. R.V. DIETRICH: Migmatites.
11. J. CHALLINOR: The Precambrian in Cambria.

Monday, August 28, 1972

Chairman: G.W. White

12. E.L. YOCHELSON: Charles D. Walcott.
13. A.M. OSPOVAT: Werner's concept of the basement complex.
14. G. REGNELL: The primaeval Earth according to Swedish 18th century naturalists.
15. A. OPARIN: Evolution of carbon compounds as a way to the origin of life on Earth.
16. V.A. MASLENIKOV, K.O. KRATZ and N.N. BARKHATOVA: Russian school in geochronological studies of the Precambrian.
17. S.S. GOWDA and R. SRINIVASAN: History of concepts of Precambrian geology as applied to Mysore Area in India.

* * *

The Volume of Abstracts of the Twenty-fourth International Congress (on pages 526-532) contains the abstracts of 16 papers of the 17 on the program. Full publication of the papers, plus another submitted subsequently to Professor Waterhouse (William A.S. Sarjeant and Anthony P. Harvey: "Uriconian and Longmyndian. A history of the study of the Precambrian rocks of the Welsh Borderland") was planned from the outset; and after the Congress the organizer of Symposium 106, Professor J.B. Waterhouse, then of the University of Toronto, Canada, attempted to find a publisher for this collection of papers. He continued his efforts after he moved from Canada to Australia, where he is now the Head of the Department of Geology and Mineralogy, University of Queensland, St. Lucia, Brisbane. Unfortunately, however, he failed to find a suitable publisher in either Canada or Australia. Professor Waterhouse arranged for the papers to be refereed and, helped by Alan Miller of the University of Toronto and Rod Banks of Brisbane, undertook some preliminary editing of them. The bundle of manuscripts then lay idle until the spring of 1976, when it was turned over to the present editors.

Of the original 17 papers three do not appear in this *Special Paper*. The paper by D.R. Bowes and A.M. Hopgood, "Significance of structural trend in Precambrian terrain" has been published in full elsewhere (*Acta Palaeont. Polon.*, vol. 26, 1976, p. 57-82). No suitable manuscript could be obtained for J.V. Batjushkova's "Evolution of ideas on the Precambrian history of continents and oceans" and no manuscript was forthcoming from Professor Oparin.

The paper by J. Challinor "The Precambrian in Cambria" was published in the Geological Magazine, 1976, vol. 113, p. 449-456, but since this area is so crucial to an understanding of the history of Precambrian studies, the author kindly agreed to prepare an expanded version for this *Special Paper*. The paper by Professor Gerhard Regnéll also differs from the one originally presented. All of the original authors have been given opportunity to bring their papers up-to-date, though not all have chosen to do so.

At the request of the Geological Association of Canada, the editors agreed to expand the material by inviting other authors to submit contributions. This provided us with the welcome opportunity to broaden the coverage of the topic by including articles on geographical areas not dealt with during Symposium 106. Over twenty invitations were sent out to geologists in different countries known to be familiar with the historical growth of the concepts of Precambrian geology. As a result, three additional papers were forthcoming – Goran Kjellström's account of the Precambrian organic-walled microfossils of Sweden; his compatriot Per Lundegårdh's more general history of the study of the Swedish Precambrian; and the review by A.H. Lang and V. Ruzicka of contributions from the search for radioactive minerals to the understanding of the Canadian Shield. However, the response to our invitations was otherwise disappointing, most replies pleading existing commitments too heavy to permit the writing of such a paper. Rather than further delaying the publication of papers that had already lain unpublished for five years, it was decided to go ahead with publication of the material already at hand. To have done otherwise would have meant further lengthy delays and the possible withdrawal of some of the papers submitted.

During the final stages of editing, the refereeing of new papers received was undertaken by Dr. W.K. Braun, Dr. Richard Forester, Dr. Fred F. Langford, Dr. David J. Mossman and Dr. Mel R. Stauffer, all of the Department of Geological Sciences, and Mr. L. Stewart, Department of History, University of Saskatchewan. The late Dr. V.A. Eyles (Great Rissington, England) courteously commented on one paper. The advice of Dr. W. Glen E. Caldwell (Department of Geological Sciences), the help in translating Russian references of Dr. William Barr (Department of Geography, University of Saskatchewan), and the assistance in editorial work of Mrs. Maire Legg, Research Assistant to the second editor, are also gratefully acknowledged. Translation into French of the abstracts was undertaken by M. Jean-Jacques Coulomb (Université de Montreal) and Dr. Jocelyne C. Legault (University of Waterloo).

The names of the editors of Special Paper 19 are listed alphabetically. Of the considerable workload involved no dominant portion should be assigned to the first-listed editor. W.A.S. Sarjeant, having a special interest in the bibliography of the history of geology, undertook the arduous task of verifying the references for all papers. In this he was ably assisted by Mrs. Maureen Dickson Czerneda, Editor for the Geological Association of Canada, who extended the task assigned her, to see Special Paper No. 19 through the printing process. Her help is gratefully acknowledged.

We should also like to express our appreciation to the Geological Association of Canada which, through their Publications Committee, not only made this *Special Paper* possible but encouraged us in several ways. The Committee recognized that, because Symposium 106 was held in Canada, some obligation rested with the principal geological organization of this country to see that the papers became part of the published literature, even though most deal with areas outside this country. It is hoped that, since the Precambrian is of such great importance to Canada, a review of the concepts that came out of Precambrian studies elsewhere will prove of interest to many Canadian geologists.

This volume does not in itself constitute a history of the study of the Precambrian system; it cannot, when such vast areas as the Precambrian shields of South Africa and Brazil and the Precambrian of China and Siberia are not dealt with at all. However, we hope it may provide a useful foundation for the eventual writing of such a history.

W. O. Kupsch
W.A.S. Sarjeant.

History of Concepts in Precambrian Geology, edited by W.O. Kupsch and W.A.S. Sarjeant, Geological Association of Canada Special Paper 19, 1979

THE CONCEPT OF PRECAMBRIAN GEOLOGY AND THE RECOGNITION OF PRECAMBRIAN ROCKS IN SCOTLAND AND IRELAND

J.G.C. Anderson
Department of Geology, University College
P.O. Box 78, Cathay's Park, Cardiff, Wales CF1 1XL, U.K.

ABSTRACT

About 95 per cent of the Precambrian rocks of the British Isles occur in the Scottish Highlands and in Ireland. This paper traces the historical development of the concept that these rocks are Precambrian and the recognition of their place in the Precambrian sequence which has gradually been established.

The use of the term Precambrian did not become possible until the term Cambrian was chosen by Sedgwick in 1835. Nevertheless, earlier views on the rocks now known to be Precambrian are discussed and attention is drawn to the stratigraphical methods first used and also to the later use of radiometric dating. Rather sparse fossil evidence is provided by the microfossils which have recently been obtained from some of the Precambrian rocks.

Although the very oldest (Katarchean) rocks have not been recognized in Scotland and Ireland, Archean, Lower Proterozoic and Upper Proterozoic rocks are all present. Within the Caledonian Fold-Belt the Upper Proterozoic sequence continues without any distinct break, into the Cambrian.

RESUME

Environ 95% des roches précambriennes des îles britanniques se trouvent dans les Highlands d'Ecosse et en Irlande. Cet article trace le développement historique du concept selon lequel ces roches sont précambriennes et la reconnaissance de leur place dans la séquence prcambrienne qui a été graduellement établie.

Il ne fut pas possible d'utiliser le terme Précambrien avant que le terme Cambrien n'ait été choisi par Sedgwick en 1835; néanmoins les premières perspectives sur les roches connues maintenant pour être précambriennes sont examinées. Nous attirons l'attention sur les méthodes stratigraphiques d'abord utilisées et sur l'usage ultérieur des datations radiométriques. Des preuves fossiles plutôt dispersées sont fournis par les microfossiles obtenus récemment de quelques roches précambriennes.

Bien que l'on n'ait pas reconnu les roches les plus anciennes (Katarchéen) en Ecosse et en Irlande, celles de l'Archéen, du Protérozoïque Inférieur et Supérieur sont toutes présentes. A l'intérieur de la ceinture plissée calédonienne la séquence du Protérozoïque Supérieur se poursuit sans coupure distincte dans le Cambrien.

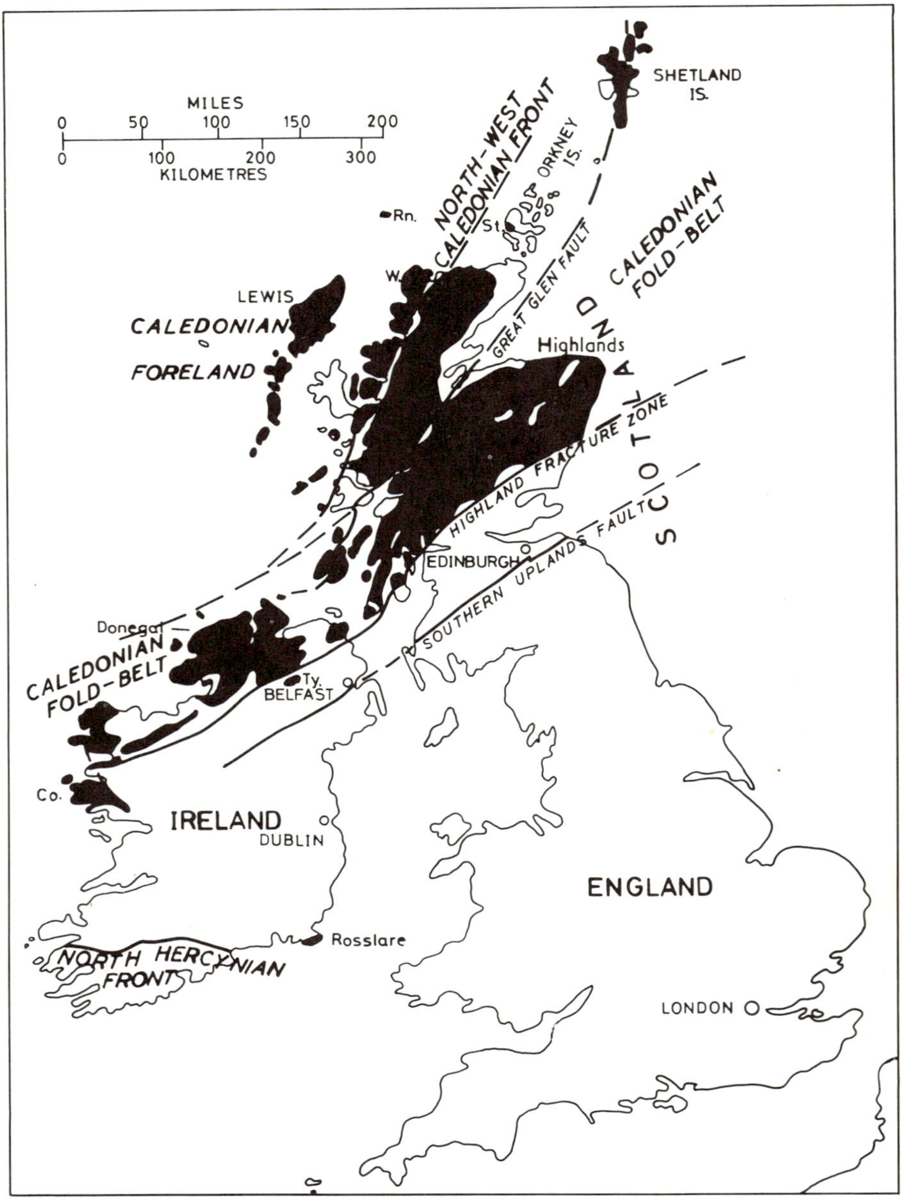

Figure 1. Precambrian outcrops of Scotland and Ireland.
Co. = Connemara; Rn. = L. Rona; St. = Stromness; Ty. = Tyrone; W = Cape Wrath.
(The whole of the Dalradian is shown as Precambrian as the top of the Dalradian, now accepted as Cambrian, cannot be satisfactorily separated on the scale used).

INTRODUCTION

Precambrian rocks occupy about 12 per cent of the British Isles. About 95 per cent of these occur in the Scottish Highlands and in Ireland. In Scotland they make up most of the Highlands, north-west of the Highland Boundary Fracture-Zone. Precambrian rocks also occur north of the same Fracture-Zone in Ireland, but are hidden to a greater extent under younger strata. Precambrian strata are present south of the Fracture-Zone in western and south-eastern Ireland.

In view of their relative abundance, it might be thought that recognition of Precambrian rocks and their naming as such would have entered Scottish and Irish geological literature at an early date. In fact, a validly-based concept of the Scottish and Irish Precambrian was reached fairly late, probably because of the sparse development of the Cambrian itself. In Scotland fossiliferous Cambrian strata are virtually confined to the North-West Highlands. In Ireland there is almost everywhere a big gap between fossiliferous Ordovician or Silurian strata and meta-sediments which are the continuation of those of the Scottish Highlands. It is only in the south-east that Cambrian rocks have been recognized, and this on rather sparse fossil evidence.

EARLY (PRE-1840) CONCEPTS

As the term Cambrian was chosen by Sedgwick in 1835 (see Sedgwick and Murchison, 1836) it is obvious that no Precambrian rocks were named as such before that date; nevertheless, it is helpful to look at early views of what we now know to be Precambrian rocks.

Hutton in his *Theory of the Earth* (read in 1785, published in 1788) referred the slates, schists, etc. of the Highlands, which he recognized as altered sediments, to the Primary, but only as the oldest so far discovered. Jameson (1800) described fossils *(sic)* from the Scottish Isles as micaceous schistus, chlorite schistus, ardesia (slate) etc. As early as 1811 Imrie, in a traverse from the edge of the south-east Grampians northwards, noted in succession clay slate, mica slate, micaceous schistus and gneiss. He thus anticipated, to some extent, Barrow's (1893) classic establishment of metamorphic zones in the same area.

On early 19th century maps (Necker de Saussure, 1808; Boué, 1820; Macculloch, 1836; Phillips, 1838; Griffiths, 1838) metamorphic strata, now known to be mainly Precambrian, are classified – following Werner – as Primary, Primitive or Transition and/or are described as clay slate, mica slate, chlorite rock, quartz rock, primary limestone, mica-schist, gneiss, etc. The term 'metamorphic' was introduced into geological literature by Lyell in 1833 and from then on was increasingly used, without, in many cases, stratigraphical implication. Nevertheless, the Wernerian concept, that all such rocks were "primitive", was slow-dying and in Scotland and Ireland, as elsewhere, led to the inclusion in the Precambrian of meta-sediments which later proved to be younger and to the exclusion of ancient unaltered sediments, e.g., the Torridonian.

CALEDONIAN FORELAND — NORTH-WEST HIGHLANDS

The development of the Precambrian concept will first be considered in the Caledonian foreland, where the relevant succession is as follows:

Cambro-Ordovician	Durness Limestone
Cambrian	Serpulite Grit
	"Fucoid Beds" (dolomitic shale)
	Pipe-rock
	Basal quartzite
(unconformity)	
Torridonian	Unaltered sediments
(unconformity)	
Lewisian metamorphic assemblage	At least three metamorphic complexes

The Lewisian, referred to by some authors as the Hebridian or Fundamental Gneiss, was regarded from the earliest days of geology as ancient. For example, Necker (1808) referred to it as primitive. It was also referred to by some authors as the Hebridean or Fundamental Gneiss. Wernerian reasons for its age were supplanted by comparison with other "Archaean" regions. After Logan and Hunt (1855) had proposed the name Laurentian for the metamorphic rocks beneath the oldest fossiliferous strata of Canada some of the Scottish gneisses were classified

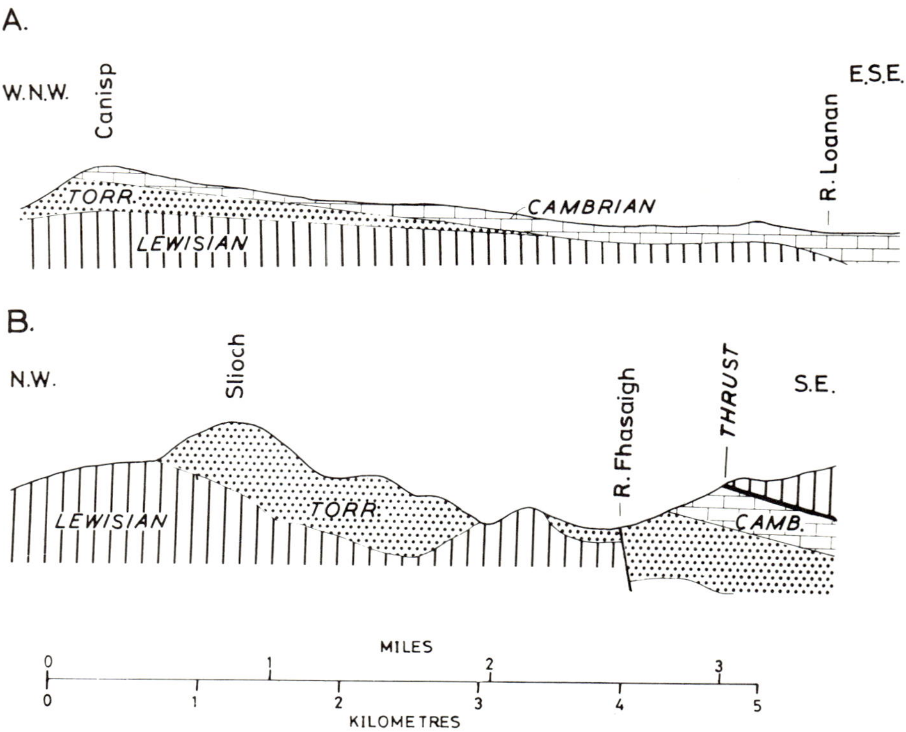

Figure 2. Precambrian relationships in Caledonian Foreland.
A — Assynt District; B — Loch Maree District.

as Laurentian (for example on Murchison and Geikie's geological map of 1862 and on Geikie's map of 1876). As early as 1860 Murchison had, however, proposed the term 'Lewisian'. For the identification of the formation as Precambrian on strict stratigraphical grounds we have, nevertheless, to consider the overlying formations.

The Torridonian (or Torridon Sandstone as it was earlier known) with its abundant red sandstones was assigned by Macculloch (1836) and a number of later geologists to the Devonian. However, in 1854 Charles Peach (1855) found fossils in the Durness Limestone. These were first considered by some authors to be Devonian, although Murchison (1856) thought they might be Lower Silurian. Nevertheless, he still placed the Torridonian sandstones of the Applecross Peninsula in the Devonian; the underlying metamorphic rocks were referred to the pre-Devonian. Nicol (1857) also put the red sandstones into the Devonian and the overlying quartzites and limestones into the Lower Carboniferous, as had been implied by Macculloch (1836) in showing three patches as "mountain limestone". The Durness Limestone fossils were, however, soon accepted as Lower Silurian and on Murchison and Geikie's map of 1862 the Torridonian is coloured as Cambrian. From then on the Lewisian was regarded as Precambrian.

In 1891 came the discovery of Lower Cambrian fossils in the Fucoid Beds (Geikie 1891b; Peach and Horne, 1892). (As early as 1821 Macculloch had discussed "Serpulites" in the quartz rock of Loch Eireboll). The 1891 find established the Precambrian age of the Torridonian and in the discussion of Peach and Horne's 1892 paper, Hudleston suggested the descriptive use of the term. In 1894 Geikie referred to the formation as a definite stratigraphical system.

Microfossils from the Lower Torridonian were reported by Naumova and Pavlovsky (1962), who suggested correlation with the Riphean, known to be Late Precambrian. Attention to these was also drawn by Sutton (1962).

Recognition of the Precambrian age of the Torridonian put the age of Lewisian well back into the Precambrian. Peach and Horne (1907) recognized that a dyke phase separated two Lewisian metamorphisms. Structural studies, and radiometric dating, available from the mid-1950s (see, for example, Giletti *et al.*, 1961), have now led to the recognition of at least three complexes in the Lewisian, the first dated at 2,700 m.y. or older and the last ending about 1,250 m.y. ago. The older part of the Torridonian has been shown to be about 935 m.y. old and to be separated by an unconformity from the Upper Torridonian dated at about 760 m.y. (Moorbath, 1969).

CALEDONIAN FOLD-BELT — NORTH OF HIGHLAND BOUNDARY — SCOTTISH HIGHLANDS AND IRELAND

Formations in the Scottish and Irish Caledonian Fold-Belt which need to be considered in arriving at the Precambrian concept are as follows:

Northern Highlands	Grampian Highlands	Ireland N. of H.B.F.	Ireland S. of H.B.F. Connemara	Ireland S. of H.B.F. S.E. Ireland
Devonian (unconf.)	Devonian (unconf.)	Devonian or L. Carb. (unconf.)	Silurian or Ordovician (unconf.)	Ordovician or Cambrian (unconf.)
	Dalradian M.A.	Dalradian M.A.	Dalradian M.A.	Bray S. Knockrath
Moinian M.A. (unconf. or tectonic contact) Lewisian M.A.	Moinian M.A.	Moinian M.A.		Cullestown Formation Rosslare Complex

H.B.F. = Highland Boundary Fracture-Zone
M.A. = Metamorphic Assemblage
S. = Series

The Moinian rocks of the Northern Highlands were once known as the Eastern or Upper Gneisses (as opposed to the Western Gneisses in the Lewisian) and it was not until 1888 that Peach used the term Moine Schists in a systematic sense, deriving it from A'Moine in Sutherland. The formation gave rise to early controversy, for along the border of the fold-belt it overlies the Durness Limestone. Murchison (1860) considered this contact to be stratigraphical and, because the Moinian is unconformably overlain by Devonian, classified it as Silurian. Nicol (1861) and Lapworth (1883) showed that the metamorphics had been thrust over the fossiliferous sediments. Although settling a major geological controversy, this proved only that the Moinian did not have to be Silurian (*sensu lato* of Murchison), to which system it had been attributed on Geikie's maps of 1876. In fact, the direct stratigraphical evidence gives only very wide brackets between unconformably underlying Lewisian and unconformably overlying Devonian.

In 1930 Peach (in Peach and Horne, 1930) emphasized a view he had probably long held that the Moinian sediments, which are dominantly psammitic, are the metamorphosed equivalent of the Torridonian. In 1951 Kennedy brought forward strong supporting petrological evidence. With the increasing use of "right-way-up" techniques after 1930 the Moinian of the Grampian Highlands was shown to be older than the Dalradian (see below). These arguments have led to a general acceptance of the Precambrian age of the Moinian.

The term Dalradian (from the ancient Scottish kingdom of Dalriada) was introduced by Geikie (1891a) for the schists of northern Ireland and parts of Scotland; on earlier maps these rocks are coloured as Lower Silurian. Until very recently most Geological Survey maps and Memoirs referred to them as "metamorphic", "of uncertain age" and "possibly Precambrian".

The Dalradian metasediments differ from those of the Moinian in the presence of limestones, quartzites and meta-volcanics. Stratigraphical successions were set up; owing to the complex tectonics there was prolonged controversy about stratigraphical order. Although, as early as 1911, J.S.G. Wilson had suggested that graded bedding indicated order of deposition it was not until 1930 that the general use of "right-way-up" techniques in the Grampian Highlands was initiated by Vogt. This had important implications for the age-concept, for it was shown that in places the stratigraphical base of the Dalradian is next to the Moinian and the top is cut off by the Highland Boundary Fracture-Zone, where it is in contact (mostly, but not entirely, mechanical) with fossiliferous Arenig. Detailed comparisons with the Norwegian Sparagmitian/Cambro-Silurian succession also became possible (e.g. Holtedahl, 1952), with the implication that the Dalradian might span the Precambrian/Cambrian boundary.

Pringle (1940) made a palaeontological breakthrough in the Highland Border with the discovery of pagetid trilobites in the Leny Limestone. This showed that the Upper Dalradian, at least, is of Lower/Middle Cambrian age, a view supported by palynological evidence (Downie *et al.*, 1971).

As increasing radiometric evidence became available in the late 1950s (see Giletti *et al.*, 1961) it became clear that part or the whole of the Dalradian *could* be Cambrian since the metamorphic dates are middle to upper Lower Palaeozoic with a statistical peak at 475 m.y.; in fact, recent work suggests that the metamorphic peak may have been of the order of 500 m.y.

An indirect stratigraphical argument involves correlations with Scandinavia.

Scotland and Ireland		*Scandinavia*
Higher Dalradian Groups.		Fossiliferous Cambrian.
Quartzitic Group.		Quartzite
Calcareous Group	(Boulder Bed	Tillite or Boulder Bed
	(Limestone	Limestone (Biri)
Moinian Groups.		Lower part of Sparagmitian

A key horizon here (and also in world comparisons) is the interstratified Boulder Bed of Lake Mjosen and the Varangian tillite further north, represented in Scotland and Ireland by very similar Boulder Beds, generally accepted as glaciogenes. On the basis of this correlation only the Lower Dalradian is Precambrian.

Since this paper was read, Johnstone (1975) has used the term Moine Succession, and Harris and Pitcher (1975) have referred to the Dalradian Supergroup spanning the Precambrian/Cambrian boundary.

Early terms for parts of the Irish metamorphics were mostly the same as those applied in Scotland, but Griffiths (1855) used Azoic, and Hull (1881) Laurentian. The term Dalradian, as in the Grampians, was introduced in 1891, but the Moinian was not recognized until 1948 (Anderson). Since all the main sub-divisions of the Dalradian have now been recognized north of the Boundary Fracture-Zone in Ireland the same arguments apply regarding age as in Scotland.

CALEDONIAN FOLD-BELT SOUTH OF HIGHLAND BOUNDARY — IRELAND

High grade metasediments appear from beneath unconformable Silurian (and possibly also Arenig) in Connemara. Modern work has made it possible to recognize the Lower Dalradian sub-divisions, including the boulder bed or glaciogene. As only the Lower Dalradian is present the Connemara Schists, as they are termed, are very probably all Precambrian.

Radiometric dates for their metamorphism are, however, Lower Palaeozoic, including the 475 m.y. peak. As the Connemara Schists appear (the field evidence is not completely conclusive) to be overlain by Arenig, this raises problems (for discussion see Anderson and Owen, 1968, p. 11).

In the south-east corner of Ireland, what is now known as the Rosslare Complex was regarded by Callaway as early as 1881 as Precambrian, but Geological Survey officers show these and other pre-Ordovician rocks of the Wexford/Rosslare area as Cambrian (1928).

The Rosslare Complex of high-grade gneisses, etc., which have undergone later cataclasis and diapthoresis, were definitely assigned to the Precambrian by Baker (1955). The Rosslare Complex was also placed in the Precambrian by Crimes and Dhonau (1967), as was the adjacent younger Cullenstown Formation.

Further north, pre-Ordovician strata come up in the Bray Anticlinorium, south of Dublin, and are shown as Cambrian on geological maps, including that of Griffiths (1855). The oldest of these rocks, the Knockrath and Bray Series, were put into the Precambrian by Tremlett (1959) but were transferred back into the Cambrian by Brück *et al.* (1974).

SUMMARY AND PRESENT CONCEPTS

The view of 19th century geologists that extensive outcrops of "primitive" or Precambrian rocks occur in the Scottish Highlands and in Ireland, although based on metamorphic state or long range correlation, has been substantiated by modern research. The Torridonian was not placed in the Precambrian until 1891, and it was as late as the 1950s or 1960s that two formations in the southeast tip of Ireland were definitely transferred from the Cambrian to the Precambrian. Research in this century has pushed the beginning of the Precambrian further back in time and has led to the recognition of additional elements.

The Scottish and Irish Precambrian may be looked at in the light of recent concepts, such as the division of geological time into the Katarchaean, the Archaean, the Lower Proterozoic and the Upper Proterozoic (Sutton, 1967). Small though it is in comparison with the great Shields of the World, the north-westerly kratogen has been shown to have the same complex history. Katarchaean rocks have not been identified, but the Archaean is represented by the two oldest units, the Scourian and the Inverian, ranging from 2,600 m.y. or more to 1,900 m.y. The youngest complex, the Laxfordian, is Lower Proterozoic. The Torridonian is certainly Upper Proterozoic, and because of the presence of an unconformity there is no difficulty in separating it from the Cambrian.

Apart from outcrops of Lewisian in the Northern Highlands and possibly also some rocks in Western Ireland (Max, 1970) and in South-East Ireland, all Scottish

Figure 3. Proof of Precambrian age of Torridonian.
Well-bedded, basal Cambrian quartzite resting unconformably on Torridonian sandstone (more massive). North of Achnashellach, Ross-shire, Scotland.

and Irish Precambrian rocks of the Caledonian Fold-Belt are Upper Proterozoic. Furthermore, it is now evident that the sequence, as in many other regions, continues without any distinct break into the Cambrian. The top of the Dalradian is certainly Cambrian but, as all the Moinian/Dalradian rocks have undergone Lower Palaeozoic metamorphism ranging from the slate/greenschist facies to the granulite/migmatite facies, recognition of the base of the Cambrian is particularly difficult.

From 190 years of study since Hutton, there has emerged the concept that the Scottish and Irish Precambrian contains elements ranging from the early Archaean (2,700 m.y.) to the base of the Cambrian.

ACKNOWLEDGEMENTS

The author gratefully acknowledges help from the late Dr. V.A. Eyles and Mrs. Eyles, Great Rissington, Cheltenham, U.K.

REFERENCES

Anderson, J.G.C., 1948, The occurrence of Moinian rocks in Ireland: Quarterly Journal of the Geological Society of London, v. 103, p. 171-190.

Anderson, J.G.C. and Owen, T.R., 1968, The Structure of the British Isles: Oxford, Pergamon Press, 162 p.

Baker, J.W., 1955, Pre-Cambrian rocks in Co. Wexford: Geological Magazine, v. 92, p. 63-68.

Barrow, G., 1893, On an Intrusion of Muscovite-Biotite Gneiss in the South-Eastern Highlands of Scotland, and its accompanying metamorphism: Quarterly Journal of the Geological Society of London, v. 49, p. 330-358.

Boué, A., 1820, Essai Géologique sur l'Ecosse: Imprimerie de Brizard-Courcier, Paris, 519 p.

Brück, P.M., Potter, T.L., and Downie, C., 1974, The Lower Palaeozoic Stratigraphy of the northern part of the Leinster Massif: Proceedings of the Royal Irish Academy, v. 74B, p. 75-84.

Callaway, C., 1881, The metamorphic and associated rocks south of Wexford: Geological Magazine, v. 8, p. 1-5.

Crimes, T.P. and Dhonau, N.B., 1967, The Pre-Cambrian and Lower Palaeozoic Rocks of South-East Co. Wexford, Eire: Geological Magazine, v. 104, p. 213-221.

Downie, C., *et al.*, 1971, A palynological investigation of the Dalradian rocks of Scotland: Institute of Geological Sciences, Report 71/9, 29 p.

Geikie, Sir. A., 1876, Geological map of Scotland: Edinburgh.

—————, 1891a, Presidential Address: Proceedings of the Geological Society of London, v. 47, p. 48-162.

—————, 1891b Discovery of the Olenellus Zone in the North-West Highlands: Nature, v. 44, p. 479.

—————, 1894, Annual Report of the Geological Survey for 1893: Memoirs of the Geological Survey of Great Britain, p. 279-282.

Giletti, B.J., Moorbath, S., and Lambert, R.St.J., 1961, A geochronological study of the metamorphic complexes of the Scottish Highlands: Quarterly Journal of the Geological Society of London, v. 117, p. 233-272.

Griffith, R., 1838, Outline of the geology of Ireland, with geological map of Ireland to accompany the Report of the Railway Commissioners: Dublin, Railway Commissioners.

—————, 1885, A general map of Ireland to accompany the Report of the Railway Commissioners, showing the principal physical features and geological structure of the country: Dublin, Railway Commissioners.

Harris, A.L., and Pitcher, W.S., 1975, The Dalradian Supergroup, *in* A Correlation of Pre-cambrian Rocks in the British Isles: Geological Society of London, p. 52-75.

Holtedahl, O., 1952, The structural history of Norway and its relation to Great Britain: Quarterly Journal of the Geological Society of London, v. 108, p. 65-98.

Hull, E., 1881, On the Laurentian rocks of Donegal, and other parts of Ireland: Scientific Transactions of the Royal Dublin Society, v. 1, (series 2), p. 243.

Hutton, J., 1788, Theory of the Earth; or an Investigation of the laws observable in the Composition, Dissolution and Restoration of Land upon the Globe: Transactions of the Royal Society of Edinburgh, v. 1, p. 209-304.

Imrie, Lt.-Col., 1811, A Description and Section of the Strata of the Grampian Mountains from the Plain of Kincardineshire to the Summit of Mount Battock: Transactions of the Royal Society of Edinburgh, v. 6, p 3-19.

Jameson, R. 1800, Mineralogy of the Scottish Isles: B. White, Edinburgh; W. Creech, London, 2 vols. 243 and 269 p.

Johnstone, G.C., 1975, The Moine Succession, *in* A Correlation of Pre-Cambrian rocks in the British Isles: Geological Society of London, p. 30-42.

Kennedy, W. Q., 1951, Sedimentary differentiation as a factor in the Moine-Torridonian correlation: Geological Magazine, v. 88, p. 257-266.

Lapworth, C., 1883, The secret of the Highlands: Geological Magazine, v. 10, p. 120-128; 193-198; 337-344.

Logan, W.E., and Hunt, T.S., 1855, Esquisse géologique du Canada . . . à l'exposition universelle de Paris: Bossange, Paris, 100 p.

Lyell, Sir C., 1830-1833, Principles of Geology: London, 3 vols.

Macculloch, J., 1821, A Geological Classification of Rocks: Longman, Hurst, Rees, Orme and Brown, London, 655 p.

————, 1836, A Geological Map of Scotland: Arrowsmith, London.

Max, M.D., 1970, Mainland Gneisses south-west of Bangor in Erris, Northwest County Mayo, Ireland: Scientific Proceedings of the Royal Dublin Society, v. 3A, p. 275-291.

Moorbath, S., 1969, Evidence for the age of deposition of the Torridonian sediments of north-west Scotland: Scottish Journal of Geology, v. 5, p. 154-170.

Murchison, Sir R.I., 1856, On the Relations of the Crystalline Rocks of the North Highlands to the Old Red Sandstone of that region, and on the recent discoveries of fossils in the former by Mr. C. Peach: Report of the British Association for the Advancement of Science (1855), p. 85.

————, 1860, Supplemental Observations on the Order of the Ancient Stratified Rocks of the North of Scotland and their associated Eruptive Rocks, with a letter from Professor Harkness: Quarterly Journal of the Geological Society of London, v. 16, p. 215-240.

Murchison, Sir R.I., and Geikie, A., 1862, First Sketch of a New Geological Map of Scotland with Explanatory Notes: W. and A.K. Johnston and W. Blanchard, Edinburgh and London.

Naumova, S.N. and Pavlovsky, E.V., 1961, The discovery of plant remains (spores) in the Torridonian Shales of Scotland (in Russian): Doklady Akademii Nauk Soyuza Sovetskikh Sosialisticheskikh Respublik, v. 141, p. 181-182.

Necker de Saussure, L.A., 1808, A Map of Scotland, coloured according to the different formations observable in that country, with an explanatory Memoir: not published. MS deposited in the Library of the Geological Society of London. Facsimile published by Edinburgh Geological Society, 1939.

Nicol, J., 1857, On the Red Sandstone and Conglomerate and the Superposed Quartz-rocks, Limestones and Gneiss of the North-West Coast of Scotland: Quarterly Journal of the Geological Society of London, v. 13, p. 17-39.

————, 1861, On the Structure of the North-West Highlands and the Relation of the Gneiss, Red Sandstone and Quartzite of Sutherland and Ross-shire: Quarterly Journal of the Geological Society of London, v. 17, p. 85-113.

Peach, B.N., *et al.*, 1888, Report on the recent work of the Geological Survey in the North-West Highlands of Scotland: Quarterly Journal of the Geological Society of London, v. 44, p. 378-441.

Peach, B.N. and Horne, J., 1892, The Olenellus Zone in the North-West Highlands of Scotland: Quarterly Journal of the Geological Society of London, v. 48, p. 227-242.

————, 1930, Chapters on the Geology of Scotland: Oxford University Press, 232 p.

Peach, C.W., 1855, Notice of the Discovery of Fossils in the Limestone of Durness in the County of Sutherland: Edinburgh New Philosophical Journal, v. 2 (new series), p. 197.

Phillips, J., 1838, Index/Geological Map of the British Isles: Weale, London.

Pringle, J., 1939, The discovery of Cambrian trilobites in the Highland border rocks near Callander, Perthshire: Report of the British Association for the Advancement of Science (1939), p. 252.

Sedgwick, A. and Murchison, Sir. R.I., 1836, On the Silurian and Cambrian Systems: Report of the British Association for the Advancement of Science (1835), p. 59-61.

Sutton, J., 1962, Torridonian microfossils: Geological Magazine, v. 99, p. 379.

————, 1967, The Extension of the Geological Record into the Precambrian: Proceedings of the Geologists' Association, London, v. 78, p. 493-534.

Tremlett, W.E., 1959, The Precambrian rocks of Southern Co. Wexford (Ireland): Geological Magazine, v. 96, p. 58-68.

Vogt, T., 1930, On the chronological order of deposition of the Highland schists: Geological Magazine, v. 67, p. 68-73.

Wilson, J.S.G., 1911, *in* The Geology of Knapdale, Jura and North Kintyre: Memoirs of the Geological Survey of Great Britain.

History of Concepts in Precambrian Geology, edited by W.O. Kupsch and W.A.S. Sarjeant, Geological Association of Canada Special Paper 19, 1979

HISTORY OF CONCEPTS OF PRECAMBRIAN GEOLOGY IN AUSTRALIA

David F. Branagan
Department of Geology and Geophysics, University of Sydney
Sydney, N.S.W. 2006, Australia

ABSTRACT

The study of Precambrian rocks in Australia can be divided into four overlapping periods. During the first period, up to the 1870s European (particularly British) ideas were dominant, and theoretical concepts about where ancient rocks should occur were influential.

After the discovery of Cambrian fossils in South Australia there followed a long period in which the basic stratigraphy and structure of the Precambrian rocks of this region were evaluated, particularly by Howchin and Mawson. This work was the basis for comparative studies in other regions.

Gold discoveries in the 1890s in Western Australia focussed attention on the huge areas of Precambrian in that state and led to the recognition of Archaean rocks. Studies by A. Gibb Maitland and E. de C. Clarke were specially important.

The final period, since the 1940s, has involved detailed stratigraphic and structural studies, combined with numerous age determinations. These have resulted in a fairly good understanding of the tectonic history of the continent during the Precambrian.

RESUME

L'étude des roches précambriennes en Australie peut être divisée en quatre périodes qui se chevauchent. Durant la première période, jusqu'en 1870, les idées européennes (particulèrement britanniques) dominèrent et les concepts théoriques, portant sur la question de savoir où devraient se trouver les occurences de roches anciennes, furent influents.

La découverte de fossiles cambriens dans le Sud de l'Australie fut suivie d'une longue période au cours de laquelle la stratigraphie et la structure de base de cette région furent évaluées, particulièrement par Howchin et Mawson. Ce travail fut à la base d'études comparatives dans d'autres régions.

Dans les années 1890, les découvertes d'or en Australie occidentale attirèrent l'attention sur les immenses surfaces de Précambrien de cet état et conduisirent à la reconnaissance de roches archéennes. Les études de A. Gibb Maitland et E. de C. Clarke furent particulièrement importantes.

Depuis les années 1940, la période finale a comporté des études stratigraphiques et structurales combinées à de nombreuses déterminations d'âges. Il en est résulté une bonne compréhension de l'histoire tectonique du continent au cours du Précambrien.

INTRODUCTION

The first settlement by Europeans was at Sydney in 1788 and for at least the next 60 years the scientific attitudes were those directly imported from Europe and particularly by the British, who were the original colonists.

It is not surprising, therefore, that early attitudes about Australian geology were essentially those of what might be called the 'British School', although the work of French and German scientists, among others, was often of great individual significance.

In the first 50 years of the Nineteenth Century before the term Precambrian had been coined, there still lingered the influence of Arduino and others, with their concepts of primitive, highly deformed, unfossiliferous mountain cores overlain by less deformed fossiliferous rocks which became progressively younger away from the cores. In 1803 the mineralogist A.W.H. Humphrey stated sagely that the explorer Francis Barrallier couldn't have penetrated the mountain ranges of Sydney to its core as he had found no granite. Much later Strzelecki (1845) wrote

> 'the best sections of the stratified masses . . . tend to prove that, of the stratified rocks, mica slate, being the nearest to the crystalline rocks, and following all the contortions of the base, forms the oldest portion of the crust, and that siliceous and argillaceous slates, which rest upon it, are the next in order of superposition'.

Even Jukes (1850) has a lingering backward glance –

> 'the east coast mountain chain . . . cannot be said to have any one well-defined axis of granitic or other rocks, . . . Large granitic masses, however, show themselves at different portions of its course . . . Upon these rocks lies a large palaeozoic formation . . . but we have at present no knowledge of the exact relation between the palaeozoic and metamorphic series'.

Both Strzelecki and Jukes produced geological maps (Figs. 1 and 2), the first attempts to synthesize the geology of large extents of the continent.

Acceptance of the newly developing European terminology is evident in Burr's (1846) comments on the rocks of South Australia which Jukes classed as metamorphic. Burr wrote 'the rocks of which one [Mt. Lofty] range is composed are those which belong to the Primary strata, probably corresponding to the Cambrian and Skiddaw systems of Sedgwick' because 'they are apparently devoid of the evidence of the existence of animal and vegetable life during their formation'.

Tenison Woods (1858) thought the same rocks 'probably of either Cambrian or Silurian Formation' but continued 'this is mere guesswork, supported by little more than resemblances in mineral character, etc.'

Selwyn (1860) believed the basal parts of the Mount Lofty chain were the equivalent of Upper Silurian (i.e., Silurian as presently used) and the metamorphic rocks of the Victorian Alps to be Lower Silurian (i.e., Cambrian to Ordovician in our present technology), but in 1861 he suggested that the metamorphic rocks at the western border of Victoria belonged to 'possibly a true Cambrian or Azoic Series'. In 1868, Selwyn went further and wrote 'perhaps the rocks of some of the large areas mapped as metamorphic represent Cambrian or Laurentian series', probably the first use of this Canadian term in Australian geology. The Rev. W.B. Clarke (1867), however, did not agree entirely. He believed there was not sufficient evidence of the existence of Azoic rocks in East Australia, and suggested that

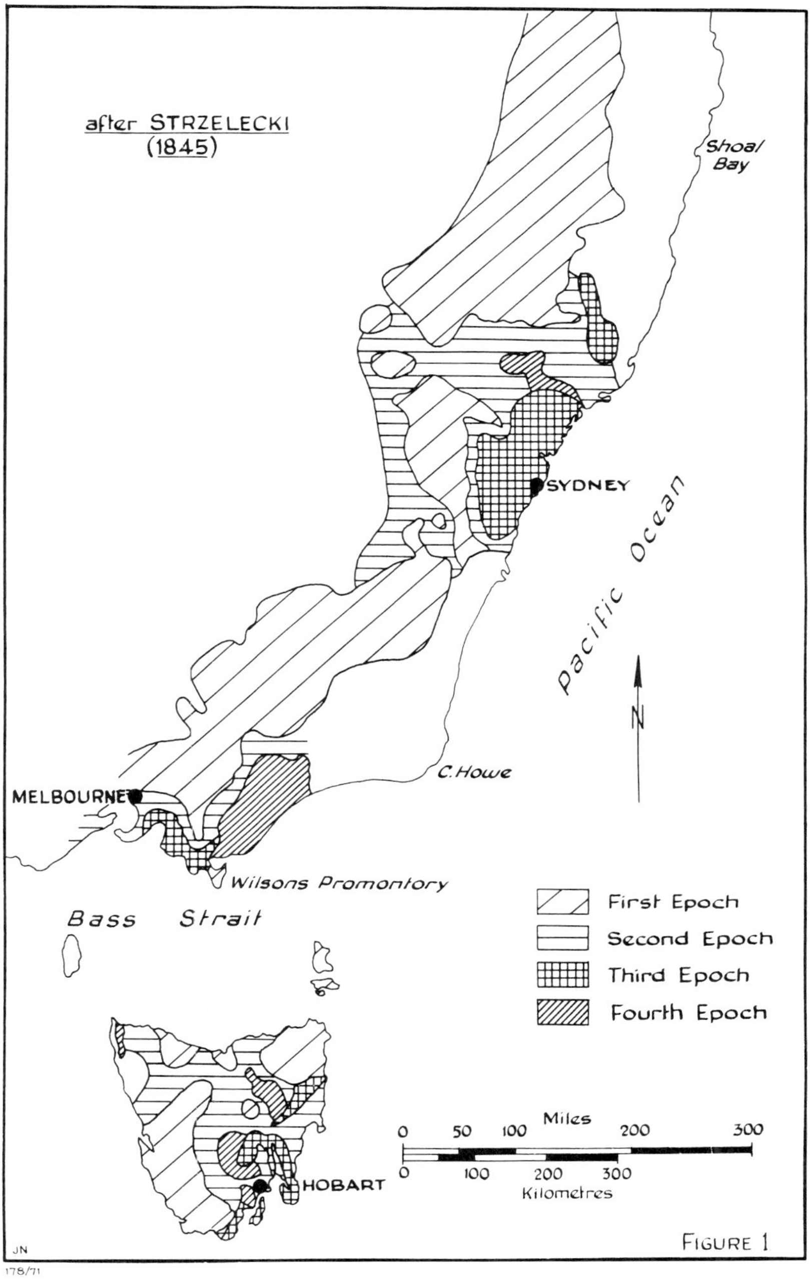

Figure 1. Distribution of Old Rocks (after P.E. Strzelecki, 1845).

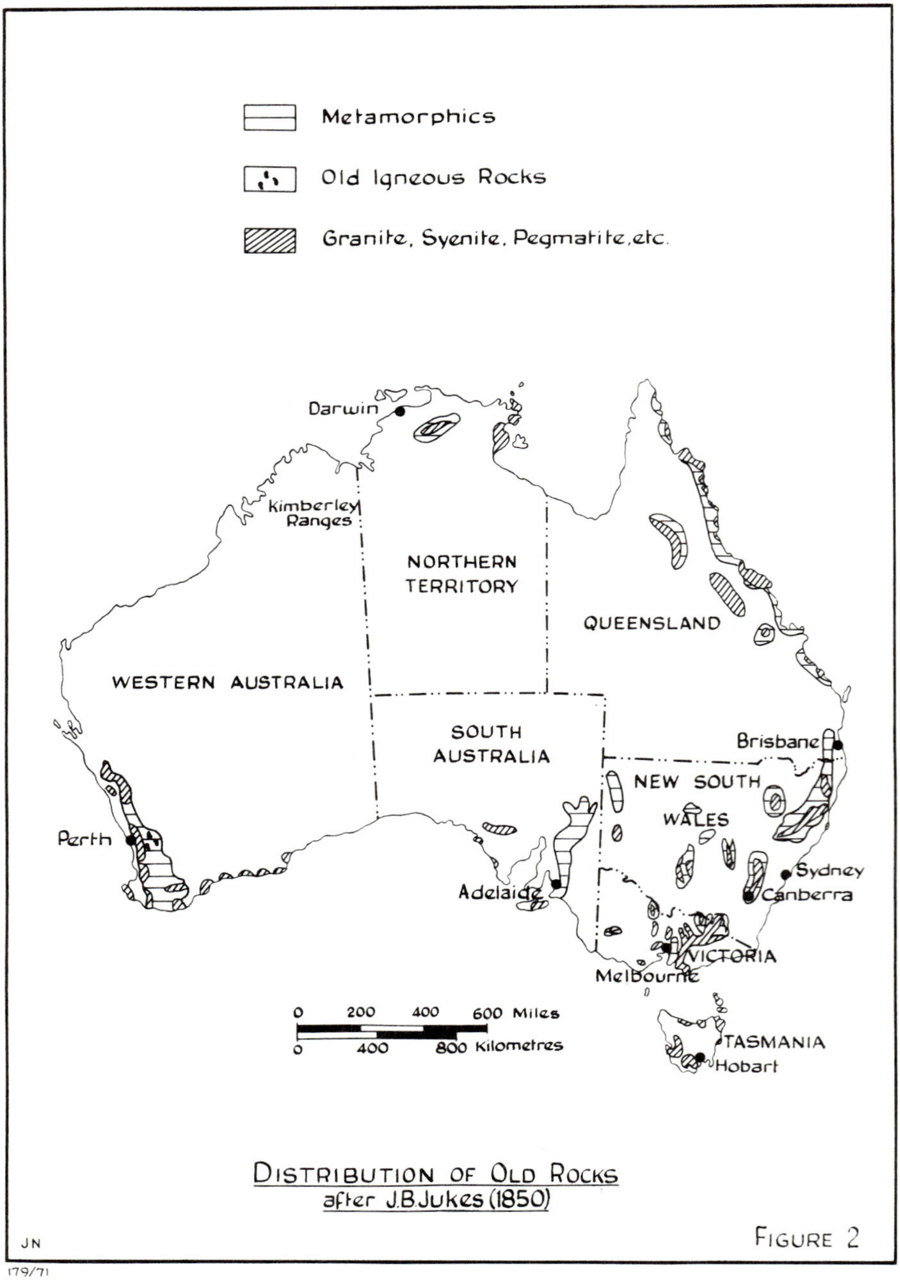

Figure 2. Distribution of Old Rocks (after J.B. Jukes, 1850).

some of the 'First Epoch' gneisses of Strzelecki were merely the products of 'transmutation'. Brough Smyth's map of 1875 (Fig. 3) summarizes the state of knowledge and opinion up to that time.

Jukes (1862, p. 434) seems to have been first to use the term 'Precambrian' with reference to Australian geology when he suggested it was highly probable that the gneiss and mica schists forming the mountain chains belonged wholly or partly to the Precambrian periods.

However, the uncertainty prevailed for some years longer. Geikie (1882) quoted the opinion of Selwyn and many others that the crystalline schists were metamorphosed Palaeozoic formations but added 'there are not improbably other areas referable to an Archaean Series', and Hardman (1884), while provisionally classing the metamorphic rocks of the Kimberley area in north-west Australia as Lower Silurian or Cambro-Silurian, wrote 'it is not improbable that these rocks, as well as similar formations in the colony and other Australian colonies, may be of Laurentian age'.

Writing of what he called 'the Fundamental Rocks or Archaean', Tate in 1893 said

> 'the generalisation which has sought to sweep all the crystalline rocks of Australia into the great Silurian net has been broken down by the discovery of unconformably superimposed Cambrian strata; and though it by no means follows that the whole of the Crystalline rock masses are of Archaean age, yet there are good reasons for the belief that those rocks which exhibit the phenomenon of regional metamorphism belong to one epoch'.

DISCOVERY OF THE FIRST CAMBRIAN FOSSILS

In 1879 the first truly Cambrian fossils (trilobites and archaeocyathids) were found at Ardrossan, South Australia by Tepper and ascribed by Tate to the Lower Silurian 'employing the term in the Murchisonian sense, though the Menevian series is implied'. Woodward (1884), Etheridge (1890a) and Tate (1892) gathered further evidence of Cambrian fossils in South Australia.

The rocks of the Mount Lofty Range underlying the recognizable Cambrian sequence consequently were regarded by Tate and others as Precambrian. Tate's belief that the Mount Lofty Range was all Precambrian and unfossiliferous was disproved by the finding of archaeocyathids by T.W.E. David and Walter Howchin in 1896.

At this time the archaeocyathids were believed to be Upper Cambrian, and the underlying (apparently unfossiliferous) beds near Adelaide were put into the Lower Cambrian. Further mapping by Howchin showed the wide extent through South Australia of these 'Lower Cambrian' unfossiliferous beds, often without any association with the 'Upper Cambrian'. There was always evidence of a time break between the Lower and Upper Cambrian even though actual unconformities were absent.

Elsewhere in the world the Archaeocyathinae horizon was being accepted as Lower Cambrian, so Howchin's 'Lower Cambrian' gradually (in the first 20 years of the 20th century) came to be regarded as late Precambrian in age.

This contention was strengthened over the years by evidence of glaciation

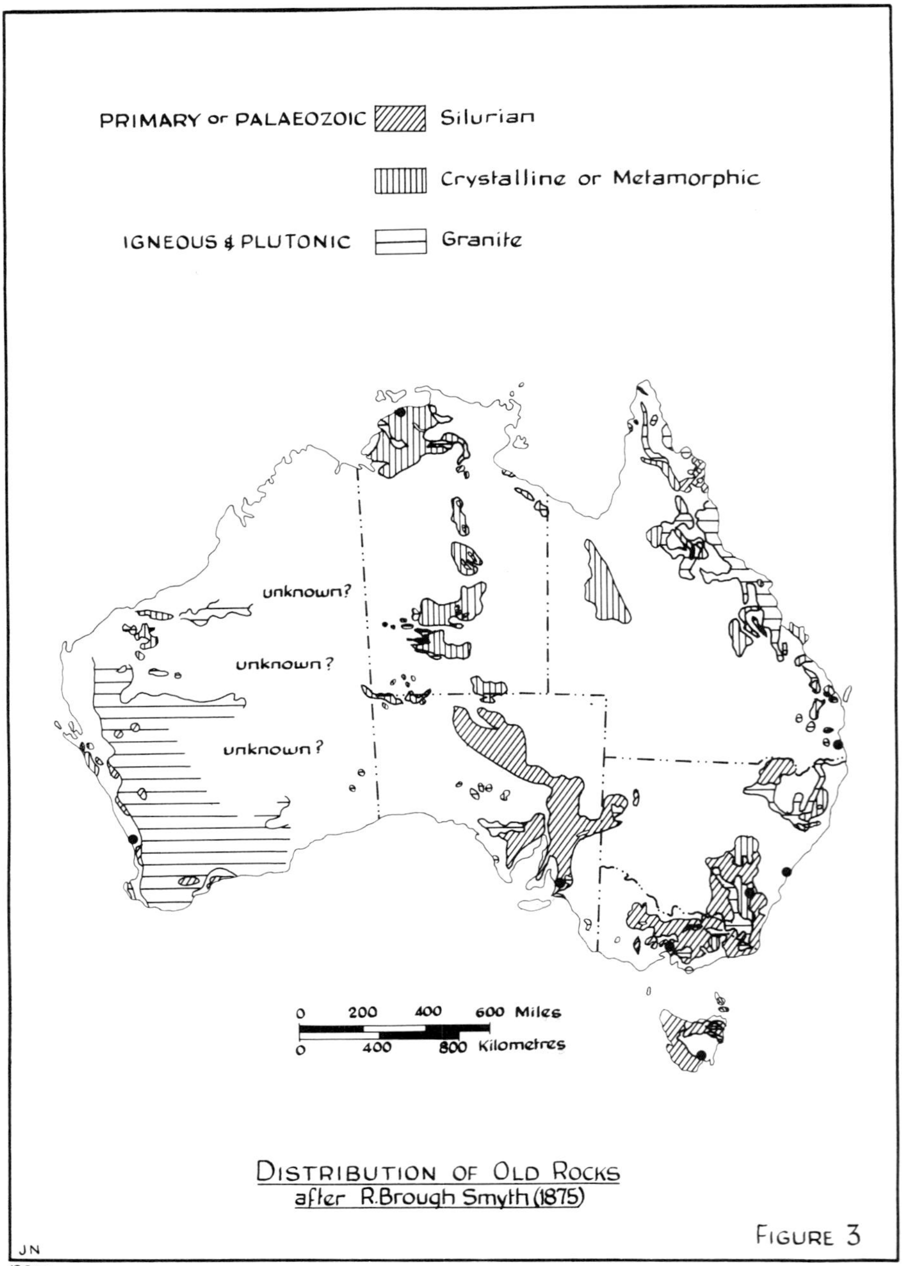

Figure 3. Distribution of Old Rocks (after R.B. Smyth, 1875).

high in this sequence of rocks, called the Adelaide Series by Mawson. H.P. Woodward (1885), and Howchin (1901) recognized glacial morainal debris at least as old as Cambrian, and Howchin traced evidence of this glaciation throughout South Australia. Mawson in 1912 extended the evidence by mapping similar beds in the Torrowangee area near Broken Hill, New South Wales.

By the first decade of this century a two fold division (albeit informal) had been established for the Precambrian of South Australia. The two major units were: (1) a 'basement' complex overlain by (2) the moderately deformed Adelaide Series. Although the Precambrian character of the oldest rocks in South Australia had been established by 1893, the opinion held by overseas geologists was that such rocks probably had limited extent. Wallace (1893) used Everitt's map of 1887 (Fig. 4) when he wrote

> 'the Blue Mountains of New South Wales, the Australian Alps, the Great Dividing Range of Victoria, as well as the Pyrenees and Grampians in the south-west, consist mainly of Silurian deposits of two periods, the lower or more ancient, perhaps corresponding to the Cambrian or even the Laurentian formations of Europe and North America, and the upper or more recent. In South Australia the same rocks form the mountainous backbone of the country . . . When we come to the west coast, however, we find the Palaeozoic rocks far less developed . . .'

SOME SUPPOSED PRECAMBRIAN OUTCROPS

The next geological map of Australia (J.W. Gregory, 1907; Fig. 5), shows an enormously increased area of the country as Archaean or ancient granite. Gregory was adamant that the Australian Alpine region (at least in Victoria) was Precambrian (he used the term 'Archaean') as he had mapped highly deformed and apparently unfossiliferous rocks lying unconformably beneath lower Palaeozoic rocks. David (1908) also suggested the rocks near Cooma in New South Wales might be Precambrian and Sussmilch (1911) continued this idea by stating that within N.S.W. there were 'two districts where Precambrian rocks are definitely known to occur – the Barrier district (i.e., the Broken Hill region) and the Cooma-Kosciusko district'. However, the finding of graptolites by Laseron (Browne, 1914) in the latter region began the change in attitude about the highly deformed rocks of the eastern highlands. Of these eastern Australian rocks only the apparently unfossiliferous Bunya Phyllites of the Brisbane Metamorphics in Queensland continued to be suggested as possibly Precambrian until recently (e.g., Houston, 1965). More recent work indicates these rocks are of Carboniferous age (Fleming *et al.*, 1974). However, there are small outcrops of Precambrian rocks in western Tasmania and King Island (Bass Strait).

Sussmilch (1911) was perhaps the first to use the term Proterozoic with reference to Australian geology and to recognize the possibility of subdividing the Precambrian rocks outside South Australia.

The important orebodies of Broken Hill (N.S.W.) attracted much attention amongst mining geologists, and in 1916, E.S. Moore compared the 'Hanging Wall gneiss' adjacent to the main orebody with Laurentian gneiss and noted the general similarity of the 'basement complex' to the Precambrain of the Canadian Shield.

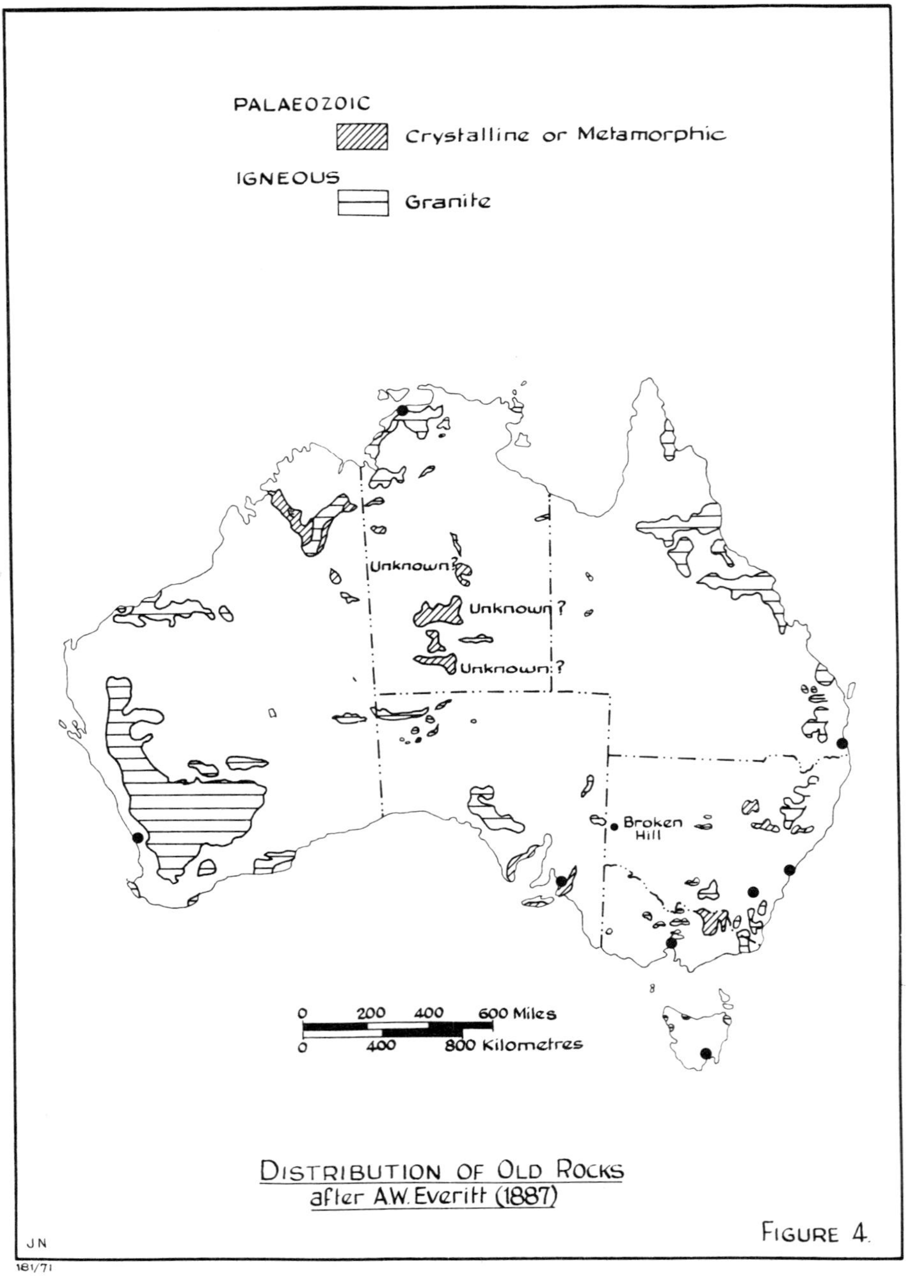

Figure 4. Distribution of Old Rocks (after A.W. Everitt, 1887).

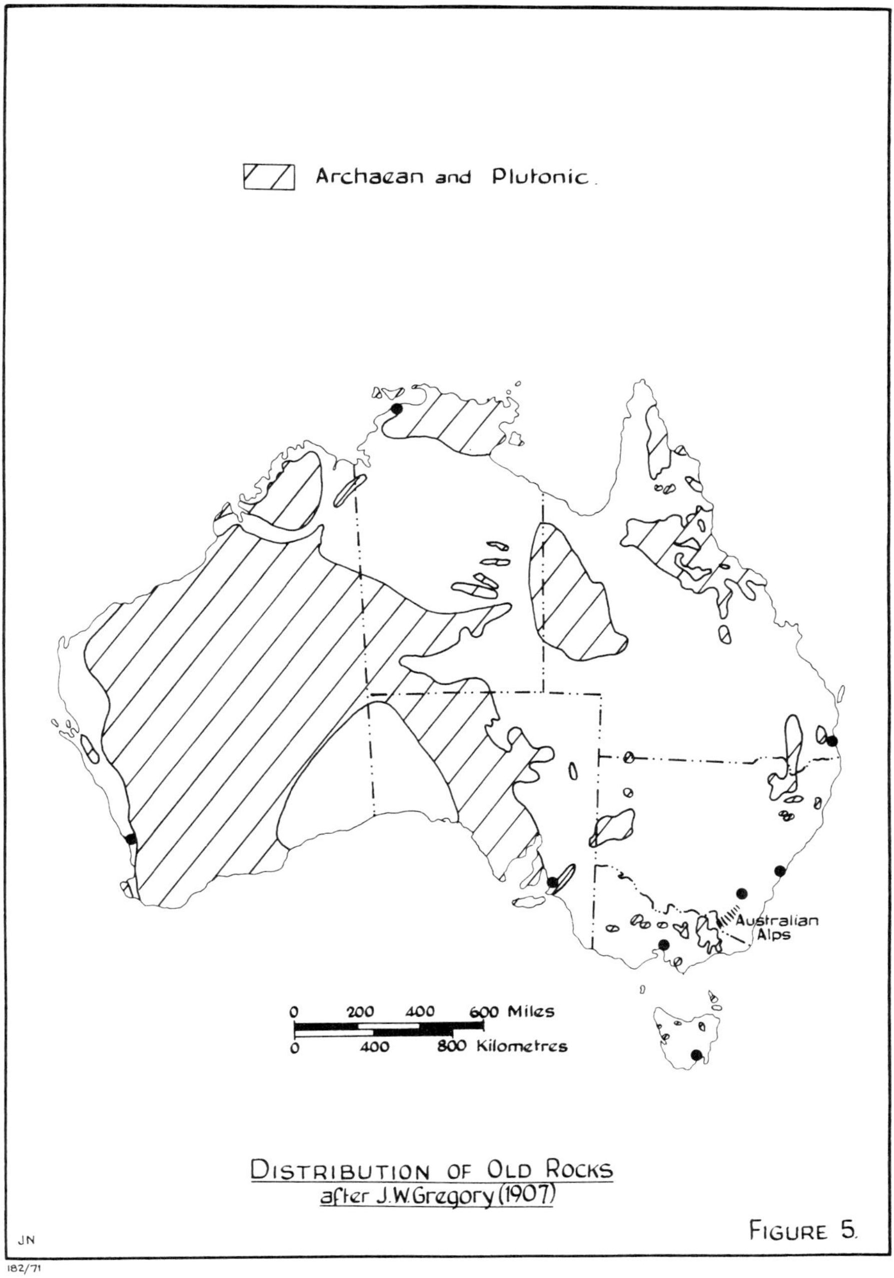

Figure 5. Distribution of Old Rocks (after J.W. Gregory, 1907).

E.C. Andrews (1922) was convinced of the Precambrian age of the Broken Hill rocks and thought them 'probably even Archaean', an indication of a growing change in the use of this word, and a tacit recognition of the significance and value of the term Proterozoic.

Meanwhile attempts to clarify the stratigraphy of both the 'Adelaide Series' and the Barossa basement complex continued in South Australia and by 1936 Mawson was able to comment on a two-fold subdivision of the Adelaide rocks and more work on the underlying basement.

THE WESTERN AUSTRALIAN PRECAMBRIAN

Considerable interest began to be directed to the rocks of Western Australia by the gold discoveries of the 1890s. Hardman in 1884 had suspected some Precambrian rocks might be present and Etheridge (1890b) had established the presence of fossiliferous Cambrian.

A. Gibb Maitland, A. Montgomery and others mapped the various mineral fields of Western Australia in the early 1900s. Although still uncertain about the actual ages of many rocks, Maitland and Montgomery (1912) presented a reasonably coherent picture of the geology (Fig. 6). Mapping had indicated the presence of several metamorphosed units with unconformable relations between them.

A definitive subdivison of the West Australian Precambrian was published by E. de C. Clarke in 1923 in which five 'units' were named: viz: (1) Yilgarn Series (oldest), (2) Kalgoorlie Series, (3) Mosquito Creek Series, (4) Later acid rocks, (5) Nullagine Series. This sequence was essentially adopted by David (1932) and was still generally acceptable in 1950 (David, 1950).

The four series were separated by unconformities. David (1932) grouped the Yilgarn and Kalgoorlie Series as Archaeozoic, the Mosquito Creek Series and the Later acid Rocks (with other minor units) as Older (?) Proterozoic and the Nullagine Series as Newer Proterozoic. Gregory and Barrett (1931) pointed out that 'the age of the Nullagine Series is not definitely established . . . The Series is similar in structure and relations to the Torridonian and Keweenawan Systems'.

David (1932) believed 'the close of Mosquito time was marked by the most stupendous intrusions of granite that have ever taken place in the geological history of Australia [see Fig. 6]. The compound granite batholith of the southern portion of W.A. is as large in area as the whole of France . . .' However David (1950) pointed out that much of this region consists of a considerable variety of gneisses (Fig. 7).

Following World War II there was an enormous expansion in geological mapping which led to a better understanding of the Precambrian rocks, specially in the northern part of the continent: The results generally confirmed the earlier subdivisions although there was by 1962 a preference for grouping all the Archaean rocks together while the Proterozoic was becoming more readily subdivisible (Fig. 8).

THE CONCEPT OF ACTUAL AGE OF THE AUSTRALIAN PRECAMBRIAN

Until the middle 1950s, the mapping of the Precambrian had been an essentially stratigraphic task, invoking the aid of palaeontology wherever possible. As

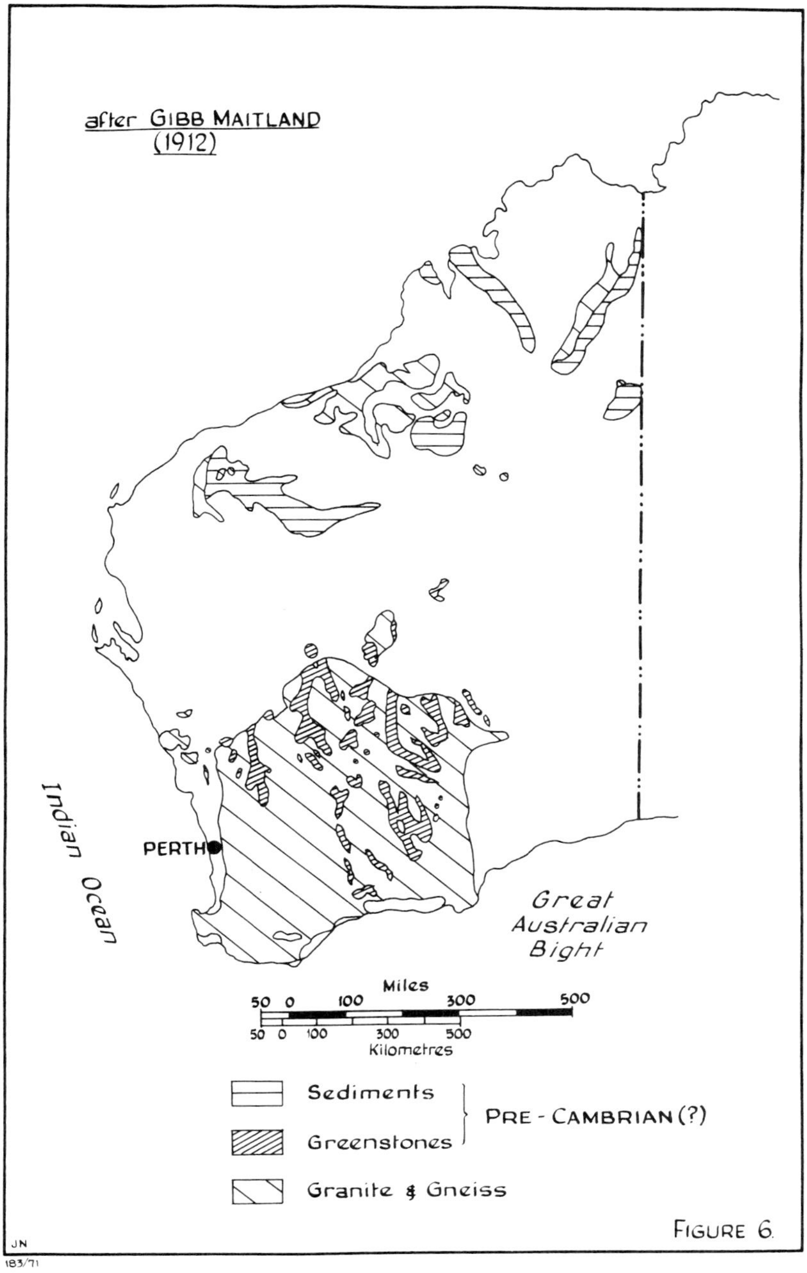

Figure 6. Distribution of Old Rocks (after Gibb Maitland and A. Montgomery, 1912).

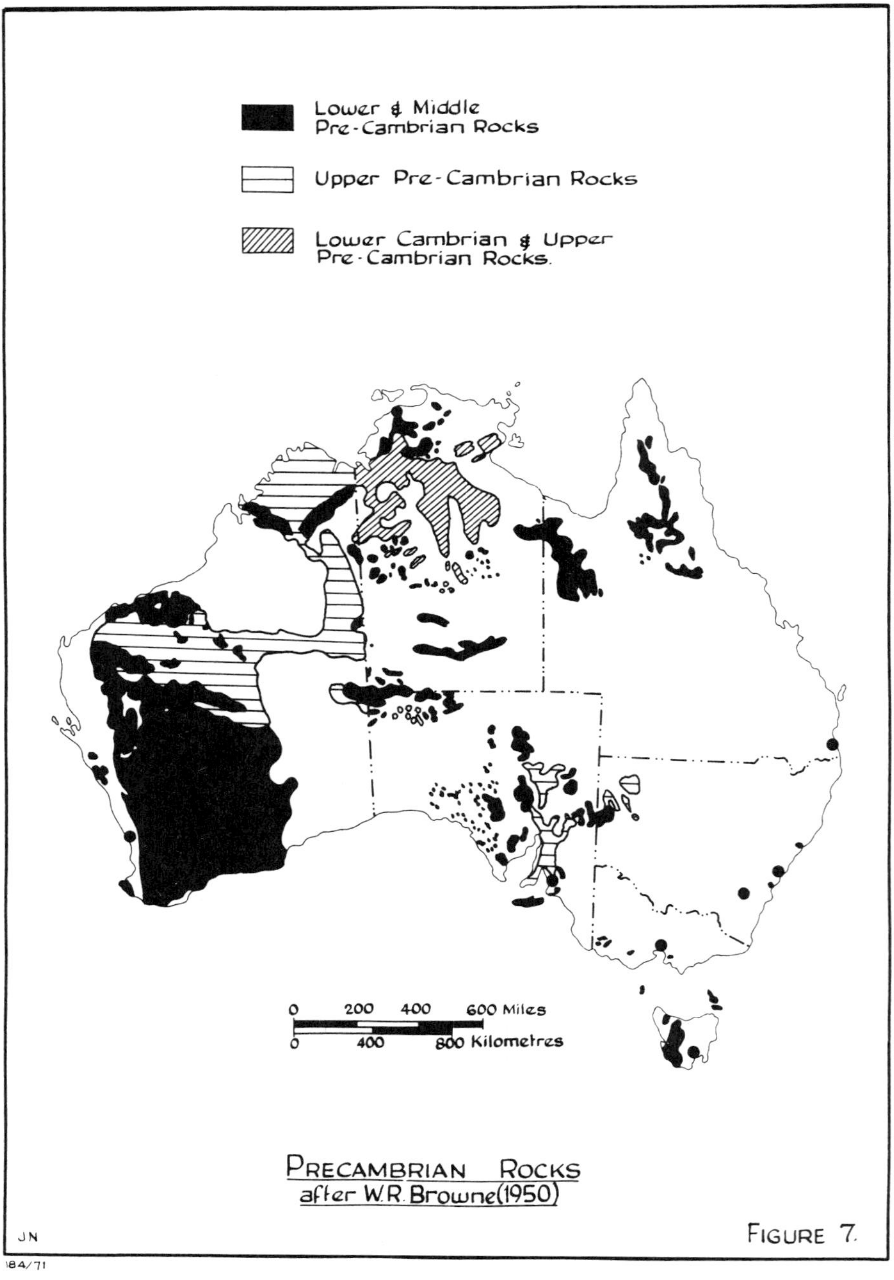

Figure 7. Precambrian rocks (after W.R. Browne *in* Browne, 1950)

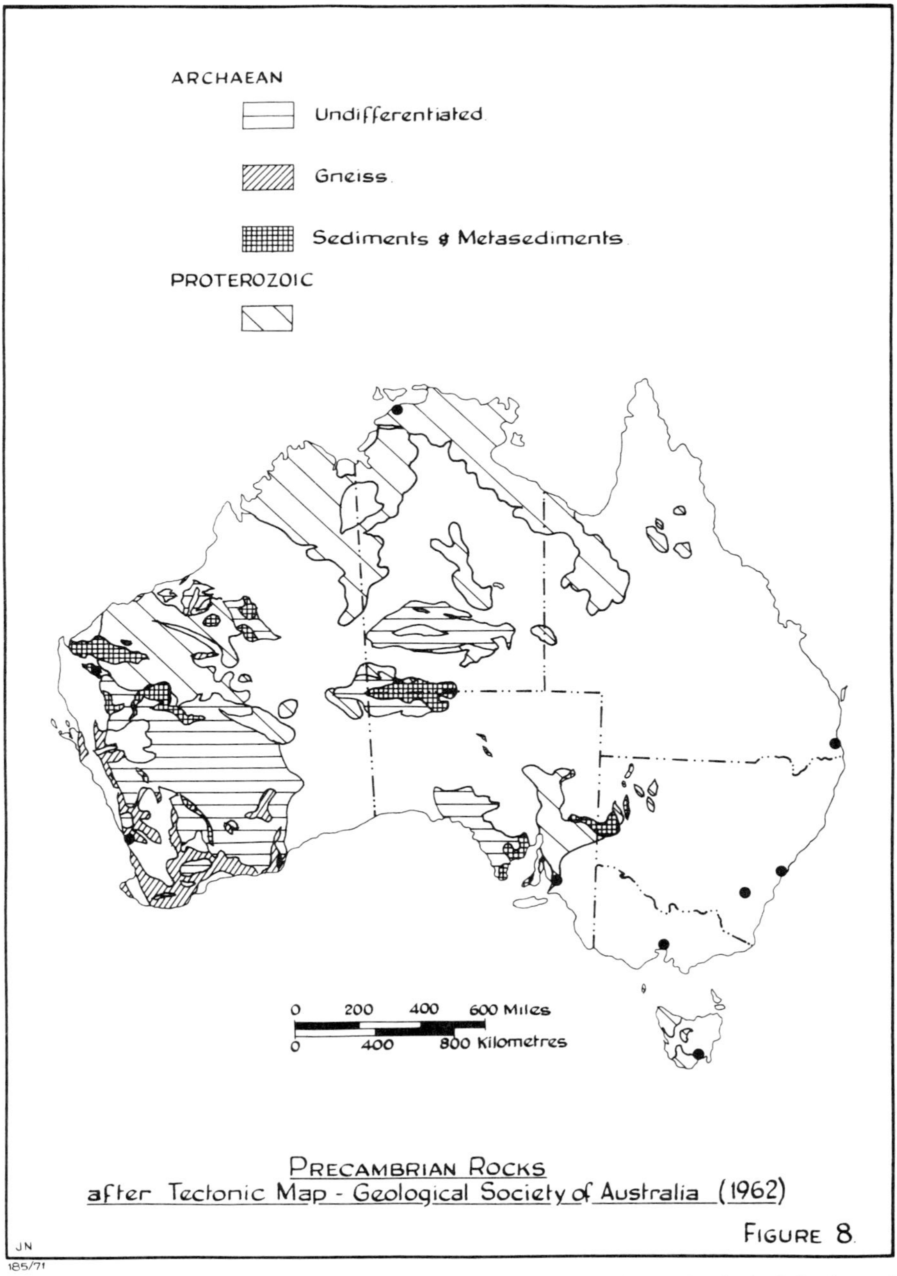

Figure 8. Precambrian rocks (after Tectonic Map of Australia – Geological Society of Australia, 1962).

early as 1896 David and Howchin referred to radiolaria and associated oolitic structures and discussed the age of some limestones in the Precambrian of South Australia. In 1936 David and Tillyard published a controversial book in which they claimed to have discovered large arthropods and giant annelids, as well as radiolaria, in the Adelaide Series. Most workers have not accepted the evidence for the large organisms and believe the features to be purely sedimentary in character.

However, in 1947 Sprigg discovered the extraordinary fauna at Ediacara which is now well-known. Elsewhere, stromatolites have proved abundant and useful in mapping (Preiss, 1973).

Guesses were made from time to time at the ages of Precambrian rocks and in general these proved underestimates, although Curran in 1899 suggested an age of more than 700 million years for the beginning of the Cambrian. David (1932, p. 32) put the beginning of 'Yilgarn time' at about 1600 million years B.P.

The first radiometric dating of Australian rocks was carried out by L.A. Cotton (1926) using the Uranium-Lead method and he obtained dates varying between 3840 m.y. (Pilbara W.A.) and 240 m.y. (Olary South Australia) which he interpreted very cautiously. Organised dating began in the late 1950s and in 1960 an important paper by Wilson *et al.* summarized the results for the Precambrian rocks examined (Fig. 9). These dates helped to strengthen the suspicion growing among those studying the tectonic history of the continent that not all the 'basement rocks' were necessarily Archaean. They also helped to show that the Upper Proterozoic sequences could be separated into two types – those essentially unaffected by post Proterozoic movements and those intimately involved with lower Palaeozoic rocks in lower to middle Palaeozoic movements.

Continuing work through the 1960s and 70s, combining detailed stratigraphic studies with age determinations has clarified the tectonic and age relationships between the various Precambrian sequences, and although there are still many problems yet to be solved, the pattern suggested by Thomson (1969) in Figure 10 still seems to be generally acceptable.

CONCLUSIONS

This essentially historical review has, of course, begged many of the philosophical questions about the Australian Precambrian which might be raised. It is clear that there has been much confusion about time, rock and time-rock terms, some of which has been clarified during the past 10 years. However, in view of the numerous attempts at classification of the Precambrian (see e.g., Willmarth 1925), it is really surprising that any sense of unity has prevailed throughout the world and that there is even some basic agreement on terms.

Today we are in a position to set up major divisions on a two-fold basis of age and tectonic environment. The recognition of tectonic environment is, of course, a still-developing process of study of lithologies and rock associations together with the metamorphic history of a sequence. The present-day attitudes defining basement and cover rocks (see Fig. 10) essentially evolved in the six years prior to 1972 in extensive discussions, led by M.J. Rickard, K.A. Plumb and H.F. Doutch of Canberra, during the preparation of the Tectonic map of Australia published by the Geological Society of Australia (1971).

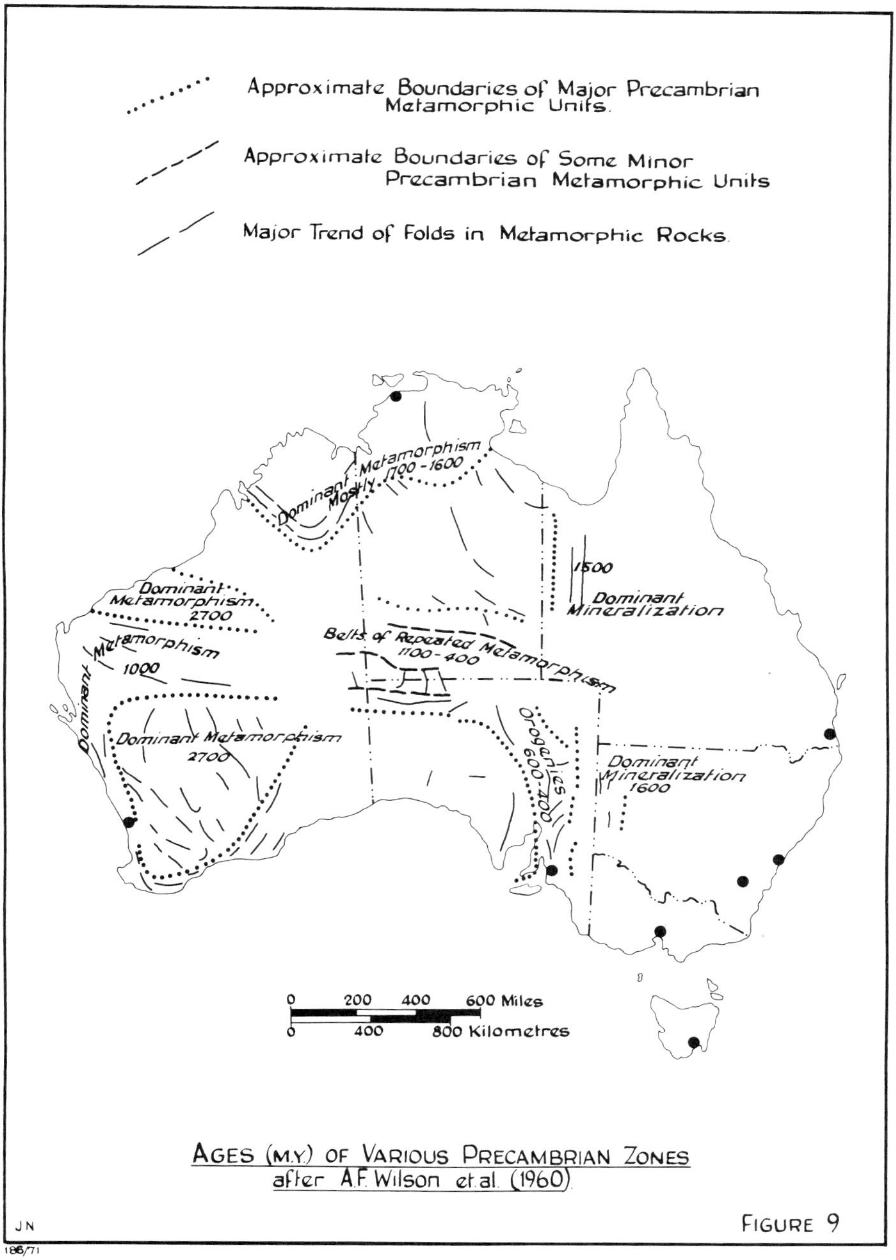

Figure 9. Ages in millions of years of various Precambrian zones (after A.F. Wilson *et al.*, 1960).

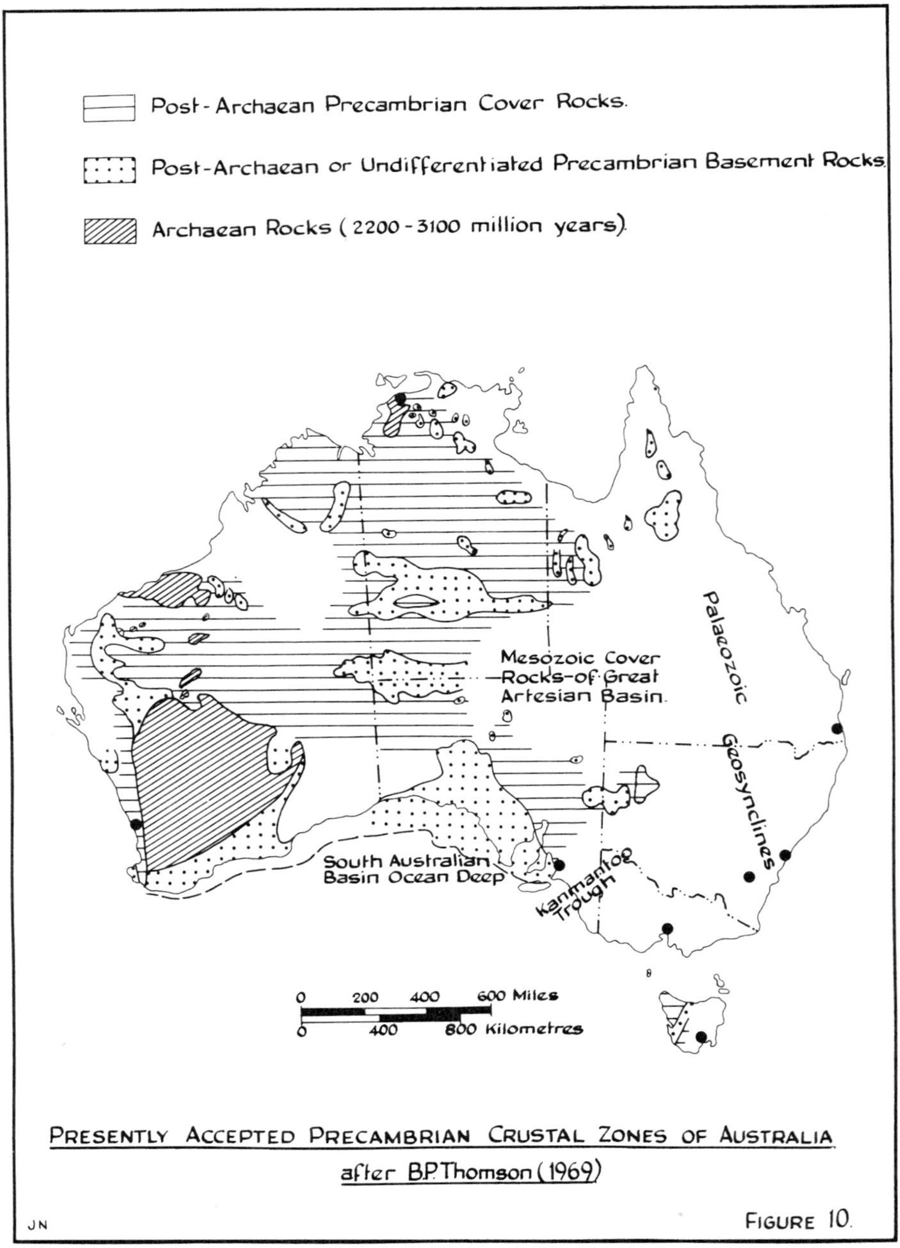

Figure 10. Presently accepted Precambrian crustal zones of Australia (after B.P. Thomson, 1969).

We are now concerned with what precisely the term Precambrian means, and how we define the limits of the Archaean, Proterozoic and particular subdivisions within these units, while Glaessner (1963), Webby (1973), and others have drawn attention to the problem of defining the base of the Cambrian succession in Australia. Crook (1966) pointed to the deficiencies in erecting a time-stratigraphic table for the Precambrian in which bodies of rock are referred, by means of radiometric dates, to periods defined by chosen points on the absolute time-scale' a method proposed by McDougall *et al.* (1965). Crook proposed instead that 'the boundaries of time-rock units be defined at selected reference points on the ground, and these reference points dated as accurately as possible by radiometric means'. This approach is preferred by Crook 'since: (a) it leads to a geological time-scale homogeneous in concept; (b) it implies that there is no difference in principle between Precambrian and Phanerozoic stratigraphy; and (c) the time-rock units so created are directly comparable with those that might be established by lithological, palaeontological, magnetometric, or climatic - lithological correlation'.

The discovery of very large ore deposits during the 1960s created intense interest in the Precambrian rocks of Australia, as is evident from the highly successful conference on Archaean Geology held in West Australia in 1970 (Glover, 1971). Perhaps of greater geological interest and future use will be a continuing study of the less-deformed Proterozoic rocks. In extent and character they offer perhaps a better opportunity for setting up useful systems and periods than elsewhere in the world.

In conclusion one interesting sidelight should be mentioned. This is the enduring opinion of the man in the street that 'Australia is the oldest continent'. It persists to the present day in numerous glossy 'coffee-table' books and even appears in more reputable literary circles. This idea has a long history in Australian and even British literature, based no doubt on the peculiar character of our present (often now nearly extinguished) fauna and flora and on the stark character of much of our landscape. No doubt this myth will endure while the geologist immerses himself in the mysteries of the Hypozoic, Pampalozoic, Azoic, Cryptozoic (Willmarth, 1925) or just plain Precambrian rocks.

ACKNOWLEDGEMENTS

I wish to thank Professor T.G. Vallance of the Department of Geology and Geophysics, University of Sydney for his helpful advice and criticism in preparing this paper. Mrs. Johann Nielsen, formerly of the same Department, also helped by her careful compilation of the figures from the original maps, several of which were made available by Fisher Library – the University of Sydney, and by the Geological Survey of New South Wales.

REFERENCES

Andrews, E.C., 1922, The Geology of the Broken Hill District: Memoirs of the Geological Survey of New South Wales, Geology 8, Government Printer, Sydney, 432 p., p. 195.

Browne, W.R., 1914, The Geology of the Cooma District – Part 1: Journal of the Proceedings of the Royal Society of New South Wales, v. 48, p. 172-222.

————, ed., 1950, Geology of the Commonwealth of Australia: London, 3 vols.

Burr, T., 1846, Remarks on the Geology and Mineralogy of South Australia: Andrew Murray, Adelaide, 32 p., p. 4.

Clarke, E. de C., 1923, The Pre-Cambrian System in West Australia: Journal of the Royal Society of Western Australia, v. 9, no. 2, p. 13-35.

Clarke, Rev. W.B., 1867, Remarks on the Sedimentary Formations of New South Wales, *in* Catalogue of the Natural and Industrial Products of New South Wales forwarded to the Paris Universal Exhibition of 1867, by the New South Wales Exhibition Commissioners: Government Printer, Sydney, p. 65-80.

Cotton, L.A., 1926, The Age of Certain Radium-bearing rocks in Australia: American Journal of Science, v. 12, July 1926, p. 41-46.

Crook, K.A.W., 1966, Principles of Precambrian Time Stratigraphy: Journal of the Geological Society of Australia, v. 13, no. 1, p. 195-202.

Curran, Rev. J.M., 1899, Geology of Sydney and The Blue Mountains, 2nd Edition: Sydney, Angus and Robertson, p. 347-348, p. 391.

David, T.W.E., 1908, Geological notes on Kosciusko, with special reference to evidence of glacial action: Proceedings of the Linnean Society of New South Wales, v. 33, p. 657-658.

——————, 1932, Explanatory notes to accompany a new geological map of the Commonwealth of Australia: The Commonwealth Council for Scientific and Industrial Research, Sydney, 117 p.

——————, 1950, *in* The Geology of the Commonwealth of Australia, Browne, W.R., ed., vol. 1, chapter 1: London, Arnold and Co., 3 vols. (1) p. 747. (2) p. 618.

David, T.W.E. and Howchin, W., 1896, Notes on the glacial features of the Inman Valley, Yankalilla and Cape Jervis District: Transactions of the Royal Society of South Australia, v. 21, p. 61-67.

David, T.W.E. and Tillyard, R.J., 1936, Memoir on Fossils of the late Precambrian (Newer Proterozoic) from the Adelaide Series, South Australia: Angus and Robertson in conjunction with the Royal Society of New South Wales, Sydney, 122 p.

Etheridge, Robert, Jr., 1890a, On Some Australian Species of the family Archaeocyathinae: Transactions of the Royal Society of South Australia, v. 13, p. 10.

——————, 1890b, Notes on the Palaeontology of Western Australia: Geological Magazine, new series, decade III, v. 7, p. 97.

Everitt, A.W., 1887, Continental Australia from the most recent information and materials supplied by the various Survey Departments of the various colonies. Constructed and engraved at the Department of Lands and Survey, Melbourne under the direction of A.J. Skene, Surveyor General of Victoria: William Slight Engraver. Geologically coloured by Arthur Everett under the direction of C.W. Langtree, Secretary for Mines and Water Supply and Chief Mining Surveyor for the Colony of Victoria. The Honourable Duncan Gillies, M.P., Minister of Mines, Melbourne.

Fleming, P.G., Day, R.W., Murray, G.C., Whitaker, W.G., 1974, Late Palaeozoic Invertebrate Macrofossils in the Neranleigh-Fernvale Beds: Queensland Government Mining Journal, v. LXXV, no. 869, March 1974, p. 105-107.

Geikie, A., 1882, Textbook of Geology: London, MacMillan and Co., 971 p., p. 640.

Glaessner, M.F., 1963, The base of the Cambrian: Journal of the Geological Society of Australia, v. 10, no. 1, p. 223-241.

Glover, J.E., ed., 1971, Symposium on Archaean Rocks held at Perth, May 23-26, 1970: Special Publications of the Geological Society of Australia, v. 3, p. 469.

Gregory, J.W., 1907, Stanford's Compendium of Geography: Australasia, v. 1. (Australia and New Zealand): London, Edward Stanford, 657 p., p. 411-412.

Gregory, J.W., and Barrett, B.H., 1931, General Stratigraphy: London, Methuen and Co. Ltd., 285 p., p. 44-45.

Hardman, E.T., 1884, Report on the Geology of the Kimberley District: Western Australia Parliamentary Paper no. 31, Perth, Government Printer.

Houston, B., 1965, Geological Map of the City of Brisbane: Geological Survey of Queensland, Publication no. 323.

Howchin, W., 1901, Preliminary note on the existence of glacial beds of Cambrian age in South Australia: Transactions of the Royal Society of South Australia, v. 25, p. 10-13.

————, 1904, The Geology of the Mount Lofty Ranges – Part L, The Coastal District: Transactions of the Royal Society of South Australia, v. 28, p. 253-280.

————, 1906, The Geology of the Mount Lofty Ranges – Part 2, The lower and basal beds of the Cambrian: Transactions of the Royal Society of South Australia, v. 30, p. 227-262.

Humphrey, A.W.H., 1803, *in* Watson, F., ed., Historical Records of Australia Series 1, v. 5, p. 589: Sydney, Government Printer.

Jukes, J.B., 1850, A Sketch of the Physical Structure of Australia, so far as it is at present known: London, T and W Boone, 95 p., p. 10-11.

————, 1862, The Students' Manual of Geology, 2nd Edition: Edinburgh, A and C Black, 764 p.

McDougall, I., Dunn, P.R., Compston, W., Webb, A.W., Richards, J.R. and Bofinger, V.M., 1965, Isotopic Age determinations on Precambrian Rocks of the Carpentaria region, Northern Territory, Australia: Journal of the Geological Society of Australia, v. 12, no. 1, p. 67-90.

Maitland, A. Gibb and Montgomery, A., 1912, The Geology and Mineral Industry of Western Australia: Bulletin Geological Survey of Western Australia, no. 50.

Mawson, D., 1912, Geological Investigations in the Broken Hill area: Memoirs of the Royal Society of South Australia, v. 2, no. 4, p. 208-319.

————, 1936, Progress in Knowledge of the Geology of South Australia: Transactions of the Royal Society of South Australia, v. 60, p. lvi-lxv.

Moore, E.S., 1916, Observations on the geology of the Broken Hill Lode, New South Wales: Economic Geology and Bulletin of the Society of Economic Geolgists, v. 11, p. 327-348.

Preiss, M.V., 1973, Palaeoecological interpretations of South Australian Precambrian Stromatolites: Journal of the Geological Society of Australia, v. 19, no. 4, p. 501-532.

Selwyn, A.R.C., 1860, Geological notes of a journey in South Australia from Cape Jervis to Mount Serle, South Australian Parliamentary Papers, no. 20, p. 1-2.

————, 1861, Geology of the Colony of Victoria: *in* Catalogue of the Victorian Exhibitions 1861: Clarson, Shallard, Melbourne, p. 175-191.

————, 1868, A Descriptive Catalogue of the Rock Specimens and minerals in the National Museum collected by the Geological Survey of Victoria with explanatory notes on their nature and mode of occurrence in place: Melbourne, Government Printer, 86 p.

Smyth, R.B., 1875, First Sketch of a Geological Map of Australia Including Tasmania: Scale 1:7,000,000, published by direction of the Hon. Wm. McLellan, M.P., Minister of Mines, November 1875, 2 sheets.

Sprigg, R.C., 1947, Early Cambrian (?) jellyfishes from the Flinders Ranges: Transactions of the Royal Society of South Australia, v. 71, no. 2, p. 212-224.

Strzelecki, P.E., 1845, Physical Description of New South Wales and Van Diemens Land: London, Longman, Brown, Green and Longmans, 462 p. p. 83.

Sussmilch C.A., 1911, An Introduction to the Geology of New South Wales: Government Printer, Sydney, 177 p., p. 8.

Tate, R., 1892, The Cambrian Fossils of South Australia: Transactions of the Royal Society of South Australia, v. 15, p. 183-189.

————————, 1893, (Presidential) Inaugural Address. Report of the Fifth Meeting, Australasian Association for the Advancement of Science, Adelaide, South Australia, 1893: Reports of Meetings, Australian Association for the Advancement of Science, Sydney, v. V, 691 p., p. 1-69.

Tectonic Map of Australia, 1962, Geological Society of Australia, Sydney, 1962.

Tenison Woods, Rev. J.E. 1858, Observations on Metamorphic Rocks in South Australia: Transactions of the Philosophical Institute of Victoria, v. ii, p. 168-176.

Tepper, O., 1879, Introduction to the Cliffs and Rocks at Ardrossan, Yorke's Peninsula: Transactions of the Philosophical Institute of Adelaide, v. 2 (1878-9), p. 71-79.

Thompson, B.P., 1969 *in* Parkin, L.W., ed., Handbook of South Australian Geology: Geological Survey of South Australia, 268 p., p. 19.

Wallace, A.R., 1893, Stanford's Compendium of Geography and Travel (New Issue) Australasia. Volume 1 (Australia and New Zealand): London, Edward Stanford, 505 p., p. 84-85.

Walpole, B.P. ed., 1962, Geological notes in explanation of the Tectonic Map of Australia: Compiled by the Tectonic Map Committee of the Geological Society of Australia, Bureau of Mineral Resources, Geology and Geophysics, Canberra, 72 p.

Webby, B.D., 1973, Trace Fossils from the Lintiss Vale Formation of New South Wales; a late Precambrian fauna: Search, Nov.-Dec. 1973, v. 4, nos. 11-12, p. 494-496.

Willmarth, M.G., 1925, The Geological Time Classification of the United States Geological Survey compared with other classifications: Bulletin United States Geological Survey, no. 769, 138 p.

Wilson, A.F., Compston, W. Jeffery, P.M., and Riley, G.H., 1960, Radioactive ages from the Precambrian Rocks in Australia: Journal of the Geological Society of Australia, v. 6, part 2, p. 179-195.

Woodward, H., 1884, Note on the remains of Trilobites from South Australia: Geological Magazine, new series, decade III, v. 1, 343 p.

————————, 1885, Mines, etc., east of Farina and Leigh's Creek: South Australian Parliamentary Paper, v. 40, no. 3.

History of Concepts in Precambrian Geology, edited by W.O. Kupsch and W.A.S. Sarjeant, Geological Association of Canada Special Paper 19, 1979

HISTORY OF THE STUDY OF THE 'PRE-CAMBRIAN' IN WALES

John Challinor
Broncastell, Capel Bangor, N. Aberystwyth, Cardiganshire, Wales, U.K.

ABSTRACT

The question of the recognition of Precambrian rocks is discussed; essentially it depends upon international agreement as to how the boundary should be defined and no hard-and-fast decision has been yet arrived at. The pioneer investigations of Sedgwick are discussed insofar as is permitted by the incomplete fashion in which they were published. The history of subsequent work in North Wales is thereafter outlined, especially attention being paid to the studies by Blake and Geikie on the rocks of the Padarn Ridge. Studies in Pembrokeshire are reviewed more briefly.

It is concluded that only the Monian rocks of Anglesey and southwestern Lleyn can truly be regarded as Precambrian and that the other so-called 'Pre-Cambrian' rocks – the Arvonian of northwestern Caernarvonshire and the Pebidian and Dimetian of Pembrokeshire – should provisionally be attributed to the Cambrian.

RESUME

Nous discutons la question de la reconnaissance des roches précambriennes, elle dépend essentiellement de l'accord international portant sur la question de savoir comment la limite devrait être définie et aucune décision ferme et définitive n'est encore intervenue. Les prèmieres investigations de Sedgwick sont discutées aussi loin que le permet la forme incomplète sous laquelle elles furent publiées. L'historique des travaux postérieurs en Galles du Nord est ensuite esquissé, une attention particulière étant accordée aux études de Blake et Geikie sur les roches de Padarn Ridge et celles de Greenly à Anglesey. Les études dans le Pembrokeshire sont revues plus brièvement.

Nous en concluons que seules les roches du Monien d'Anglesey et du Sud-Ouest de Lleyn peuvent vraiment être considérées comme précambriennes, et que les autres roches soi-disant précambriennes – l'Arvonien du Nord-Ouest du Caernarvonshire et le Pebidien et le Dimetien du Pembrokeshire – devraient être attribuées à juste titre au Cambrien.

AREAS OF OUTCROP

The rocks which are currently classified as Precambrian in Wales (*Cambria*, the Latin name, derived from the Celtic *Cymry*) occur in relatively small outcrops in the extreme north and the extreme south-west of the Principality. It will be convenient to group these areas of outcrop into three regions: (1) the island county of Anglesey and the Lleyn peninsula of Caernarvonshire, (2) north-western Caernarvonshire, and (3) western Pembrokeshire.

(1) Rather more than half the surface of Anglesey is occupied by the *Monian* group of rocks. These are more or less highly metamorphosed sedimentary, with some metamorphosed volcanic, rocks (schists and gneisses) and intruded granites. Similar rocks occur along a strip of land bordering the coast in the west of Lleyn. (*Mona*, the Latin name of Anglesey, from the Celtin *Môn*; the name Monian was first used by J.F. Blake in 1887 and 1888a.)

(2) There are two areas of outcrop in north-western Caernarvonshire, each elongated in a north-east to south-west direction. One forms a low-lying ridge about 13 km long, extending from Bangor to Caernarvon parallel to and near the Menai Strait (which separates Anglesey from the Caernarvonshire mainland); this has been named the 'Bangor ridge'. The other, farther inland, forms higher ground, about 22 km long, extending from Bethesda to Penygroes, passing through the north-western part of the lake of Llyn Padarn; this has been named the 'Padarn ridge'. (These two names will be used throughout the following account for the two outcrops of the *Arvonian* rocks.) The rocks of both outcrops have been described as porphyries, rhyolites, felsites, and tuffs: they are now interpreted as, largely, ignimbrites. A granitic rock near Caernarvon is included. (*Arvon*, the Celtic name for the region between, roughly, the Menai Strait and the higher mountains of Snowdonia. The name *Arvonian* was originally used by Henry Hicks in 1878 and 1879; he included in it some of his Pembrokeshire rocks. It was redefined and used exclusively for the occurences in Caernarvonshire, by Edward Greenly in 1930 and 1945.)

(3) In the extreme western promontory of Pembrokeshire are several small outcrops of the *Pebidian* and *Dimetian* rocks. The former are volcanic extrusions, the latter granitic intrusions. (*Dimetia*, or *Demetia*, the Latin name for, roughly, what is now Pembrokeshire; *Pebidiauc*, or *Pepitiauc*, the Celtic name for the district about St. David's. The names *Pebidian* and *Dimetian* were first used by Hicks in 1877.)

The outcrops and characters of all these rocks are described, in detail or in summary, in a number of more recent works. We give these (which are in the list at the end of this paper) in chronological order, with numbers in brackets to indicate the regions with which they deal: 1908, Green (3); 1912, Thomas and Jones (3); 1919, Greenly (1); 1928, Matley (1); 1930, Green (3); 1945, Greenly (2); 1961, George (1, 2); 1965, Anderson (1, 2, 3); 1969, Shackleton (1, 2); 1970, George (3). In addition, there are the three recent textbooks covering the geology of the British Isles: 1967, Rayner; 1968, Anderson and Owen; 1969, Bennison and Wright.

It should be mentioned that the counties of Wales have recently been regrouped for offical purposes but, for other purposes, they will doubtless continue to be referred to by their original names, as in the present case.

THE GENERAL BASE OF THE CAMBRIAN

The recognition of Precambrian rocks naturally depends on the recognition of the base of the Cambrian. The questions as to how this base might be defined and how, when defined, it could be recognized in various regions, have often been considered; among the more recent general discussions those by Glaessner (1963) and Harland (1974) are especially important. For Britain the matter has been examined by Stubblefield (1956). These questions remain unsettled. There are, however, some general considerations that may be emphasized, as they apply with particular force to the relevant rocks in Wales.

Empirically and practically, though not logically, the one most important division of the whole geological column is this one, the base of the Cambrian; but it is peculiarly elusive. In 1930 Chadwick proposed the name *Phanerozoic* for all the rocks, and all the time, from Cambrian to Present, and *Cryptozoic* for everything below and before the Cambrian; these names referred to the fact that, in the former, evidences of the forms of life are 'plainly visible' while in the latter they are 'hidden'. Thus on the William Smithian principle – 'strata identified by organized fossils' – the rocks of the Phanerozoic division can be set in order and correlated, while those of the Cryptozoic remain unsorted. This division results from the exceedingly curious fact that when the fossil succession, representing the history of life, is traced downwards through the rocks, and thus backwards in time, it comes to a sudden and obviously inconclusive 'loose end'.

The first definite and plainly visible fauna to appear in the rock-succession, starting now from below upwards, is that characterized by the *Olenellus* group of trilobites; the presence of this fauna is taken as determining the Lower Cambrian series. Thus it seems, at first sight, that the base of the Cambrian might be defined as the stratigraphical level at which this *Olenellus* fauna first appears; but here there are several difficulties, of which three may be stated briefly:

(1) Despite a number of proposals for local and regional time-divisions depending on fossils of animal life, particularly trilobites, no universally satisfactory zonal subdivision of this series has so far been established, so that what constitutes the lowest (earliest) sub-fauna of the *Olenellus* fauna is unknown other than locally.

(2) There are in some regions strata containing faunas determining a Lower Cambrian, a Middle Cambrian, an Upper Cambrian, or even a later, age which are conformably underlain by a downward extension of unfossiliferous strata. As, therefore, fossils first appear in the Cambrian succession at various horizons in various regions, it would not be reasonable to expect the earliest appearance of the *Olenellus* fauna, where this happens at all, to occur everywhere at the same time, because the vagaries and accidents of the conditions necessary for fossilization naturally account for the presence or absence of fossils in the rocks. (See the cases of the Llanberis Slates, and the Cambrian of South Wales, as detailed in later sections of this paper.)

(3) Beds with the *Olenellus* fauna may occur immediately, or at some remove, overlying a surface of unconformity, the beds below the unconformity being unfossiliferous. In that case the local base of the Cambrian is usually taken at the unconformity by reason of these two separate features, the one structural, the other palaeontological, though neither is in itself a satisfactory stratigraphical boundary. Such a separation between the 'Cambrian' and 'Pre-Cambrian' in any district must be merely convenient and provisional. To equate the strata immediately overlying such unconformities, in separated and perhaps far-distant regions, as being isochronous, and thus as providing a universal base to the Cambrian, is quite impermissible. But it has been done, and was advocated by no less an authority than C.D. Walcott. On the common-sense principle of a general 'uniformitarianism' throughout geological time it would indeed be surprising to find anywhere in the stratal succession

(Cryptozoic or Phanerozoic) a universal unconformity immediately above which all the rocks were isochronous or even approximately so. In fact, any such unlikely structural state at the beginning of Cambrian times has been disproved, particularly in North America. [See especially the two papers, by Wheeler (1947) and Snyder (1947)].

The question as to whether a certain rock-group is Precambrian or Cambrian seems usually to be posed on the presumption that there is some criterion, some outside standard of reference, by which this question can be judged or settled. Whatever stratigraphical horizon is taken as the base of the Cambrian must necessarily also be the top of the Precambrian. An horizon defined at one specified locality and thenceforward requiring its time-equivalent to be sought in all other places where the question of assignment arises, is definitely very desirable. Unfortunately, no horizon has yet been defined in a continuous sequence in any part of the world which has been universally agreed on as dividing what is to be called Pre-Cambrian below from the Cambrian above.

SEDGWICK'S LOWEST ROCKS

In this paper we shall be chiefly concerned with the second of our three areas, that of north-western Caernarvonshire, as here lies the core of the problem of the 'Pre-Cambrian' in Wales.

Nothing was known about the geology of this area until Adam Sedgwick (Fig. 1) began his researches:

> "I commenced my task in Wales in 1831, accompanied for a short time by my friend Charles Darwin – a name well known and honoured in the whole world of science – but other engagements soon drew him away from North Wales. We laid down the northern boundary of the Carboniferous Limestone together, and I at first purposed to examine the rocks in descending order. But I found in Denbighshire the same interrupted broken masses of Old Red Sandstone which I had so many times noticed between the Carboniferous and the Silurian and Cambrian rocks in the North of England. I therefore despaired of establishing a good base to work upon in that district; and after a short examination of the rocks on both side of the Menai Straits, I resolved to fix a provisional base-line on the Caernarvon side of the Straits" (Sedgwick, 1863, p. xv; and see p. xvi and xvii).

In a letter to Murchison, dated from Llanllyfni near Caernarvon, 13 September 1831, Sedgwick said that he "finally entered North Wales on the 5th August" (Clark and Hughes, 1890, v. 1, p. 378).

Murchison had begun his 'Silurian' researches in June; Sedgwick had been invited to accompany him, but decided to work out these old rocks independently in a separate part of the Wales and Welsh Borderland region. It is interesting to note that Sedgwick had intended to work downwards from the stratigraphical datum of the base of the Old Red Sandstone, as Murchison did with such success; but he encountered the great unconformity at the base of the Carboniferous, as he had done in the Lake District of northern England. (His 'Old Red Sandstone' in both regions is in fact basal Carboniferous.)

Sedgwick's whole geological life was one of fits and starts, acute perception and deplorable failure to see a matter through; and, in particular, a fatal negligence as

Figure 1. The Rev. Adam Sedgwick when aged 82. From a crayon drawing by Lowes Dickinson.

to publication.

After finding that he could not work downwards through the rocks, from the known to the unknown, he decided that he must work upwards from the lowest rocks he could find. In the letter to Murchison mentioned above he writes:

"I spent some days in the Isle of Anglesea in the hopes of learning my lesson for Snowdonia. Henslow's paper is excellent but the lesson worth next to nothing; for Anglesea is almost as distinct in structure from Snowdonia, as if they had been separated by the Atlantic sea rather than the straits of Menai. I have now been at real hard work, cracking the rocks of Caernarvonshire for rather more than three weeks; and can report progress. . . . I have no room for a Snowdonian transverse section. The structure is on the whole regular; I have nearly completed one *base line* to work upon; the rest must be done by traverses" (the italics are Sedgwick's).

Thus by a sure instinct, guided by Henslow's paper (1822), Sedgwick went straight to the lowest (oldest) rocks that were visible in this region of the western part of southern Britain. In these three weeks he had made out something of the general structure of Snowdonia and had fixed a base-line for his researches into the succession of those rocks which in 1835 he was to name 'Cambrian' (Sedgwick and Murchison, 1836). From his reminiscent account (1873) and his contemporary letter (1831) it may be said that this base-line was the south-eastern shore of the Menai Strait. From the brief and inadequate reports of the highly important papers that Sedgwick read to the Geological Society in 1838 and 1843, and from the paper printed in full in 1847, he evidently realized from the start that the Anglesey rocks (with which he rightly came to include a similar strip in the south-western part of the Lleyn Peninsula of Caernarvonshire) were older than any of those forming the country between the Menai Strait and the Snowdon mountains.

In his second season (1832), Sedgwick resumed his work in North Wales towards the end of May, and on July 23 wrote a long letter to Murchison describing an itinerary through the county of Merioneth and ending with the words "I will fill up this sheet with one or two sections for your amusement" (Clark and Hughes, 1890, v. 1, p. 394). The three very rough sections on the folded-sheet inserted opposite this letter in Clark and Hughes's biography seem to be the only ones of Sedgwick's ever to be published. Such sections were displayed at several meetings of the Cambridge Philosophical Society and the Geological Society during the period 1831 to 1838 and must have formed the most important part of Sedgwick's expositions. It is extraordinary that there should have been such carelessness that nothing of these sections was published at the time; it was indeed a very lucky chance that the scrawled sketches, sent with the letter to Murchison, survived.

For our present purpose we note on this sheet 'Section 1', from Caernarvon to Bala, where at the extreme north-west end we have, in ascending order, (1) 'porphyry of the Menai' (the Bangor ridge), (2) 'slate and porphyry', (3) 'porphyry' (the Padarn ridge). Reports of the Geological Society meetings of May 1838 and June 1843 indicate the same succession. In that of 1838 we have: "The rocks are grouped in the ascending order as follows. (1) Chlorite slate, quartz rock, and mica slate of Anglesea and SW Caernarvonshire . . . (2) The old slate series of Caernarvonshire and Merionethshire, alternating indefinitely with bands of porphyry and felspare rock. . . ." He thus gives the two groups which are, in modern nomenclature, (1) Monian and (2) Arvonian and Llanberis Slates combined. Sedgwick himself writes in his paper in the *Quarterly Journal* for 1847: "In short the slates near Bangor and Caernarvon are among the very oldest rocks of North Wales; of course excepting the crystalline and hypozoic groups of Anglesea and the south-western shore of Caernarvonshire, which I do not at present wish to notice". Finally, in his 1852 paper, Sedgwick gives a table of which the following is the lower part:

Cambrian Series	Lower	Festiniog group	Arenig slates and porphyry Tremadoc slates Lingula flags
		Bangor group	Harlech grits Llanberis slates
Metamorphic Granite			

The question arises: what is Sedgwick including in his 'Llanberis slates'; that is, where is he drawing the base of his Cambrian? From the sketch-section of 1882 it is

Figure 2. Andrew Crombie Ramsay. From a photograph by D. Hains.

clear that he takes the 'slates' at least as far down as the Bangor ridge ('porphyry of the Menai'), and as he is including the upper 'porphyry' (Padarn ridge) it is reasonable to suppose that he also includes the lower (Bangor ridge). The report of the 1838 meeting confirms this view. The use of the name 'Bangor group' is a further confirmation. The beds between the two Arvonian ridges, which Sedgwick was including in his 'Llanberis slates', are all now taken to be of Ordovician (Arenig) age; that is, they, belong to the lower part of Sedgwick's 'Arenig slates and porphyry' series. In Caernarvonshire and Anglesey the Arenig rocks overstep the several divisions of the lower rocks,. As to the Anglesey and south-west Lleyn 'metamorphic' rocks (and presumably the associated 'granite'), these are placed definitely below the Cambrian.

Sedgwick, being concerned solely with Lower Palaeozoic rocks, altogether ignores the Carboniferous outcrop along the south-east side of the Menai Strait.

THE GEOLOGICAL SURVEY AND OTHERS IN NORTH WALES

The work of the Geological Survey under Ramsay (Fig. 2) in North Wales was begun about 1849 and culminated in the publication of the large memoir in 1866, of which a second edition, very little altered in essentials, was issued in 1881. As regards Anglesey, this work has been summarized by Greenly (1919, v. 1, p. 6-7), with frequent reference to the revealing letters and diary given in Geikie's *Memoir*

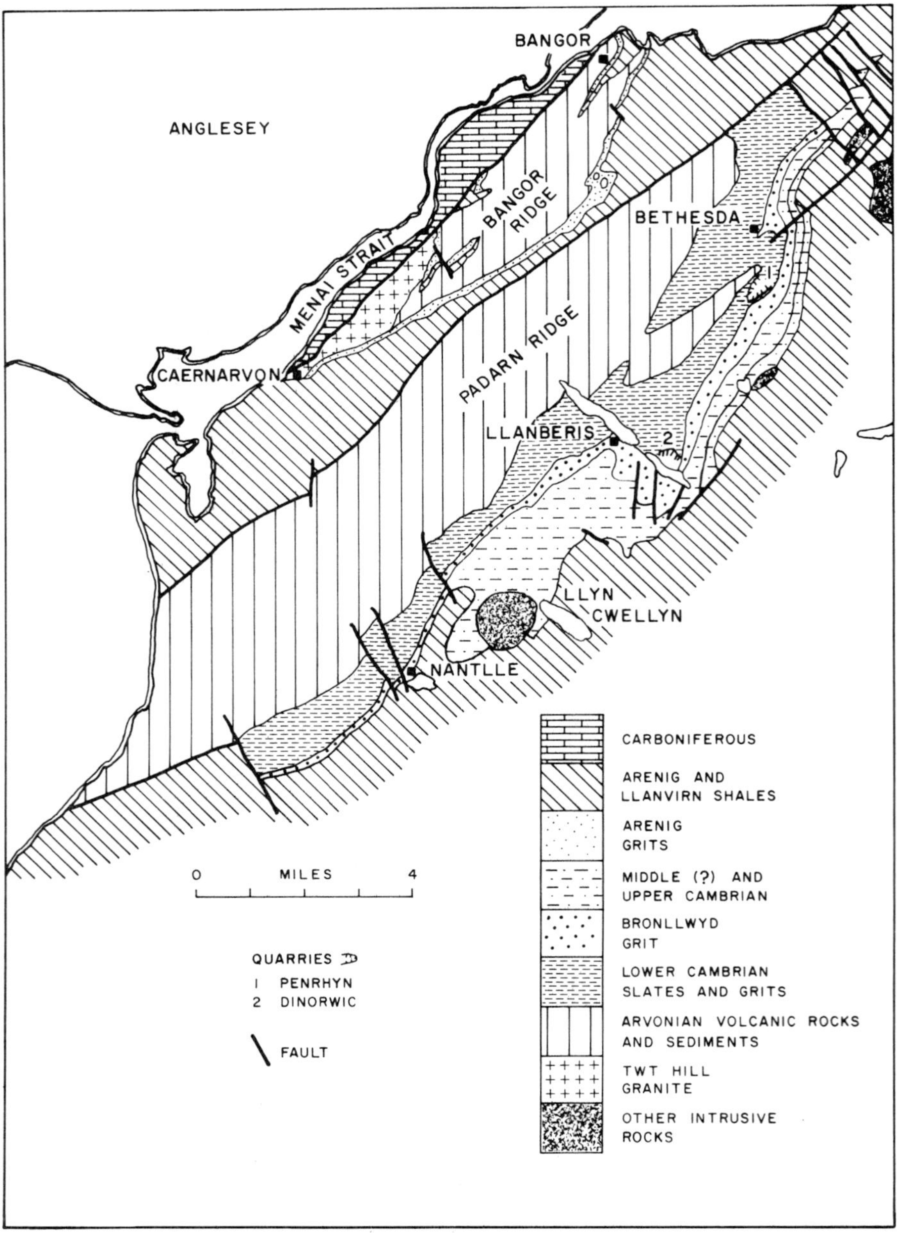

Figure 3. Geological Map of the area south of Bangor, Wales.

of Ramsay (1895). In a letter to Aveline (October, 1849), Ramsay remarks: "We had a short rap at Anglesey at very old rocks – older than the Cambrian" (Geikie, p. 154). However, by 1853, Ramsay was considering the Anglesey schists to be metamorphosed Cambrian and this assignment was firmly stated in the Survey memoir of 1866 and repeated unaltered in 1881. In the accompanying map covering the whole of Wales and the Welsh Borderland, on the scale of ten miles to the inch, there are no rocks shown as being older than the Cambrian.

This map is a reduction from the one-inch sheets of north-west Caernarvonshire (1850, 1852). The rocks of the Bangor ridge and the south-easterly part of what is at the present day included in the Padarn ridge are shown as 'intrusive felspathic porphyry etc; chiefly of Lower Silurian age'. Between these two outcrops the ground is shown as occupied partly by 'Cambrian' outcrop elliptically enclosing the Padarn ridge. This arrangement is shown on sections 2 and 3 which are reductions from the six-inch 'horizontal sections' (1853 and 1854, revised 1880).

It should be noted that the only rocks included in the 'Cambrian' at this time by the Survey were those now placed as Lower Cambrian. The Survey 'Lower Silurian' comprised all the rocks between the 'Lingula beds' (and the 'Menevian beds' inserted in the 1881 edition) and the 'Lower Llandovery beds'; that is, all the rocks ranging, in modern nomenclature, from the Middle Cambrian, through the Ordovician, to the lowest division of the Silurian.

An interesting idea is detailed with considerable persuasion in the text of the Memoir (1866, p. 140-142; 1881, p. 173-176). To quote one graphic passage:

> "Beneath the grits rise the bluish purple and purple slates of the larger slate quarries of Penrhyn and Llanberis, and below these are other thinner bands of slate, grit, and conglomerate, the lowest of which, by Llyn Padarn and elsewhere, *passes by insensible gradations into* the great mass of quartz-porphyry that for 13 miles strikes along the midst of the Cambrian strata, between Llanllyfni and where St. Ann's chapel, now buried beneath slate quarry rubbish, once stood near the river Ogwen. So closely does the matrix of the altered rock resemble the adjoining typical porphyry in colour, texture and even in porphyritic character, and by such insensible gradations do they melt into each other, that the suspicion or rather the conviction constantly recurs to the mind that the solid porphyry itself is nothing but the result of the alteration of the stratified masses carried a stage further than the state of porcellanite into the condition of that kind of absolute fusion that in many other regions seem to have resulted in the formation of granites, syenites, and other rocks, commonly called intrusive" (1881, p. 173. The italics are Ramsay's. The passage in the 1866 edition, p. 140, is the same, except for an insignificant word or two.).

This is clearly a reference to the process now called 'granitization' which is here taking the form of what might be called 'porphyritization'. Ramsay applies the same idea to the rocks of the Bangor ridge (1886, p. 165; 1881, p. 199); indeed he interprets the aplitic granite of Twt Hill at Caernarvon as due to this process of granitization.

It may be interpolated that the Llanberis Slates are mineralogically reconstituted by metamorphism. "Any highly cleaved slates, such as those of North Wales, composed essentially of new minerals with a general parallel arrangement, may be regarded as crystalline schists of very fine texture" (Harker, 1950, p. 162). The group is about 5,000 feet in stratigraphical thickness. It is completely unfossiliferous except in the uppermost beds, where the trilobite *Pseudatops* [*Concoryphe*] *viola* occurs, regarded as indicating a late Lower Cambrian age for those beds (Howell and Stubblefield, 1950).

Blake (1888b) and, especially, Geikie (1891, 1897) maintained that the conglomerates overlying the volcanic rocks of the Padarn ridge did not indicate any significant unconformity and that all the rocks of the two ridges should be classed as Cambrian.

> "The quartz-porphyry, felsite or rhyolite of Llyn Padarn is not a metamorphic but an eruptive rock, as has been demonstrated by Professors Hughes and Bonney. . . . But most of the critics of the work of the Geological Survey have proceeded to certain further deductions. They have maintained that the presence of fragments of the poryhyry in the overlying conglomerate marks an unconformability between the two rocks, that the conglomerate shows the base of the Cambrian system, and that the porphyry is therefore pre-Cambrian. These assertions and inferences do not seem to me to be warranted. They have already, in my judgement, been disproved by Mr. Blake in an excellent memoir read before the Society in 1888, with the main conclusions of which I agree. Mr. Blake shows, in my opinion conclusively, that there is no break in the Cambrian series, that the various porphyries and their accompaniments are parts of that series, and that there is no certain proof of the existence of any pre-Cambrian rocks in the whole district" (Geikie, 1891, p. 92-93).

With regard to the actual character of the rocks of the two ridges; Ramsay's particular view of their metamorphic character was not supported by later observers. Several of these, especially Bonney (1879), showed that the rocks were truly of volcanic origin.

The Geological Survey quarter-inch map of North Wales (1910) gives the outcrops, over the area we are considering, as shown on the map accompanying the 1881 memoir except that the north end of the Cambrian outcrop surrounding the Padarn ridge is extended to Bangor, obliterating the 'Lower Silurian' of that part in the 1881 map and leaving only a small tongue south and east of Caernarvon. This tongue is now, on this map, labelled Ordovician. (Charles Lapworth, in 1879, had proposed and defined the Ordovician system.) The 'schists and associated rocks (Anglesey)' together with the 'purple slates, conglomerates etc. (Longmynd)' (Shropshire) are tabulated as 'Longmyndian etc.' and placed below the Cambrian, but the name 'Pre-Cambrian' is not used. The rocks of the Bangor and Padarn ridges are placed in a quite separate category of 'igneous rocks'.

A second edition of the 1910 quarter-inch map was published in 1930. This map reverts to the outcrops shown in the 1881 map, all Ramsay's 'Lower Silurian' being labelled Ordovician. The name 'Pre-Cambrian' is now used in the index for the Angelsey schists (etc.) and for the 'Longmyndian' and 'Uriconian' of Shropshire, the rocks of the Bangor and Padarn ridges being separately tabulated as igneous rocks. There is an unaltered reprint of this 1930 map, dated 1968; this unfortunately does not incorporate the important revision of the mapping of the Padarn ridge and its immediate surroundings.

This revision was carried out during the period 1935 to 1937 by David Williams, and I am greatly obliged to Professor Williams for giving me some details. He was at the time collaborating with Edward Greenly. Greenly read his papers on 'The Ordovician rocks of Arvon' in 1943 and 1944 respectively; his sketch-maps show the boundary of the Arvonian of the Padarn ridge extended to the north-west, but do not otherwise include the Padarn ridge outcrop. (Both these papers were published in 1945.) Greenly, while expressing some doubts, was inclined to accept the interpretation that there was an unconformity between the Llanberis Slates and the Arvonian and to assign the Arvonian to the Precambrian; however, at the reading of Greenly's second paper, Williams said in the discussion that, during his mapping of the country north-east of Llyn Padarn (that is, on the south-east side of Padarn ridge), he had been unable to draw a line of demarcation between the Arvonian and Cambrian formations because of the gradational passage between them, this gradation being chiefly a matter of repeated alternations and intercalations of volcanic and sedimentary rock-types. He felt there was, notwithstanding Greenly's opinion, great doubt as to whether the Arvonian as a whole should be separated structurally from the overlying Cambrian and therefore a doubt as to whether it should be called Precambrian. He was thus confirming the opinions of Blake and Geikie in the nineteenth century. In the Geologists' Association *Guide to Snowdonia* (Williams and Ramsay, 1968) the 'Arvonian rhyolite' is placed as '?Pre-Cambrian'.

Williams's maps, which he tells me were far from complete, were never published; but his and Greenly's findings were incorporated in the ten-mile map of Great Britain (1948). The Padarn ridge outcrop is widened to the north-west, in accordance with Greenly's published sketch-map (1945), absorbing what had been called 'Cambrian' on that side, and leaving a narrow outcrop of Ordovician sedimentaries separating it from the Bangor ridge. The rocks of these two ridges (that is, the Arvonian rocks) are included among those listed in the index as 'rhyolitic and trachytic lavas; tuffs, ashes, and agglomerates', with 'granite' at Caernarvon. These, together with the Anglesey schists and granite, are placed as 'Pre-Cambrian'.

In the tectonic map (1966) the rocks of the ridges are shown as 'acid and intermediate rocks folded by Pre-Cambrian movements'. The Padarn ridge is shown on the map as an anticlinal structure though this is not apparent in the accompanying section.

The most recent work is the illuminating essay by Wood (1969), in which a detailed map and discussion is presented of the critical outcrops of the Arvonian and the Llanberis Slates groups just north of Llyn Padarn. This confirms the view that there is no significant structural break between these two groups.

PEMBROKESHIRE

The 'Pre-Cambrian' rocks of the main district where outcrops occur in south-western Pembrokeshire – the St. David's-Hayscastle district, north of St. Bride's Bay – were mapped by the Geological Survey. The first edition of the one-inch geological map was published in 1845, and a later one (and, even now, still the latest one) in 1857. These rocks are indicated as intrusive granites and porphyries of 'Lower Silurian' age. During the period 1871 to 1886 the district was carefully examined by

Henry Hicks, the results being published in a series of papers in the *Quarterly Journal of the Geological Society* during these years. Hicks maintained that a conformable series of sedimentary rocks unconformably overlay igneous rocks which were partly intrusive (Dimetian) and partly extrusive (Pebidian). He drew the base of the Cambrian at his unconformity, thus relegating all the igneous rocks to the Precambrian.

Meanwhile, in 1883, Archibald Geikie wrote a paper in the *Quarterly Journal* 'On the supposed Pre-Cambrian rocks of St. David's' in which he denied the existence of any such unconformity and stoutly supported the view of the Geological Survey. He maintained this opinion in his later reviews of volcanic activity in Britain (1891, 1897). In subsequent papers, particularly several by Green (e.g., 1908, 1930) and that by Thomas and Jones (1912), it was maintained that there was some unconformity between the 'Cambrian' and the 'Pre-Cambrian', but the published evidence appears slight. The series of rocks underlying the rocks with a Middle Cambrian fauna are reasonably assigned to the Lower Cambrian; but no undoubted fossils directly indicating a Lower Cambrian age have been found in them.

It thus appears that here, in Pembrokeshire, we have very much the same state of affairs as we have found in Caernarvonshire.

CONCLUSION

The conclusion at which the present thesis arrives is, therefore, that the Monian, of Anglesey and south-western Lleyn, is the only rock-group in Wales to be placed as Precambrian, the other so-called 'Pre-Cambrian' rocks – the Arvonian of north-west Caernarvonshire, and the Pebidian and Dimetian of Pembrokeshire – being placed in the Cambrian. These assignments are advocated on both historical and practical grounds.

The whole question has been neglected for many years because it has been prematurely supposed to have been solved. An examination of the history of such matters is thus not merely interesting, purely as history, but may reveal substantial flaws, or at least doubts, about our present assumptions. The study of the history of geology is thus vital to the prosecution of knowledge at the present day.

Note on the Usage of the Term Pre-Cambrian and Some Other Related Terms

The terms, forming a set, of which the following are examples – Cambrian, Carboniferous, Cretaceous (all with capital initial letters) – are adjectival in form but are used as either adjectives or nouns. They refer primarily to particular groups of rocks, the stratigraphical systems, but refer also to the periods of time during which the systems were respectively accumulated.

The term Precambrian is, in its primary sense, obviously a time adjective meaning 'before the Cambrian', and in the form Precambrian that sense is the sole, exclusive meaning. In practice, however, in the form Pre-Cambrian or Precambrian, it refers not only to a portion of time but also to a material entity of rocks: all those lying below the Cambrian in the stratigraphical table. A more direct term for Precambrian rocks is Infra-Cambrian, though this usually refers to rock-series immediately below what is taken to be the base of the Cambrian.

The divisions between the particular systems and between the corresponding particular periods are defined arbitrarily. The Precambrian, as a course of time, is now known to have been some seven times as long as all the later periods put together. As a rock-succession it has an indeterminate base. If there were sufficient knowledge about the rocks formed during Precambrian time, and if they could be correlated throughout the world, no doubt Precambrian rocks and Pre-Cambrian time would be divided into generally applicable particular systems and periods.

It is found that as a matter of convenience, and purely on the basis of the character of the material evidence of the passage of geological time, geological history falls into two portions of time, the Precambrian and the post-Precambrian. These have been named the Cryptozoic (hidden evidence of life) and the Phanerozoic (clear evidence of life) portions of time. These terms, proposed in 1930, have recently been revived and have immediately been carried over into the material sphere to apply to rock as well as to time (as in the first few sentences of the Introduction to the Geological Society of London's volume *The Phanerozoic Time-scale*).

The long-familiar terms Palaeozoic (or Paleozoic), Mesozoic, and Cainozoic (of the ages of ancient, middle and recent life respectively) are, again, primarily time-terms, applied to the successive eras of Phanerozoic time. The way in which all these terms we are here considering are used for rock and time interchangeably (and as adjectives or nouns) is shown in the terms Lower Palaeozoic and Upper Palaeozoic, each constructed half of a word referring purely to rock-position and half of a word referring primarily to a portion of time. These two particular compound names are, however, more properly and usually applied to rock rather than to time.

(I am indebted to Mr. Anthony Wyatt for suggesting a look at this matter.)

ACKNOWLEDGEMENTS

Subsequent to the presentation of this paper in Montreal, the sections dealing with North Wales were published, under the title "The 'Pre-Cambrian' in Cambria", in the *Geological Magazine* (1976); they are here presented in a modified and updated form.

The two portraits included here are reproduced with the permission of the Publications Committee of the Geological Society of London.

REFERENCES

Anderson, J.G.C., 1965, The Precambrian of the British Isles :*in* Rankama, K., ed., 1965, The Precambrian, Vol 2: New York, Interscience Publishers, p. 25-111.

Anderson, J.G.C. and Owen, T.R., 1968, The structure of the British Isles: Oxford University Press, 162 p.

Bennison, G.M. and Wright, A.E., 1969, The Geological History of the British Isles: London, Edward Arnold, 406 p.

Blake, J.., 1887, Introduction to the Monian system of rocks: British Association for the Advancement of Science, Report of the 56th Meeting, Birmingham 1886, p. 669.

Blake, J.F., 1887, Introduction to the Monian system of rocks: British Association for the Advancement of Science, Report of the 56th Meeting, Birmingham 1886, p. 669.

————————, 1888a, On the Monian system of rocks: Geological Society of London, Quarterly Journal, v. 44, p. 463-547.

————————, 1888b, The Cambrian and associated rocks in north-west Caernarvonshire: Geological Society of London, Quarterly Journal, v. 44, p. 271-290.

Bonney, T.G., 1879, The quartz-felsite and associated rocks at the base of the Cambrian

series in north-western Caernarvonshire: Geoogical Society of London, Quarterly Journal, v. 35, p. 309-320.
Chadwick, G.H., 1930, Subdivision of geologic time: Geological Society of America, Bulletin, v. 41, p. 47-48.
Challinor, J., 1976, The 'Precambrian' in Cambria: Geological Magazine, v. 113, p. 440-456.
Clark, J.W. and Hughes, T.M., 1890, The life and letters of Adam Sedgwick: Cambridge University Press, 2 vols., 539 p.; 640 p.
Cox, A.H. *et al.*, 1930, The Geology of the St. David's district, Pembrokeshire: Geologists' Association, London, Proceedings, v. 41, p. 241-273.
Geikie, A., 1883, The supposed Pre-Cambrian rocks of St. David's: Geological Society of London, Quarterly Journal, v. 39, p. 261-333.
——————, 1891, History of volcanic action in the British Isles from the Devonian to the Tertiary (Anniversary Address of the President): Geological Society of London, Proceedings, p. 63-162.
——————, 1895, Memoir of Sir Andrew Crombie Ramsay: London, Mcmillan, 397 p.
——————, 1897, The ancient volcanoes of Great Britain: 2 vols., London, Macmillan, 447 p; 492 p.
George, T.N., 1961, British regional geology: North Wales, 3rd edition. (Based on previous editions by B. Smith and T.N. George, 1935, 1948): London, H. M. Stationery Office, 96 p.
——————, 1970, British regional geology: South Wales, 3rd edition. (Based on previous editions by J. Pringle and T.N. George, 1937, 1948): London, H. M. Stationery Office, 155 p.
Geological Survey maps and sections (separately published):
Maps on the scale of 1 inch to 1 mile. Old series.
(Pembrokeshire) Sheet 40 (1845, revised 1857). (Caernarvonshire) Sheets 75NW (1850), 78SW (1852), 78SE (1852).
Sections on the scale of 6 inches to 1 mile.
(Caernarvonshire) Sheets 28 (1853, revised 1880), 31 (1854, revised 1880).
Maps on the scale of ¼ inch to 1 mile. New series.
North Wales Sheet 9 and 10 (1910, revised 1930, reprinted 1968).
Map on the scale of 10 miles to 1 inch.
Great Britain (1948). Sheet 2.
Map on the scale of 25 miles to 1 inch.
Tectonic map of Great Britain and Northern Ireland (1966).
Glaessner, M.F., 1963, The base of the Cambrian: Geological Society of Australia Journal, v. 10, p. 223-241.
Green, J.F.N., 1908, The geological structure of the St. David's area (Pembrokeshire): Geological Society of London, Quarterly Journal, v. 64, p. 363-383.
——————, 1930, Pre-Cambrian rocks: *in* Cox *et all.*, The geology of the St. David's district, Pembrokeshire: Proceedings of the Geologists' Association, London, v. 41, p. 247-250.
Greenly, E., 1919, The geology of Anglesey: Memoirs of the Geological Survey of the United Kingdom, 2 vols., 980 p.
——————, 1930, A question of nomenclature: Geological Magazine, v. 67, p. 287-288.
——————, 1945, The Ordovician rocks of Arvon: Geological Society of London, Quarterly Journal, v. 100 (for 1944), p. 75-83.
——————, 1945, The Arvonian rocks of Arvon: Geological Society of London, Quarterly Journal, v. 100 (for 1944), p. 269-287.

Harker, A., 1950, Metamorphism: a study of the transformations of rock-masses: 3rd edition, London, Methuen, 362 p.
Harkness, R. and Hicks, H., 1871, On the ancient rocks of the St. David's promontory, South Wales, and their fossil contents: Geological Society of London, Quarterly Journal, v. 27, p. 384-404.
Harland, W.B., 1974, The Pre-Cambrian – Cambrian boundary, *in* Holland, C.H., ed., 1974, Cambrian of the British Isles, Norden and Spitzbergen: London, J. Wiley, p. 15-42.
Henslow, J.S., 1822, Geological descriptions of Anglesea: Transactions of the Cambridge Philosophical Society, v. 1, p. 359-452.
Hicks, H., 1873, On the Tremadoc Rocks in the Neighbourhood of St. David's, South Wales, and their fossil contents: Geological Society of London, Quarterly Journal, v. 29, p. 39-52.
————————, 1875, On the Successon of the Ancient rocks in the vicinity of St. David's, Pembrokeshire, with special reference to those of the Arenig and Llandeilo Groups, and their fossil contents: Geological Society of London, Quarterly Journal, v. 31, p. 167-195.
————————, 1877, On the Pre-Cambrian (Dimetian and Pebidian) Rocks of St. David's: Geological Society of London, Quarterly Journal, v. 33, p. 229-241.
————————, 1878, Additional Notes on the Dimetian and Pebidian Rocks of Pembrokeshire. With an appendix by Thomas Davies: Geological Society of London, Quarterly Journal, v. 34, p. 153-169.
————————, 1878, On some new Pre-Cambrian areas in Wales: Geological Magazine, dec. II, v. 5, p. 460-461.
————————, 1879a, On the classification of the British Pre-Cambrian rocks: Geological Magazine, dec. II, v. 6, p. 433-436.
————————, 1879b, On a new group of the Precambrian rocks (the Arvonian) in Pembrokeshire: Geological Society of London, Quarterly Journal, v. 33, p. 285-294.
————————, 1884, On the Pre-Cambrian rocks of Pembrokeshire, with special reference to the St. David's district. With an appendix by Thomas Davies: Geological Society of London, Quarterly Journal, v. 40, p. 507-560.
————————, 1886, Further proofs of the Pre-Cambrian age of certain Granitoid, Felsitic, and other rocks in N. W. Pembrokeshire: Geological Society of London, Quarterly Journal, v. 42, p. 351-356.
Holland, C.H., ed., 1974, Cambrian of the British Isles, Norden and Spitzbergen: London, J. Wiley, 300 p.
Howell, B.F., and Stubblefield, C.J., 1950, A revision of the fauna of the North Welsh *Conocoryphe viola* beds implying a Lower Cambrian age: Geological Magazine, v. 87, p. 1-16.
Lapworth, C., 1879, On the Tripartite classification of the Lower Paleozoic rocks: Geological Magazine, v. 6 (v. 16), p. 1-15.
Matley, C.A., 1928, The Pre-Cambrian complex and associated rocks of southwestern Lleyn (Caernarvonshire): Geological Society of London, Quarterly Journal, v. 84, p. 440-504.
Rankama, K., ed., 1965, The Precambrian. Vol. 2: New York, Interscience Publishers.
Ramsay, A.C., 1866, The geology of North Wales, 1st edition: Memoirs of the Geological Survey of the United Kingdom, 381 p.
————————, 1881, The geology of North Wales, 2nd edition: Memoirs of the Geological Survey of the United Kingdom, 611 p.
Rayner, Dorothy H., 1967, The stratigraphy of the British Isles: Cambridge University Press, 453 p.
Rodgers, J., ed., 1956, El Sistema Cambrico, su paleogeografia y el problema de su base.

Part 1: Europe, Africa, Asia (435 p.). Part 2: Australia, America (762 p.): Mexico, International Geological Congress Report, 20th session.

Salter, J.W., 1873, A catalogue of the collection of Cambrian and Silurian fossils contained in the geological museum of the University of Cambridge: Cambridge University Press, 204 p.

Sedgwick, A., 1831, letter to R.I. Murchison dated Llanllyfni, near Caernarvon, 13 September 1831; *in* Clark, J.W. and Hughes, T.M., 1890, The life and letters of Adam Sedgwick:Cambridge University Press, v. 1, p. 377-379.

——————, 1832, Letter to R.I. Murchison dated Barmouth, 23rd July 1832. Enclosing sketch-section of the structure of N. Wales: Published *in* Clark, J.W., and Hughes, T.M., 1890, The life and letters of Adam Sedgwick. Cambridge University Press, v. 1, p. 319-394, with a facsimile of the sketch.

——————, 1938, Report of a meeting held 23 May 1838. "A memoir entitled a Synopsis of the English series of stratified rocks inferior to the old red sandstone – with an attempt to determine the successive natural groups and formations . . . . commenced on 21st March was concluded": Proceedings of the Geological Society of London, v. 2, p. 675-685.

——————, 1843, Report of a meeting held 21 June 1843. "Outline of the geological structure of North Wales": Proceedings of the Geological Society of London, v. 4, p. 212-224.

——————, 1847, On the classification of the fossiliferous slates of North Wales, Cumberland, Westmoreland and Lancashire (being a supplement to a paper read to the Society, March 12, 1845): Geological Society of London, Quarterly Journal, v. 3, p. 133-164.

——————, 1852, On the classification and nomenclature of the Lower Palaeozoic rocks of England and Wales: Geological Society of London, Quarterly Journal, v. 8, p. 136-168.

——————, 1873, Preface to Salter, 1873.

Sedgwick, A. and Murchison, R.I., 1836, On the Silurian and Cambrian Systems, exhibiting the order in which the older sedimentary strata succeed each other in England and Wales: British Association for the Advancement of Science, Report of the 5th Meeting, Dublin, 1835, Pt. 2, p. 59-61.

Shackleton, R.M., 1969, The Pre-Cambrian of North Wales, *in* Wood, A., ed., 1969, The Pre-Cambrian and Lower Palaeozoic rocks of Wales: Cardiff, University of Wales Press, 461 p; p. 1-22.

Snyder, F.G., 1947, The problem of the Lipalian interval: Journal of Geology, v. 55, 146-152.

Stubblefield, C.J., 1956, Cambrian palaeogeography in Britain, *in* Rodgers, J., ed., El Sistema Cambrico, su paleogeografia y el problema de su base: Mexico, International Geological Congress Report, 20th session, Pt. 1, p. 1-43.

Thomas, H.H., and Jones, O.T. 1912, The Pre-Cambrian and Cambrian rocks of Brawdy, Haycastle, and Brimaston (Pembrokeshire): Geological Society of London, Quarterly Journal, v. 68, p. 374-401.

Wheeler, H.E., 1947, Base of the Cambrian system: Journal of Geology, v. 55, p. 153-159.

Williams, D., 1945, Discussion *in* Greenly, E., 1945, The Arvonian rocks of Arvon: Geological Society of London, Quarterly Journal, v. 100 (for 1944), p. 285.

Williams, D. and Ramsay, J.G., 1968, Snowdonia, Second Edition (First Edition, 1959): Geologists' Association Guide.

Wood, A., ed., 1969, The Pre-Cambrian and Lower Palaeozoic rocks of Wales: Cardiff, University of Wales Press, 461 p.

Wood, D.S., 1969, The base and correlation of the Cambrian rocks of North Wales, *in* Wood, A., ed., 1969, The Pre-Cambrian and Lower Palaeozoic rocks of Wales: Cardiff, University of Wales Press, p. 47-56.

History of Concepts in Precambrian Geology, edited by W.O. Kupsch and W.A.S. Sarjeant, Geological Association of Canada Special Paper 19, 1979

HISTORY OF CONCEPTS: MIGMATITES

R.V. Dietrich
Department of Geology, Central Michigan University, Mount Pleasant, Michigan 48859

ABSTRACT

The idea that mixed rocks, now called migmatites, may be formed as the result of partial melting of pre-existing rocks may be traced to Hutton. Other early expounders of the concept include Lyell, Durocher, and Delesse. In the early 1900s Sauer called such rocks 'amphoteric' gneisses and Sederholm named them migmatites. The later designation has gained general acceptance since the classical localities of the Precambrian Fennoscandian Shield were visited in 1910 by many of the geologists who attended the International Geological Congress meetings held in Stockholm.

Suggested origins for migmatites and migmatite-appearing rocks include formation of their igneous or igneous-appearing components by injection, transformation *in situ*, partial anatexis, chemical metamorphic differentiation, isochemical metamorphism, and physical metamorphic differentiation. Both field occurrences and laboratory investigations indicate that most of these rocks have been formed under conditions generally attributed to the high amphibolite facies of metamorphism.

RESUME

L'idée selon laquelle des roches mixtes, appelées maintenant migmatites, puissent résulter de la fusion partielle de roches préexistantes peut être retracée jusqu'à Hutton. Lyell, Durocher et Delesse comptent parmi les premiers interprètes du concept. Au début des années 1900 Sauer appela de telles roches des gneiss "amphotères" et Sederholm des migmatites. Cette dernière désignation a gagné l'approbation générale depuis que de nombreux géologues, qui participaient au Congrès Géologique International tenu à Stockholm, ont visité les localités classiques du bouclier précambrien fenno-scandien.

Les origines qui ont été suggérées pour les migmatites et les roches d'apparence migmatitique comprennent la formation de leurs composants ignés ou d'apparence ignée par injection, transformation *in situ,* anatexie partielle, différenciation métamorphique chimique, métamorphisme isochimique, et différenciation métamorphique physique. Les occurences sur le terrain et les investigations en laboratoire indiquent que la plupart de ces roches ont été formées sous des conditions généralement attribuées à la partie supérieure du faciès des amphibolites.

INTRODUCTION AND DEFINITION

Migmatites exhibit classical development and constitute thousands of square miles of bedrock in many Precambrian terranes. The type locality for these 'mixed rocks' is a group of glaciated exposures on the islands of Spikarna just east of

Hangö near the southern coast of Finland.

The term migmatite — from the Greek $\mu\iota\gamma\mu\alpha$, meaning mixture — was introduced by the famous Finnish Geologist, Jakob Johannes Sederholm (1907, p. 88-89). When he introduced the term, he noted "this rock group occupies in a sense a transitional position between the granites and the crystalline schists . . . and has no sharp boundary with any of them."

With subsequent modifications, the definition now followed by most petrologists active in the field has both descriptive and genetic aspects – a migmatite is a "megascopically composite rock that once consisted of geochemically mobile and immobile, or less mobile, parts, – i.e., it consists of igneous or igneous-appearing and/or metamorphic materials" (Dietrich and Mehnert, 1961, p. 62).

As can be seen, this definition includes three important requirements: 1) that the two main components are distinguishable with the naked eye; 2) that one of the components was once relatively mobile while the other was relatively immobile; and 3) that one of the two main components looks igneous (e.g., like a granite) and the other metamorphic (e.g., like an amphibolite) – see Figure 1.

HISTORY

Depending upon one's prejudices and interpretation of apparently ambiguous statements published more than one and three quarters of a century ago, considera-

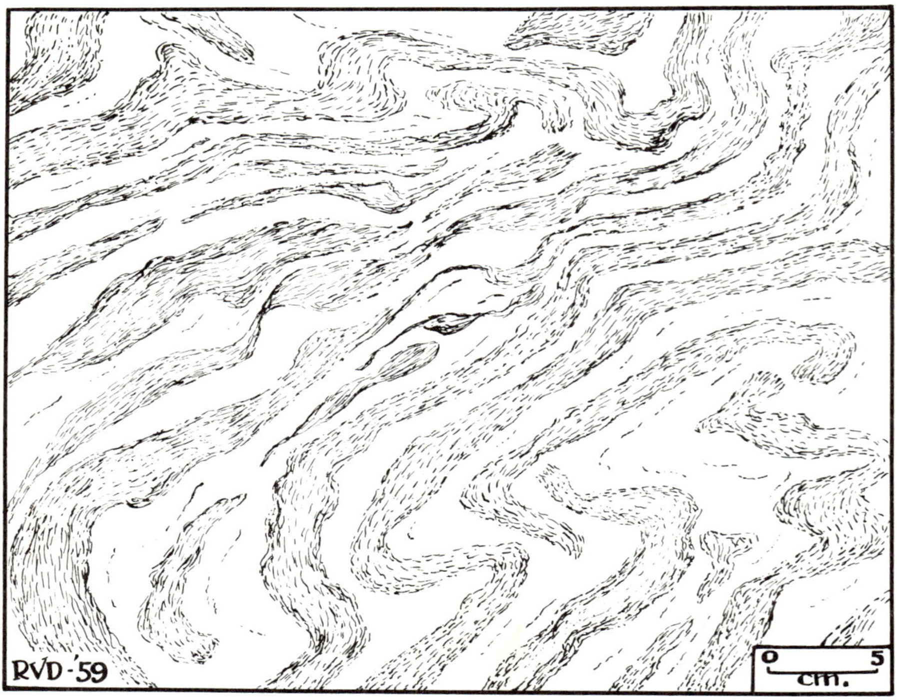

Figure 1. (see next page)

Figure 1. Schematic diagram of a migmatite with two chief petrographic components – one "granitic" (light); the other, e.g., "amphibolitic" (patterned). Alternative hypotheses of origin for migmatites (1-4) and migmatitic-appearing rocks (5-6) plus the suggested derivations for at least a major portion of each of the components may be tabulated as follows:

Hypothesis	"Granitic" Component	"Amphibolitic" Component
1. Injection	injected magma	metamorphosed rock
2. Transformation	introduced fluids and/or ions	metamorphosed rock
3. Partial anatexis	melted rock constituents	metamorphosed rock (not melted) rock
4. Chemical metamorphic differentiation	mobilized ("exuded") phase	less- or non-mobilized phases
5. Physical metamorphic differentiation	segregated mineral grains	segregated mineral grains
6. Isochemical metamorphism	treptomorphosed rock	treptomorphosed rock
	(other commonly applied terms)	
Any of above	leucosome	melasome
1 or 2 above	neosome	paleosome
4 or possibly 3 of above	mobilizate	restite

tion of at least some of the basic aspects of migmatites may or may not be attributed to Hutton (1795). In those statements, he mentions

> "the waved structure of the stratified alpine stone ['granite'] . . . has been brought to a great degree of softness, so as to have the original streight lines of its stratification changed to those undulated or waving lines which are in some cases extremely much incurved." (p. 318)

and

> "the collection of different materials, and the subsequent fusion of the compound mass" (p. 319).

In any case, Lyell (1855), Durocher (1856), and Delesse (1861) were aware of these mixed rocks and the concepts now termed anatexis and palingenesis, and several other geologists had described occurrences of representative rocks before they were named 'amphoteric' gneisses by A. Sauer in 1903. Roth (1871) has summarized the early literature; Sederholm (1907 and 1923) expanded the historical coverage through about 1920. The list of pre-1920 contributors is a veritable 'Who's Who' of the geologists of the period. Along with those already mentioned, it includes: F.D. Adams, E.C. Andrews, A.E. Barlow, J. Barrell, C. Barrois, C.P. Berkey, R.A. Daly, J.D. Dana, L. Duparc, O.H. Erdmannsdörffer, P. Eskola, C.N. Fenner, V.M. Goldschmidt, U. Grubemann, A.G. Högbom, P.J. Holmquist, Th. Kjerulf, J. Koenigsberger, A. Lacroix, A.C. Lawson, J. Lehman, F. Loewinson-Lessing, A. Michel-Lévy, H. Reusch, G. Steinmann, J.J.H. Teall, P. Termier, and A. and N.H. Winchell.

Strictly speaking, however, these rocks did not gain truly widespread attention until after Sederholm introduced the term migmatite and more particularly until after geologists from all over the world visited some of the now classical localities in southern Finland and near Stockholm, during excursions held in conjunction with the International Geological Congress meetings held in Stockholm in 1910. Since then, migmatites have been one of the primary foci of essentially all of the controversies dealing with the igneous versus metamorphic origin of granite rocks (e.g., Gilluly, 1947). This is true because most of the development of hypotheses of origin for migmatites has dealt with possible modes of formation of the more mobile, typically granitic, component.

In essence all of the suggested modes of origins for migmatites may be reduced to derivation of the granitic portion either by rearrangement of constituents of parent rock materials or by introduction of some or all of the granitic constituents from without, or of course, by some combination of these activities (see the caption of Fig. 1).

When Sederholm (1907) introduced the term migmatite, he indicated that he believed that much, if not all, of the granitic material was *injected* from without. Holmquist (1916) disagreed and became the chief protagonist advocating derivation by local exudation from within – that is, derivation as the result of *chemical metamorphic differentiation*. The metamorphic concept for the formation of migmatites has subsequently undergone several modifications – most stemming from geologists working in the Pyrenees and Fennoscandia – and now includes formation by '*transformation in situ*'. This, of course, includes long- as well as short-distance transfer of ions and molecules as a consequence of either metasomatism or solid diffusion. Such origin is generally thought to have taken place under the influence or orogenic stresses (see e.g., Wegmann, 1935). Since the early 1930s, *partial anatexis* (partial melting of pre-existing rock) has gained the support of many geologists and has been suggested as a mode of genesis for some migmatites (e.g., see Eskola, 1933). Although most such melting of pre-existing rocks has been attributed to deep burial, frictional heating has also been suggested (e.g., see Reitan, 1969) and a few petrologists have advocated alternatives involving reduction of pressure (e.g., see Bowen, 1928, p. 314 and Dietrich, 1964, p. 57). With regards to alternatives such as these, special attention is directed to two recent papers by Waldbaum (1971a and 1971b) and the ensuing discussions by Dodson (1971) and Ramberg (1971). Waldbaum described isenthalpic and isentropic temperature changes as being possibly associated with adiabatic decompression in accompaniment to certain geological processes. Evidence should be sought for adiabatic migmatization.

Additionally pertinent to considerations of the origins of migmatites are units that may be termed *migmatite-appearing rocks*. These are composite rocks that look like migmatites but are not migmatites *sensu stricto* according to any of the currently accepted definitions (including the one subscribed to in this paper). Examples are composite gneisses that have been formed as a result of essentially isochemical metamorphism with or without physical differentiation or disruption (numbers 5 and 6 in caption for Fig. 1). Wenk (1937) described migmatitic-appearing banded gneisses which he then interpreted as having been formed as the result of chiefly physical metamorphic differentiation. Dietrich (1963) favored formation of similar, migmatitic-appearing banded gneisses by treptomorphism – that is, essentially isochemical metamorphism resulting from recrystallization with little if any chemical or physical differentiation as those processes are generally conceived.

So far as the true formation of migmatites (and migmatite-appearing rocks), current consensus apppears to be that each occurrence must be carefully studied and considered from the standpoint that it may have originated as a result of any one or diverse combinations of the processes just noted.

MIGMATITE-BASED CONCEPTS

In 1935, C.E. Wegmann introduced a comprehensive concept in which migmatization was interrelated to kinematics and depth zones (see Fig. 2). His concept includes the hypothesis that some essentially homogeneous plutonic rocks are really migmatites because they consist of both paleosome material of local derivation and neosome material introduced from remote sources. Extended to the nth degree, such 'homogeneous migmatites' could have essentially every molecule of every mineral with paleosome and neosome components. As a consequence of this hypothesis, Wegmann has maintained that the term migmatite should be redefined to include such homogeneous-appearing rocks. Although his hypothesis is widely recognized as theoretically possible and evidence for 'homogeneous migmatites' is diligently sought, the suggested redefinition of the term has rather generally been considered untenable in that such a change would render the term migmatite essentially useless for anything other than theoretical discussions. Therefore, a very large

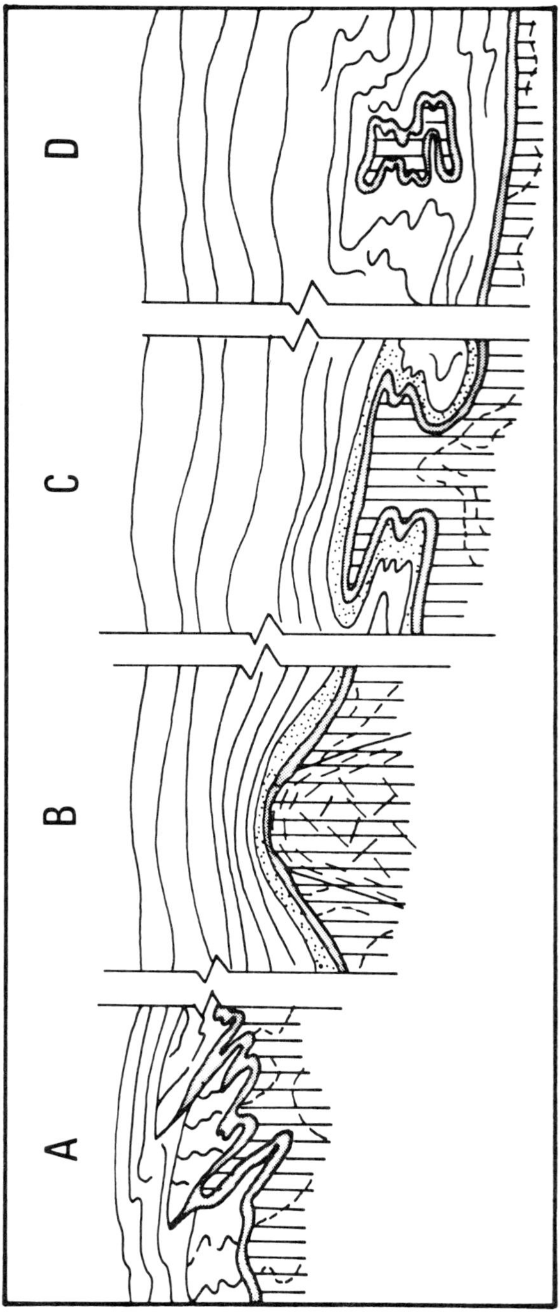

Figure 2. (see next page)

Figure 2. Broken schematic cross-section of continental crust showing major domains of migmatite formation (vertical dimension exaggerated).

Stippled areas represent 'transitional zone of detachment', the boundary zone between migmatized (vertically lined on diagram) and non-migmatized rocks (plain on diagram). From place to place, the zone may range greatly both in thickness and in style of dynamic and chemical metamorphism.

The higher grade side of this zone, often designated the 'migmatite front', has been correlated by several petrologists with the boundary between Wegmann's 'Oberbau' and 'Unterbau'. Actually, the 'front' is a conditional boundary the position of which may differ in crustal level because of differences in such aspects as original lithology and thickness of cover. The 'front' may be wholly within either 'basement' or 'cover' or it may transgress or essentially correspond to the contact between them.

The relative thickness of the cover indicated on the diagram would be as follows: "A" would be characteristic of an area of thin to moderate cover –possibly representing platform deposits; "B", a region of moderate cover; and "C" and "D", regions of moderate to thick cover, possibly geosynclinal deposits. Section A shows partial remobilization of basement and fracturing of overlying rocks (cf. Wegmann, 1935). In Section B there is a 'mantled gneiss dome' of the type hypothesized by Eskola (1948) – that is, the beds conformable with the top of the dome were originally deposited unconformably on the exposed surface of a granitic mass, the remobilized material of which now constitutes the dome. Section C shows a migmatite massif (cf. Haller, 1956 and 1958). Section D depicts an apparently 'rootless' mass of migmatite; it would appear that such masses have gained their present positions as the result of diapiric movements or even lateral flow; some of the masses so interpreted, however, may be portions of masses like the one shown in "C" which only appear to be rootless because of the lack of third dimensional data.

---

percentage of geologists who now work in migmatite terranes retain the megascopic aspect of the designation for mapping and descriptive purposes even though they fully accept and appreciate the implications of this part of Wegmann's concept. [Haller's (1971, p. 161-162) use of the designation 'synorogenic granite' for essentially homogeneous migmatites in eastern Greenland exemplifies methods that have been used in the literature to deal with the dichotomy.]

Another important migmatite-based concept involves the term *migma*, which was first used by Reinhard (1934-1935). Migma is generally employed to describe rock material, with either real or potential mobility, which upon losing its mobility becomes a migmatite. Today, such mobility is usually, but by no means exclusively, attributed to migmatites the granitic constituents of which are interpreted as once having been molten. Mobility during migmatization can hardly be left without mentioning three rather closely akin tectonic features: the 'stockwerk' of Wegmann (1935) – that is, the migmatitic infrastructure (Unterbau) versus the overlying non-migmatitic suprastructure (Oberbau); the mantled gneiss domes of Eskola (1958); and the migmatite massifs of Haller (1956 and 1958). A brief description of each of these is included in the caption of Figure 2.

Also with regards to migmas, it is noteworthy that the migma: migmatite and the magma: igneous rock relationships are *not* completely analogous. Whereas an igneous rock requires by definition the preëxistence of a parent magma, a migmatite cannot be considered *prima facie* evidence for the pre-existence of a migma.

RECENT INVESTIGATIONS

During the last few decades, hardly a year has passed without the publication of some maps or reports indicating the presence of newly discovered or more comprehensively investigated migmatities. Nonetheless, no 'new' kinds of migmatitic features have been reported for several years. On the other hand, a number of laboratory techniques have only recently been utilized in the study of migmatites. For example, Büsch (1966) has reported a petrochemical investigation in which he was able to relate possible silica-rich 'mobilizates' (neosomes) and mafic-rich 'restites' (paleosomes) to intermediate composition 'parent rocks' (see Fig. 3).

Also important to considerations relating to migmatites are the laboratory experiments dealing with solid state diffusion, phase equilibria (e.g., Tuttle and Bowen, 1958; Grant, 1973), and melting of natural rocks (e.g., Piwinskii, 1973). It has already been shown, for example, that rock melts which would consolidate to give the granitic components of migmatites should be expected to form from

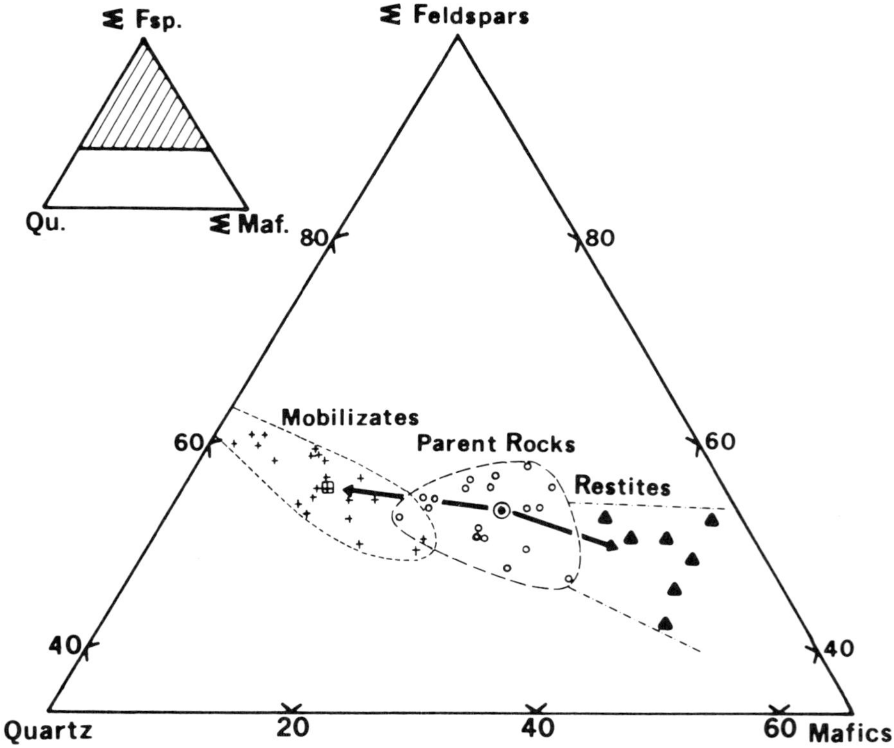

Figure 3. Example of the interpreted breakdown of intermediate composition parent rocks (O) to form quartz plus feldspar-rich mobilizates (+) and mafic-rich restites (△). In the general case, this might represent either metamorphic differentiation or partial anatexis. (Lined area within upper left 100 per cent triangle indicates area of main triangle.) Diagram is modified after Mehnert and Büsch (1966, p. 257).

rocks of appropriate compositions under conditions generally thought to obtain within or above the high amphibolite facies (see Fig. 4).

NOMENCLATURE

A regrettable chapter in the history of the growth of concepts about migmatites is tied up with the fact that both the rocks and the concepts about them have been so enmeshed in the coining of new terms. Indeed, a cursory review of the pertinent literature can rather logically be interpreted to suggest that the petrologists who have dealt with the rocks were as much interested in etymological exercises as they were in the rocks and their geneses – see, for example, Sederholm (1907), Holmquist (1921), Scheumann (1936), Jung and Roques (1936 and 1952), Huber

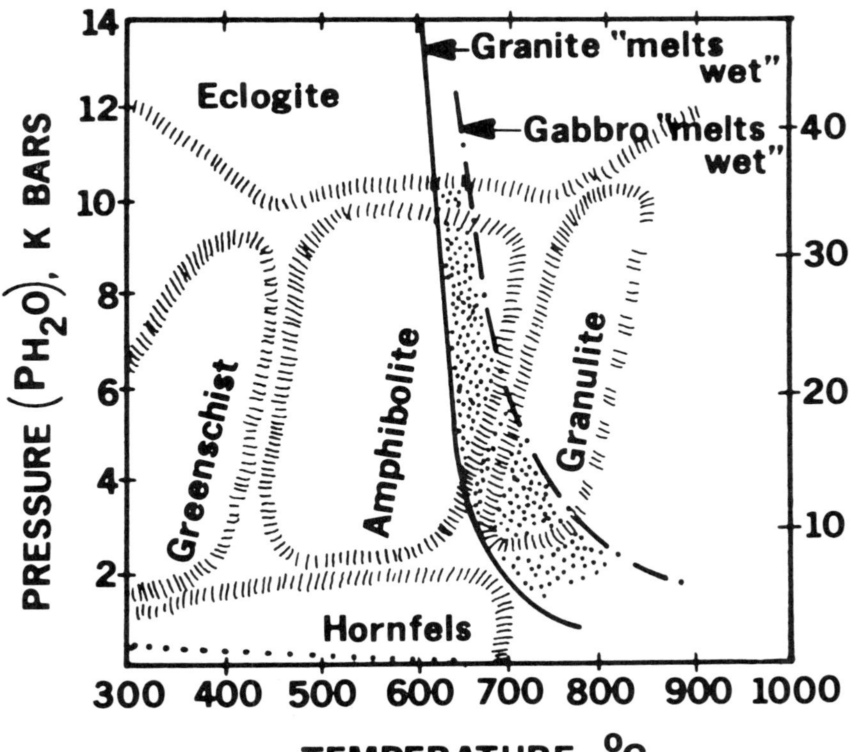

Figure 4. Conditional domain for formation of migmas, indicated by stippling, correlated with Turner's "tentative scheme of metamorphic facies . . ." and to curves indicating the conditions required for the beginning of melting for granite and for gabbro (of approximate amphibolite composition), both in water-saturated systems. Metamorphic facies 'gradational' boundaries (including the dotted line in lower left part of diagram which indicates the "lower limit of metamorphism") are modified after Turner, 1968, p. 366. Granite and gabbro 'melts wet' curves are plotted from data compilation of Merrill *et al.*, 1970.

and Niggli (1953), Piccoli (1958), Dietrich and Mehnert (1961), and Mehnert (1968). Is it not ridiculous, for example, that

> "one and the same outcrop, depending upon the observer's experiences and prejudices [might] be called such diverse things as agmatite, anatectic breccia, arterite, chorismite, diabrochite, ectexite, ectinitic breccia, embrechite, entexite, granitized breccia, injection breccia, intrusion breccia, metatexite, migmatite, rheomorphic breccia, and venite" (Dietrich and Mehnert, 1961, p. 56).

This dilemma was brought into special focus at a symposium held during the 21st International Geological Congress meetings held in Copenhagen – quite coincidentally, the second such congress held in Scandinavia and just a half century after the one during which migmatites gained their first large international audience. The resulting publication (Sörensen, 1961) affords little but a possibly useful dictionary of terms. At least some of us who are currently investigating migmatites, however, strongly believe that the best guideline for the future is to avoid most of the terms previously used and to use instead unambiguous basic language plus widely accepted petrologic terms – for example, use "intrusive breccia" instead of "mictite", "partially melted" for "diatexis', etc., etc. A slightly modified and expanded version of Mehnert's (1968, p. 232-233) suggestions, whereby migmatites would be described on the basis of their correlation with metamorphic facies plus or minus some references to their internal geometric relationships, would then serve for essentially all purposes. Furthermore, it must not be forgotten that a good picture may save many words (Fig. 5).

EPILOGUE

Anyone who reviews the main concepts relating to migmatites becomes impressed with the fact that very few new ideas have been introduced since Sederholm and his co-workers made migmatites one of their chief concerns during the first third of this century. Even the few hypotheses that have been subsequently introduced as 'new' can be shown to have been conceived, if not articulated as such, by Sederholm and his associates. Equally impressive is the fact that nearly all of the concepts have been based on empirical field observations plus induction and thus were little more than conjectures, albeit some of which required great foresight as well as insight. Further, it becomes readily apparent that geologists' first encounters with migmatites have frequently served as veritable 'eye-openers'. Even the master, himself, J.J. Sederholm (1907, p. 11) related his early experiences with these rocks to his becoming an 'actualist' and to his gaining the 'thread of Ariadne'!

ACKNOWLEDGEMENTS

This is a brief summary of the paper given at the Montreal Symposium. Much of the additional information the author presented has been incorporated in his paper *Migmatites – A Résumé* which appeared as CEGS Short Review Number 24 in the September 1974 issue of the *Journal of Geological Education.*

I again wish to thank Drs. John Haller, D.G. Jorgenson, M.E. McCallum, J.L. Weitz, and J.L. Younker for reviewing the manuscript of that paper and also the geologists who so graciously guided me and discussed various aspects relating to migmatites with me during my visits to localities in Canada, Finland, France, Germany, New Zealand, Norway, Rhodesia, South Africa, Sweden, Switzerland, and different parts of the United States of America.

Figure 5. Typical piece of migmatitic Morton Gneiss from the early Precambrian exposed in the Minnesota River Valley of southwestern Minnesota. (Height of block is about 1.15 m.)

REFERENCES

Bowen, N.L., 1928, The Evolution of the Igneous Rocks: Princeton, N.J., Princeton University Press, 334 p.

Büsch, W., 1966, Petrographie und Abfolge der Granitisation im Schwarszwald-V: Neues Jahrbuch für Mineralogie: Abhandlungen, 104, p. 190-258.

Delesse, A.E., 1861, Etudes sur le métamorphisme de roches: Imprimerie Imperiale, Paris (p. 87 – not seen by R.V.D., cited by Sederholm, 1923, p. 4).

Dietrich, R.V., 1963, Banded gneisses of eight localities: Norsk Geologisk Tidsskrift, v. 43, p. 89-119.

————, 1964, Igneous activity in the Southern Appalachians, *in* Lowry, W.D., ed., Tectonics of the Southern Appalachians: Virginia Polytechnic Institute, Geological Sciences, Memoir 1, Blacksburg, Virginia, p. 47-61.

————, 1974, Migmatites – A Résumé (CEGS short review 24): Journal of Geological Education, v. 22, p. 144-156.

Dietrich, R.V. and Mehnert, K.R., 1961, Proposal for the nomenclature of migmatites and associated rocks: International Geological Congress, 21st session (Norden), Part XXVI, Supplement vol., Section 1-21, p. 56-67.

Dodson, M.H., 1971, Isenthalpic flow, Joule-Kelvin coefficients and mantle convection: Nature, v. 234, p. 212.

Durocher, J., 1856, Etudes sur la structure orographique et la constitution géologique de la Norvège, de la Suède et de la Finlande: Mémoires de la Société Géologique de France, v. 6, (p. 1 – not seen by R.V.D., cited by Sederholm, 1923, p. 4).
Eskola, P., 1933, On the differential anatexis of rocks: Bulletin de la Commission Géologique de la Finlande, No. 103, p. 12-25.
——————, 1947, The problem of mantled gneiss domes: Geological Society of London Quarterly Journal, v. 104, p. 461-476.
Gilluly, J., ed., 1948, Origin of granite: Geological Society of America, Memoir 28, 139 p.
Grant, J.A., 1973, Phase equilibria in high-grade metamorphism and partial melting of pelitic rocks: American Journal of Science, v. 273, p. 289-317.
Haller, J., 1956, Probleme der Tiefentektonik. Bauformen im Migmatit-Stockwerk der ostgrönlandischen Kaledoniden: Geologische Rundschau, v. 45, p. 159-167.
——————, 1958, Der "Zentrale metamorphe Komplex" von NE-Grönland, Tiel II: Meddelelser om Grönland, v. 154 (3), 153 p.
——————, 1971, Geology of the East Greenland Caledonides: Interscience Publishers, New York, 410 p.
Holmquist, P.J., 1916, Swedish Archaean structures and their meaning: Bulletin of the Geological Institution of the University of Uppsala, 15, p. 125-148.
——————, 1921, Typen und Nomenklatur der Adergesteine: Geologiska Föreningens i Stockholm Förhandlingar, v. 43, p. 612-631.
Huber, J.M. and Niggli, P., 1943, Uber die Namengebung bei grobgemengten Gesteinen: Schweizerische Mineralogische und Petrographische Mitteilungen, v. 23, p. 72-160.
Hutton, J., 1795, Theory of the Earth, with Proofs and Illustrations: Vol. 1: Cadell, Junior and Davies; London, and William Creech; Edinburgh (p. 316-319).
Jung, J. and Roques, M., 1936, Zonéographie des schistes cristallins du Massif Central Français: Revue des Sciences Naturelles d'Auvergne, v. 1 (4), p. 7ff.
——————, 1952, Introduction à l'étude zonéographie des formations cristallophylliennes: Bulletin du Service de la Carte Géologique de France, 235, p. 1-62.
Lyell, C., 1855, A Manual of Elementary Geology (5th edit.): London, J. Murray, 655 p. (p. 603 – not seen by R.V.D., cited by Sederholm, 1923, p. 4).
Mehnert, K.R., 1968, Migmatites and the origin of granitic rocks: Amsterdam, Elsevier, 393 p.
Mehnert, K.R. and Büsch, W., 1966, Dioritbildung durch Remobilisation: Tschermaks Mineralogische und Petrographische Mitteilungen, v. 11 (3 - 4), p. 246-265.
Piccoli, G., 1958, Il problem delle migmatiti attraverso mezzo secolo di recerche: Periodico di Mineralogia, vol. 27, p. 1-87. [Also, Spanish translation of this article: Piccoli, G., 1960, El problema de las migmatites a través de medio siglo de investigaciones: Revista de la Facultad de Ciencias Exactas, Fisicas y Naturales, Cordoba, Communicacions del Museo de Mineralogia y Geologia, no. 37, p. 5-104.]
Piwinskii, A.J., 1973, Experimental studies of igneous rock series, central Sierra Nevada batholith, California, Part II: Neues Jahrbuch für Mineralogie, no. 5, p. 193-215.
Ramberg, H., 1971, Temperature changes associated with adiabatic decompression in geological processes: Nature, vol. 234, p. 539-540.
Reinhard, M., 1934-1935, Ueber die Gesteinmetamorphose in den Alpen: Jaarboek van de Mijnbouwkundige Vereeniging, Delft, vol. 39.
Reitan, R.P., 1969, Temperatures with depth resulting from frictionally generated heat during metamorphism, *in* Igneous and metamorphic geology . . . : Geological Society of America, Memoir 115, p. 495-512.

Roth, J., 1871, Ueber de Lehre von Metamorphismus und die Enstehung der crystallinischen Schiefer: Abhandlungen der Königlichen Akademie der Wissenschaften zu Berlin, (not seen by R.V.D., cited by Sederholm, 1923, p. 4).

Sauer, A., 1903, Das alte Grundgebirge Deutschlands: Compte Rendu. IX Congrès Géologique International, Vienna (not seen by R.V.D., cited by Sederholm, 1907, p. 4).

Scheumann, K.H., 1936, Zur Nomenklatur migmatitischer und verwandter Gesteine: Tschermaks Mineralogische und Petrographische Mitteilungen, v. 48, p. 297-302.

Sederholm, J.J., 1907, Om granit och gneis: Bulletin de la Commission Géologique de la Finlande, No: 23, p. 1-110 [An English translation of this article appears in Sederholm, J.J., 1967, Selected Works: Granites and Migmatites: Wiley, New York, 608 p.]

————————, 1923, On migmatites and associated Pre-Cambrian rocks of southwestern Finland. Part I, The Pellinge Region: Bulletin de la Commission Géologique de la Finlande, no. 58, 153 p.

Sörenson, H., ed., 1961, Symposium on Migmatite Nomenclature: International Geological Congress, 21st Session (Norden), Part XXVI, Supplement volume, Section 1-21, 78 p.

Turner, F.J., 1968, Metamorphic Petrology — Mineralogical and Field Aspects: New York, McGraw-Hill, 403 p.

Tuttle, O.F., and Bowen N.L., 1958, Origin of granite in the light of experimental studies in the system $NaAlSi_30_8$-$KAlSi_30_8$-$Si0_2$-H_20: Geological Society of America, Memoir 74, 153 p.

Waldbaum, D.R., 1971a, Temperature changes associated with adiabatic decompression in geological processes: Nature, v. 232, p. 545-547.

————————, 1971b, Adiabatic decompression coefficients of minerals (Abstract): Transactions of the American Geophysical Union, v. 52, p. 348-349.

Wegmann, C.E., 1935, Zur Deutung der Migmatite: Geologische Rundschau, v. 26, p. 304-350.

Wenk, E., 1937, Zur Genese der Bändergneisse von Ornö Huvud: Geological Institution of Uppsala Bulletin, v. 26, p. 53-89.

History of Concepts in Precambrian Geology, edited by W.O. Kupsch and W.A.S. Sarjeant, Geological Association of Canada Special Paper 19, 1979

THE HISTORY OF THE STUDY OF THE PRECAMBRIAN ROCKS OF CHARNWOOD FOREST, ENGLAND

Trevor D. Ford
Department of Geology, University of Leicester,
University Road, Leicester LEI 7RH U.K.

ABSTRACT

Used in ancient times for the manufacture of Neolithic axe-heads and from Roman-times onwards for roofing slates, the Charnian rocks are now amongst the most intensively quarried rocks in England, being extracted for concrete aggregate and roadstone from five major quarries.

The rocks of Precambrian age, now known as the Charnian system, outcrop as a group of partially exhumed mountain top inliers surrounded by Triassic sediments in an area of some 30 square miles (78km^2) in Leicestershire in central England.

The pioneer geologist, John Farey, did not appreciate the true nature of the relationship of Charnian to Triassic and regarded the former as a gigantic nodule in Triassic sediments in 1811. Conybeare and Phillips in 1822 still did not appreciate the relationship, but showed the Triassic terminating abruptly against the Charnian. By the time of the primary Geological Survey, in the mid 19th century, the unconformable relationship was understood and the probable correlation with other Precambrian regions was being discussed. Investigations by Bonney and Hill in 1877 to 1878 and later by Lapworth established the broad outlines of a stratigraphic subdivision of the Charnian, but it was W.W. Watts who, in a series of publications from 1895 to 1947, formulated the stratigraphic sequence as used until recently. He also worked out the basic structural history.

The Charnian System comprises some 12000 feet (3900m) of bedded pyroclastic rocks and sediments directly or indirectly derived from them, together with dacitic and dioritic intrusives. Radiometric dating suggests that the Charnian is at least 684 million years old; this provides support for correlation at least in part with the English Longmyndian and Uriconian, and with the Conception Group of Newfoundland.

The discovery of elements of the Ediacaran fauna in the upper part of the Maplewell Group of the Charnian supports the correlation with the Conception Group and with other late Precambrian sequences elsewhere.

RESUME

Des roches d'âge précambrien, connues en tant que Système Charnien, affleurent dans une région de quelques 30 miles carrés (78 km^2) du Leicestershire, Angleterre, sous la forme d'un groupe de pointements montagneux témoins partiellement dégagés, entourés de sédiments triassiques. Le Système Charnien comprend environ 12000 pieds (3900 m) de roches pyroclastiques litées et de sédiments en dérivant directement ou indirectement, cet ensemble étant associé avec des intrusifs dacitiques et rhyolitiques. Des corrélations avec le Longmyndien et l'Uriconien anglais et avec le groupe de Conception de Terre Neuve ont été suggérées, et sont partiellement étayées par des évidences paléontologiques et des datations radiométriques. Utilisées jadis pour la fabrication de tête de hâche, et à partir de l'époque romaine pour les toitures d'ardoise, les roches du Charnien sont maintenant en Angleterre, parmi les plus intensivement exploitées en carrière; elles sont extraites pour la fabrication de béton et de matériel d'empierrage de route à partir de cinq carrières des plus importantes.

En 1811 John Farey, pionnier en géologie, considérait les roches du Charnien comme un gigantesque nodule dans les sédiments triasiques. Les travaux postérieurs de Bonney et Hill en 1877-78 et de Lapworth en 1898 établirent les grandes lignes de la stratigraphie et de la structure. Watts publia de 1895 à 1947 l'histoire détaillée. Ces dernières années l'auteur a décrit un ensemble fossile de type Ediacaran dans les Lits Woodhouse de la séquence du Charnien. La datation radiométrique suggère pour le Charnien un âge minimum de 684 millions d'années.

INTRODUCTION

The study of the history of investigation of any area of ancient rocks, such as Charnwood Forest, requires answers to certain questions: 1). When were the rocks recognised as something distinctive? 2). When were they first utilized? 3. When were they first studied as a geological entity? 4). When was their age established? 5). What is the detailed story told by these rocks? When were the stages in that story worked out and who did the research? 6). What is the present state of knowledge? 7). What are the gaps in the present state of knowledge?

The answers to the first two questions when applied to Charnwood Forest are simple, since the Precambrian rocks there are virtually the only 'hard' rocks which outcrop in a wide area of the English Midlands. The outcrops cover an area of some seven miles (12 km) wide and long and they are mostly prominent crags on hill-tops, so that they must have been obvious in Prehistoric times. Early man made good use of such hard rocks and Shotton (1959) has noted that Group XX of his petrographic classification of Neolithic axe-heads is a tuff of Charnian type, though the exact locality of axe manufacture is not known. Thus Charnian rocks were known in Neolithic times as something distinctive and useful in axe-making.

Subsequently Swithland Slate was used for roofing material on buildings in Roman Leicester, having been transported some five miles (8 km) from the outcrop. The stone and rubble-fill of walls in Roman Leicester includes many blocks of Charnian rock, but no evidence has yet been obtained that stone was deliberately quarried at that time. From that time on Charnian rocks, particularly the more coarse-grained igneous rocks, were utilized for building purposes and early small quarries were enlarged to be swallowed up in the present vast excavations. The Swithland Slate has been used for roofing ever since Roman days, though extraction ceased almost a century ago. Swithland Slate does not cleave into slabs as thin as those of the Welsh slates, necessitating strong timber supports. Slate on buildings being demolished today is much in demand for re-use on new buildings. Swithland Slate has also been used for tombstones at least since the 16th century: its resistance to weathering has meant that inscriptions survive almost intact (Herbert, 1945).

Another use has been for butter troughs.

In the late 18th and more particularly, the 19th centuries the coarse-grained igneous rocks, such as the "syenite" (really diorite) of Markfield, were increasingly in demand as trimmed rectangular blocks or setts for the streets of London and other cities, and the arrival of the railway about 1830 gave a boost to this trade for Charnwood is the nearest source of suitable hard-rock to London. Kerb-stones have been made by hand at Markfield until the last few years. Recently the demand has been more for crushed stone for concrete aggregate and for road-making and it is today extracted from five major quarries.

Figure 1. Geological Map of Leicestershire (1821) by William Smith, topography by John Cary.

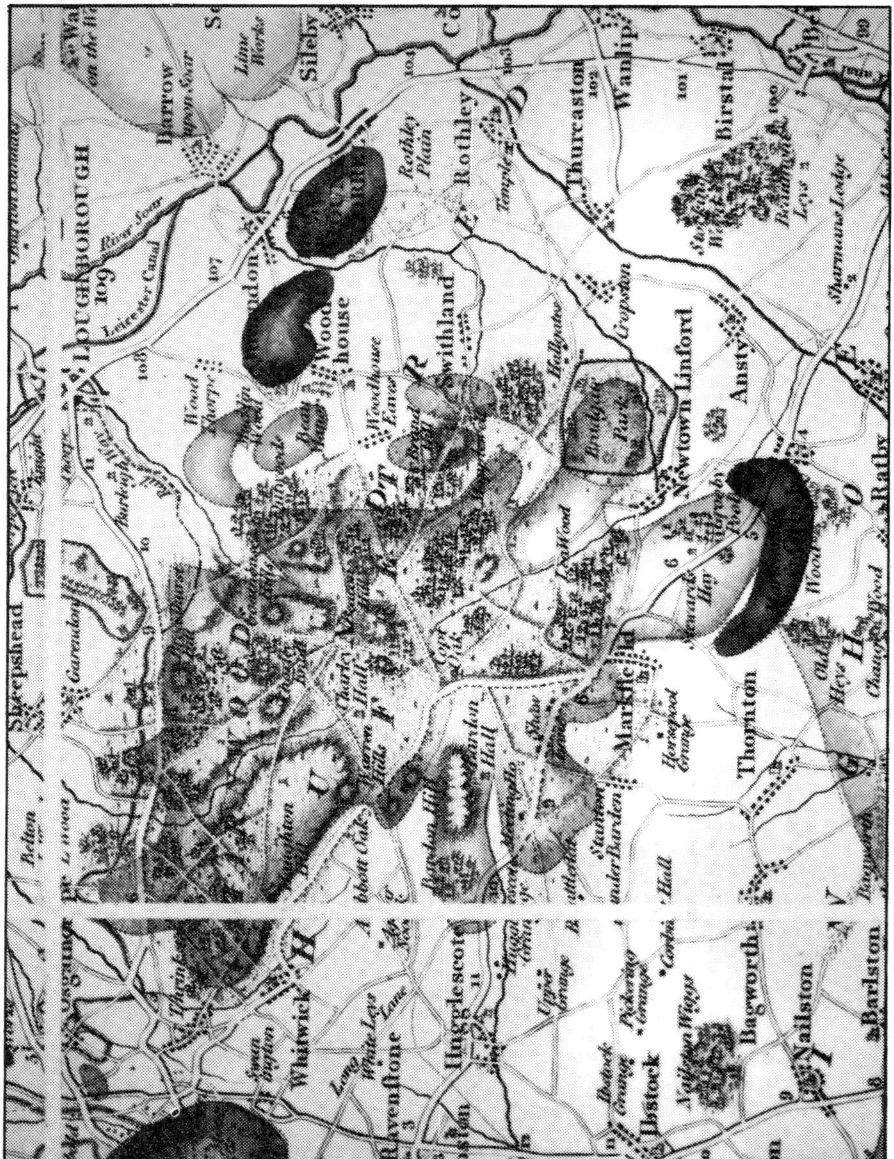

Figure 2. The Charnwood Forest region as shown on the Geological Map of Leicestershire (1821) by William Smith (see Fig. 1). Dark grey shading of Grooby and Mount Sorrel, is "Sienite on Mount Sorrel Stone". Medium grey shading is "Swithland Slate and other argillaceous rocks". Surrounding the lighter area is gypsiferous marl. Part of the Leicestershire coalfield can be seen in dark shading at the extreme northwest.

HISTORY OF GEOLOGICAL STUDY

Charnwood Forest and its rocks were known to the pioneer geologists, though their age and relationships were in doubt. Thus William Smith, the "father of stratigraphy", noted slaty rocks on his map of England and Wales as early as 1799 and indicated "Swithland Slates and other argillaceous rocks" on his county map of Leicestershire in 1821. This showed the rough outline of Charnwood Forest, but Smith made no comment on it in his writings.

His disciple John Farey was a mineral surveyor who strayed over the county boundary into Leicestershire in his classic *General Review of the Agriculture and Minerals of Derbyshire* in 1811. In this he outlined Smith's scheme of succession of strata and applied it in the Derbyshire area in what was effectively the first regional geological memoir. In it Farey listed most of the summits of hills in Charnwood Forest as "coarse slate or syenite", but he believed that "Charnwood Forest . . . was a huge nodular concretion in the red earth or gypsiferous marl, underlying the lias clay". In thus regarding the Charnwood Forest rocks as a "nodule" in the Trias he appears to have been following the Wernerian system of classifying rocks into Primary, Transition, Secondary and Tertiary divisions, and so he found it difficult to understand how Primary rocks could occur in such intimate association with the higher Secondary rocks. Farey also wrongly regarded the slates as being without stratification.

Whatever its shortcomings, this account of Farey's was probably the first of any British area now known to be Precambrian. Both Smith and Farey, and many subsequent writers, included the granite mass of Mountsorrel as part of Charnwood Forest, though it has little petrographic affinity and is now known to be of Late Silurian age about 405 million years old (Miller and Podmore, 1961). Mountsorrel will not be considered in detail in this paper.

G.B. Greenough, politician, geographer and one of the founders (and first President) of the Geological Society of London in 1807, up-dated Smith's map in 1819, though at a smaller scale. On it he showed Charnwood Forest as "porphyry and clay slate", the first recognition of the presence of porphyritic rocks.

Six years later, W.D. Conybeare and W. Phillips (1822) – an unlikely pair of pioneer geologists since Conybeare was Dean of Llandaff and Phillips was a printer and bookseller – produced the first textbook of British stratigraphy in which they gave the first section across Charnwood Forest. It was shown as a mass of unstratified rock against which the Triassic rock terminated abruptly. Their book was Volume 1, going as far back as the Old Red Sandstone only, so if Volume 2 for the older rocks had ever appeared it might have dealt with Charnwood in more detail.

Within a year or so, Phillips had revised this view and, with the assistance of Luck Kent (Phillips and Kent, 1824), produced a map discriminating between slate, greywacke, porphyry, syenite and Mountsorrel granite, and noting that the Triassic New Red Sandstone overlapped into the hollows in the Forest. They noted the dip of the stratified slates: "If these layers be taken as the order of deposit, and therefore regular stratification, they tend to show it having taken place in a mantle-shaped form". This, with their observations of the consistent NW-SE strike and their distinction of bedding from cleavage, shows that the basic structure of Charnwood

Forest was recognized as early as 1824, only 13 years after Farey's work; it also shows that the fundamental distinction of bedding and cleavage in "Primary" rocks was by then clearly established. However, this did not preclude confusion with much younger fold structures in the adjacent coalfield!

The nature of the Charnwood Rocks very soon attracted Adam Sedgwick, the first professor of geology in the University of Cambridge, whose studies in the slates of North Wales founded the Cambrian system. Unfortunately, Sedgwick (1834) wrote little on Charnwood and it was his friend, J.B. Jukes (1842), of the Geological Survey (in Potter's *History and Antiquities of Charnwood Forest,* 1842) who confirmed much of the above, adding that the folding was earlier than the New Red Sandstone (Trias), though he still confused it with the post-Carboniferous folding seen in the Carboniferous inliers of Breedon-on-the-Hill. However, Sedgwick (1834) did present a comparison of the Charnian slates with those associated with Borrowdale Volcanic Series of the Lake District (Ordovician) and of the Charnwood syenites with the intrusives of the Lake District.

Anonymous writings in White (1846 and 1863), subsequently attributed to the Rev. W.H. Coleman (1846, 1863) by Watts (1947), compared the Charnwood rocks with other ancient rocks in Britain. "The Slate Rocks of Charnwood are amongst the oldest in the British Isles . . . exceed in antiquity the Longmynd and Llanberis . . . the only possible older rocks are the Gneissic Series about Cape Wrath in the North of Scotland". Coleman also opined that the porphyries were clay-slate heated to melting point and crystallized on cooling, but that they must have flowed to some extent and enclosed fragments of less fusible rocks. This statement indicates that certain peculiar gradational and xenolithic textures known in the porphyroids had already been noticed but could not be explained; they are still problematical today. Coleman also deduced that since there were no Devonian rocks present, Charnwood Forest must have stood above sea-level then.

THE FIRST GEOLOGICAL SURVEY

The Geological Survey officers moved into the area in the late 1850s though at first they only mapped the geology on a 1 inch : 1 mile scale. Later Hull (1860) added to the detail, though he still did not establish a stratigraphic succession within the Charnian. This was an era of petrographic investigation and the Surveyors noted that Charnwood was a miniature mountain range partly made of lavas poured out on the bed of the ocean. They described every grade of alteration from common unaltered slate to "rocks which in hand specimen seem to be igneous but on a large scale show traces of stratification, showing them to be sedimentary . . . and thus pass into so-called igneous rocks of deep-seated kinds". . . . "In Charnwood Forest it would be easy to arrange a suite of specimens showing a perfect passage from stratified to igneous rocks". This was at the time of the classic experiments of melting and cooling rocks at different rates and of Sorby's application of the microscope to observe the results.

Holl (1865) first linked Charnwood rocks with the Precambrian by comparing them with the Malverns, Donegal and the St. Lawrence area of Canada. About this time James Plant (1865), one of a succession of geological curators of Leicester City Museum, drew attention to the NW-SE strike of Charnwood being the same as that in the N.W. Highlands, though this is now known to be fortuitous. Plant also commented on the variety of rocks in Charnwood in contrast with the monotony of the Longmynd rocks in Shropshire.

A number of workers filled in details during the 1860s, and 1870s, e.g. Baden-Powell (1868), Mott (1868) and Maw (1868), who gave the first correct interpretation of the relationship of the unconformable Trias to the valleys eroded in the Charnian surface. W.J. Harrison (1877), curator of Leicester City Museum and a shrewd observer of glacial lake deposits and of the Mount Sorrel granodiorite, summarized all previous work and gave a very useful description of many exposures, but he added little that was new in concept and so cannot be said to have advanced knowledge greatly. A Middle Pleistocene pro-glacial lake covering much of Leicestershire was subsequently named Lake Harrison in his honour.

THE WORK OF HILL AND BONNEY

A series of papers by C.E. Hill and T.G. Bonney, spread over the period from 1877 to 1891, attempted to give a full description of the petrography and to deduce the origin of the Charnian rocks. Both were local men, Hill being born in Leicester and Bonney in Uppingham; both became tutors at St. John's College, Cambridge, where Bonney assisted the ageing Sedgwick, and subsequently taught Watts. Hill later entered the church, while Bonney became professor at University College, London. They mapped in detail at a time when the Geological Survey were still using the 1 inch : 1 mile scale, but their final map was crudely presented and difficult to interpret. Even so their work was the foundation of all subsequent research in spite of its mistakes, such as regarding some of the felsites and porphyries as clastic. (This was, however, in the very early days of microscopic petrography and perhaps may be excused). They deduced that the Charnian rocks included lava-streams, bomb-deposits (including some bombs which had fallen into flowing lava), dust tuffs, and volcanic breccias. Stratigraphically they laid the basis for the subsequent three-series division of the sequence by Watts.

The work of Hill and Bonney stimulated much local investigation and many notes by Horace Brown and Montagu Browne, another curator, appear in the *Transaction of the Leicester Literary and Philosophical Society* during the 1880s and 1890s. Earlier contributions by Society members were summarized by Mott (1878). About this period the Geologists' Association held one of their annual excursions to Charnwood Forest and the Report contains the following delightful extract regarding one of the slate quarries in Swithland Wood: "It was already too dark, however, for much accurate observation. Two members of the party volunteered to descend the chasm and were, unfortunately, left behind; a large portion of the remainder missed their way in the big wood and after wandering about for some time, ultimately regained the carriages" (Hudleston, 1875).

SUBDIVISION OF THE CHARNIAN SEQUENCE

The Geological Survey returned to the area for the primary six inches to one mile survey about the turn of the century, but the surveyor, C. Fox-Strangways, to a large extent relied on the mapping of that eminent geologist W.W. Watts. For a time Watts was petrographer to the Geological Survey, and after a spell as assistant to

Professor Lapworth at Birmingham he became Professor at Imperial College, where he was for many years leader of his profession. While at Birmingham he was a very active University extramural course teacher all over the Midlands. During his period with the Geological Survey he mapped in Charnwood (1900) and the results were incorporated in the Loughborough and Atherstone (1905) 1 inch : 1 mile sheets (nos. 141 and 155) and in Fox-Strangways' descriptive Memoirs. The six inches to one mile maps (sheets Leics. 16 SE, 17 SE, 17 SW, 18 SW, 24, 25 NW and 30 NW) were published by the Geological Survey between 1892 and1905.

Watts (in Lapworth *et al.*, 1898, p. 335) introduced the term "Charnian System" and divided the sequence into three divisions, usually referred to as Series, and several subdivisions:

Brand Series:	Swithland Slate Brand Grits Hangingstones Conglomerate
Maplewell Series:	Woodhouse Beds (Slates) Slate Agglomerate Beacon Beds (Slates) Felsitic Agglomerate
Blackbrook Series:	Blackbrook Beds (Slates)

In addition he noted that the porphyries (for which he used the term porphyroid) were partly extrusive lava flows from one or more volcanic vents and partly intrusions of indeterminate form. He deduced that volcanoes in northwest Charnwood had contributed "bombs" to the "bombrock" agglomerate and that this was of equivalent age to the Beacon Beds. Since pebbles eroded from the porphryoids were to be found in the Hangingstones Conglomerate of the Brand Series, it established that the porphyroids were largely contemporary with the Maplewell Series. By contrast, the large coarse-grained intrusive masses of Markfield and Groby, later described under the name "markfieldite" as a variety of diorite by Hatch in his Petrography textbook, were shown by Watts (1947) to post-date the Swithland Slates.

Although markfieldite is now known to be a diorite, Watts and most of his contemporaries consistently referred to it as a syenite. The intense alteration of the feldspars made them difficult to determine in thin section, and it was not until 1952 that Snowball (Keeper of Geology in Leicester Museum) drew attention to the dioritic composition of these rocks.

Watts wrote a number of short accounts of his mapping around 1900 to 1910, particularly in the Geologists' Association's *Geology in the Field,* but his full version did not appear in print until just after his death in 1947. It still remains the standard work on Charnwood and is a classic in its own right. Apart from a detailed description of the Forest it contains a much more comprehensive review of previous geological investigation than is possible in the limited space available herein. By present-day standards, some of Watts' findings need reinterpretation and his petrographic nomenclature is in need of revision.

A few points on his maps were challenged by the Leicester Literary and Philosophical Society members between the World Wars, notably F.W. Bennett, E.E. Lowe, H.H. Gregory and F. Jones (1928). Brought together through the Leicester Literary and Philosophical Society, these four collaborators were a curious mixture. Bennett was a medical man but a distinguished amateur geologist as well and benefactor of the infant University College of Leicester, where the Chair of Geology and one of the buildings bear his name; Lowe was Director of Leicester City Museum and Gregory the Curator of Geology therein, while Jones was a lecturer in the University of London. They noted that some distinctive rock types, such as the Felsitic Agglomerate, were found to be repeated in some sections and not in others. We appreciate today that it is a normal phenomenon for repetitions of the same type of volcanic activity but 40 years ago it led to a stratigraphic controversy! Bennett published a number of sketch maps and short descriptions of various parts of Charnwood Forest while Lowe was more concerned with Mountsorrel. Jones carried out detailed structural work at a time when this aspect of geological science was little known. Gregory provided a brief but useful review for the British Association's "Scientific Survey" in 1933.

RECENT STUDIES

Watts' Series and other units were converted into two groups and a number of formations by Dunning (1975) without discussion of their definition on current stratigraphic principles; and it has been left to Moseley (1978) to attempt to sort out the stratigraphic controversies by a detailed remapping of the whole of Charnwood Forest. Moseley has redefined Watts Charnian as a Supergroup, subdivided into three Groups named as for Watts' Series, with further subdivision into a number of formations and members (Table I).

In particular Moseley has noted that slump breccias occur at a number of separate horizons, none of them of more than very local importance in correlation, so that earlier use as marker horizons was unjustified.

PALEONTOLOGICAL AND RADIOMETRIC DATING

The work of Watts (1910) and Bennett *et al.* (1928) made it clear that Charnwood is a complex inlier of mildly metamorphosed volcanics and sediments surrounded and partly covered by Triassic rocks. Its Precambrian age was still occasionally disputed until the discovery of Charnwood-type igneous rocks at Nuneaton, 15 miles (24 km) away, clearly covered unconformably by Lower Cambrian quartzites (Wills and Shotton, 1934). Even then there was doubt as to where in the Precambrian the Charnian rocks should be placed. Were they older or younger Precambrian? The usual key of comparison of fossils could not be used and radioactive dating was still in its infancy and inapplicable.

It was not until 10 years after Watt's death that frond and disc-like fossils were found by Roger Mason (a Leicester schoolboy, now Lecturer in Geology at University College, London) in 1957 and named *Charnia masoni* and *Charniodiscus concentricus* by the writer (Ford, 1958, 1968). Together with *Cyclomedusa davidi,* subsequently discovered at a different locality, these are members of what has become known as the Ediacaran fauna, well known in South Australia through the work of Glaessner (1962, 1966) and his colleagues in the University of Adelaide.

TABLE I
THE CHARNIAN SUPERGROUP

GROUPS	FORMATIONS	THICKNESS (METRES)	MEMBER
Brand Group	Swithland Fm.	260	None Differentiated
	Brand Hills Fm.	102	Stable Pit Quartz Arenite Member Hanging Rocks Conglomerate Member
Maplewell Group	Bradgate Fm.	c. 643	Hallgate Tuff Member Sliding Stone Slump Breccia Member
	Beacon Hill Fm.	1454	Fine-Grained Tuffaceous Pelites Member Outwoods Conglomerate and Slump Breccia Member Fine-Grained Tuffs Member Buck Hills Member Sandhills Lodge Member Beacon Fine-Grained Tuffs Member Benscliffe Member Charnwood Lodge Volcanic Breccia Member
Blackbrook Group	Blackbrook Reservoir Fm.	c. 612	Upper Fine-Grained Tuffaceous Pelites Member Coarse-Grained Tuffs Member Lower Fine-Grained Tuffaceous Pelites Member
	Ives Head Fm.	c. 823	South Quarry Slump Breccia Member Lubcloud Greywackes Member Morley Lane Tuffs Member

The Charnwood fossils place the Charnian firmly in the very late Precambrian when metazoan animal fossils evolved.

Among the late Precambrian fossils found in the Conception Group of South-eastern Newfoundland (Misra, 1969) were frond-like forms which subsequent studies by M.M. Anderson and by the writer have shown to be closely comparable with *Charnia masoni*.

The palaeontological assignment to the late Proterozoic has been supported to some extent by Meneisy and Miller's (1963) radiometric dating in the Geophysical Laboratories of the University of Cambridge of a porphyroid at Bardon Hill as 684 ± 29 m.y. and of a diorite at Markfield at 547 ± 24 m.y. They found that most Charnwood rocks show evidence of re-heating and argon loss, particularly as a result of the intrusion nearby of the late Silurian Mountsorrel granodiorite, so these are both regarded as minimum dates; on stratigraphical grounds the Charnian should be at least 600 m.y. old.

Rubidium-Strontium methods of dating have been applied to the igneous rocks of Leicestershire by Cribb (1975) with results which are not entirely consistent with previous studies. The northern and southern diorities of Charnwood yielded dates of

311 ± 92 and 552 ± 58 million years respectively. While the latter date for the markfieldite intrusions would place them on the Precambrian-Cambrian boundary and would be consistent with the geological history outlined above, the date for the northern diorites seems low, in view of the petrographic similarity of the two groups of intrusions. The low figure may reflect alkaline metasomatism associated with Hercynian mineralization. Too few results were obtained from the porphyries to construct an isochron, but the strontium isotope ratios suggest an age between 500 and 700 million years, broadly confirming the K-Ar date of Meneisy and Miller (1963).

THE TECTONIC SETTING

With the age of the Charnwood rocks reasonably well established, attention has turned to placing them in a plate tectonic setting or in an associated sedimentary and volcanic environment. J.H. Taylor (1958), late Professor of Geology at King's College, London, regarded the later Charnian rocks, the Brand Series, as of post-orogenic molasse type, that is, the erosional products from denudation of a nearby mountain range, but Baker (1971) in his general review of the Proterozoic history of southern England, thought that the Charnian rocks were more reminiscent of a mobile belt owing to the amount of volcanic material. He also regarded the diorite intrusions as post-orogenic, so that he tentatively re-assigned the Charnian to a much earlier age in the Middle Proterozoic – without considering the palaeontological implications and disregarding the radiometric dates as being too low. Baker thus placed the Charnian as older than the Uriconian volcanics of Shropshire, but Wright (1969) correlated these with the Charnian, and this view seems much more reasonable in view of the palaeontological and radiometric evidence of a late Proterozoic age. A corollary of the difference in the Wright (1969) and Baker (1971) interpretations of the nature of the Charnian sequence is that Wright correlates the Charnian at least in part with the Caldecote Volcanic Series at Nuneaton whilst Baker considers the latter to be considerably younger. Wright's view is thus much closer to that of Taylor and is consistent with palaeontological and radiometric dating evidence. Thus the Charnian rocks can be regarded as a volcanic-sedimentary association formed as part of a fold-belt close to a plate-margin which could have been part of the Proto-Atlantic rift in late Precambrian times. However, Dunning (1975) appears to regard the Charnian as part of a volcanic and sedimentary cover of a Midland Craton. Correlation of the Lower and Middle Charnian with the Uriconian is suggested and the Upper Charnian is regarded as a possible equivalent of the Longmyndian sediments. The nature and form of the Midland Craton are still highly speculative.

Within a broad reconstruction of the North Atlantic in Lower Palaeozoic times which suggested correlation of the English Midlands Precambrian with that of Southeast Newfoundland (Kay, 1969), Papesik (1970) drew petrological comparisons between the Harbour Main volcanics of Newfoundland and the Uriconian and, by inference, the Charnian, while Anderson (1972) compared the Conception Group of Newfoundland (at least partly equivalent in age to the Harbour Main Volcanics) with both Longmyndian and Charnian. Comparison of radiometric dates and Late Precambrian fossils in both areas strengthens the correlation and the dating of at least part of the Charnian at around 684 million years. The Charnian and Conception Group strata were interpreted as a partly volcano-genic greywacke sequence filling an early phase of the Proto-Atlantic Rift, between around 700 and 600 million years ago.

On geochemical grounds Thorpe (1972, 1974) has argued that although the Charnian porphyries are similar to and of broadly the same age as the Uriconian-Longmyndian sequence, there are significant trace-element differences. They are, however, of calc-alkaline type and probably originated at or near a plate-margin. Cribb (1975) has noted that the initial strontium isotope ratios suggest that the origin of the igneous rocks is probably partial melting of mantle and/or lower crustal material in a subduction zone between oceanic and continental plates.

Rast and Crimes (1969) have added to the confusion of the placement of the Charnian in relation to Precambrian episodes in the history of the Caledonian orogenic cycle by noting Meneisy and Miller's (1963) Precambrian date for Charnian porphyries and then placing the Charnian with the Uriconian, as an immediately post-Laxfordian volcanic phase around 1400 million years. No evidence for this was presented.

However, Rast *et al.* (1976) have subsequently reviewed the whole southeast flank of the Caledonian orogenic belt. In this they have argued that the Charnian is the broad correlative of the Longmyndian and the upper part of the Uriconian in the Welsh Borders, and the correlative of the Conception Group with some part of the Harbour Main Volcanics in Southeast Newfoundland, thereby supporting the deductions of Kay, Thorpe and Papezik. They regard the Charnian as a volcano-genic sequence resting on a metamorphic basement (visible in other parts of the belt but not in Charnwood) which was part of a Precambrian continent. Rast *et al.* distinguish an Avalonian orogeny in Late Precambrian times (post-Charnian) as a separate continental collision phase from that of the Caledonian, with the Brand Series at the top of the Charnian as a molasse facies, as noted by Taylor (1958), although it is involved in the same folding as the rest of Charnwood Forest. The Caldecote Volcanic Series of Nuneaton are regarded by Rast *et al.* as post-orogenic and thus younger than the Charnian.

The wider extent of Charnian rocks in the sub-surface outside the area of Charnwood remains a problem. Rocks "of Charnian type" have been encountered in a number of boreholes sunk in the search for oil in the English Midlands and recorded by B.P.'s Chief Geologist, now Sir Peter Kent (1967), but direct correlation has so far proved impossible. The South Leicestershire diorites, generally thought to be of Palaeozoic age and a part of the Caledonian orogenic cycle (Le Bas, 1968, 1972) were thought by Cribb (1975) on the basis of Rb/Sr dates to be of roughly the same age as the markfieldite intrusions in Charnwood, i.e., about at the base of the Cambrian.

Regarding structural detail, Evans (1963) has shown that the NW-SE trend of the cleavage is not consistent over the whole area and that it is not quite coincident with the fold axis, as claimed by Watts and Bennett *et al.* Detailed mapping by Moseley (1978) has shown that the cleavage is slightly oblique to and post-dates the main fold, but that there is also some minor folding both contemporary with cleavage and flexural slippage later than the cleavage.

No mineral deposits of consequence are known in Charnwood Forest which are demonstrably of Precambrian age. A few specks of gold were, however, found in a quartz vein on Bardon Hill (King, 1967) and they may represent either a Precambrian or a subsequent phase of mobilization.

CONCLUSIONS

Thus, in the present state of knowledge, the Charnian Supergroup consists of some 12,000 feet (3900 metres) of fine and coarse volcano-genic sediments, volcanic breccias, intrusive and extrusive rhyo-dacites, fine and coarse arenites and greywackes, intruded by slightly younger diorites. Post-Charnian folding has given rise to an asymmetrical anticlinal structure plunging gently southeast, but truncated to the north by a major east-west fault. Cleavage is slightly oblique to the folding and affects all except the coarsely crystalline igneous rocks. Primitive fossils and the few radiometric dates suggest a Late Proterozoic age. Much detailed work on the mode of eruption of the volcanics and on the sedimentation processes still remains to be done. The petrogenesis of the intrusive rocks is far from being understood, and correlation with other British Precambrian areas is still in some doubt.

There is little doubt that as geological science advances the story of the Ancient Rocks of Charnwood Forest will become even more clear, and the generations of geological students who visit the area will learn even more from our rocks – provided the "No Hammering" rule is obeyed and some rocks are left for inspection!

ACKNOWLEDGEMENTS

The substance of this article was presented verbally at the History of Geology committee's Symposium on the History of Research in Precambrian Geology held in Montreal in 1972. The comments of members of the committee have been incorporated as far as possible. The constructive criticism of Dr. R.J. King, Dr. W.A.S. Sarjeant and Mr. J.B. Moseley is gratefully acknowledged.

POSTSCRIPT

Since writing the above Helen Boynton (1978) has recorded some further occurrences of fossils and pseudo-fossils in the Charnian, while Jenkins and Gehling (1978) have redefined *Charniodiscus* Ford and have discussed its relationship to other Ediacaran frond-like fossils elsewhere in the World. Ford (1978) has also discussed these fossils in a review of Precambrian life.

The Institute of Geological Sciences have recently re-issued the Loughborough geological map (Sheet 141) on a scale of 1:50000 with but minimal revision of Precambrian details.

REFERENCES

Anderson, M.M., 1972, A possible time span for the Late Precambrian of the Avalon Peninsula, Southeastern Newfoundland, in the light of Worldwide correlation of fossils, tillites and rock units in the succession: Canadian Journal of Earth Sciences, vol. 9, p. 1710-1726.

Baden-Powell, 1868, On the igneous rocks of Charnwood Forest and its neighbourhood: Geological Magazine, vol. 5, p. 111.

Baker, J.W., 1971, The Proterozoic History of Southern England: Proceedings of the Geologists' Association, London, vol. 82, p. 249-266.

Bennett, F.W., Lowe, E.E., Gregory, H.H. and Jones, F., 1928, The Geology of Charnwood Forest: Proceedings of the Geologists' Association, London, vol. 39, p. 241-298.

Boynton, H.E., 1978, Fossils from the Pre-Cambrian of Charnwood Forest, Leicestershire: Mercian Geologist, v. 6, p. 291-296.

[Coleman, W.H.], 1846, Sketch of the Geology of Leicestershire, *in* White's History, Gazetteer and Directory of Leicestershire: Sheffield; White, 704 p., (p. 24-27).

——————, 1863, Geology in Leicestershire, *in* White's History, Gazetteer and Directory of the counties of Leicestershire and Rutland, 2nd edition: Sheffield and London; White, p. 77-111. (Geology section issued separately in 1862.)

Conybeare, W.D. and Phillips, W., 1822, The Geology of England and Wales: London; Phillips.

Cribb, S.J., 1975, Rubidium-strontium ages and strontium isotope ratios from the igneous rocks of Leicestershire: Journal of the Geological Society of London, vol. 131, p. 203-212.

Dunning, F.W., 1975, Precambrian craton of central England and the Welsh Borders: Geological Society of London Special Report No. 6, Precambrian, p. 83-95.

Evans, A.M., 1963, Conical folding and oblique structures in Charnwood Forest, Leicestershire: Proceedings of the Yorkshire Geological Society, vol. 34, p. 67-79.

Farey, J., 1811, A General View of the Agriculture and Minerals of Derbyshire: vol. 1, London, 532 p.

Ford, T.D., 1958, Precambrian fossils from Charnwood Forest: Proceedings of the Yorkshire Geological Society, vol. 31, p. 211-217.

——————, 1968, The Precambrian palaeontology of Charnwood Forest, *in* P.C. Sylvester-Bradley and T.D. Ford (eds.), Geology of the East Midlands: Leicester University Press, 400 p., p. 12-14, Chapter 1 on Charnwood Forest.

Ford, T.D., 1978, Precambrian fossils and the origin of the phyla, *in* Symposium on the Origin of the Phyla: Systematics Assoc. meeting in Hull, April 1978. Volume in press.

Fox-Strangways, C., 1900, The Geology of the Country around Atherstone and Charnwood Forest: Memoirs of the Geological Survey of England and Wales, 102 p.

——————, 1905, The Geology of the Country between Derby, Burton, Ashby and Loughborough: Memoirs of the Geological Survey of England and Wales, 83 p.

Glaessner, M.F., 1962, Precambrian fossils: Biological Reviews, vol. 37, p. 467-494.

——————, 1966, Precambrian palaeontology: Earth Science Review, vol. 1, p. 29-50.

Gregory, H.H., 1933, Geology. Chapter 2 in A Scientific Survey of Leicester and District: Leicester; British Association for the Advancement of Science, 100 p., p. 17-25.

Harrison, W.J., 1877, A sketch of the geology of Leicestershire and Rutland: Sheffield, 67 p. (also in the Directory of the County by W. White, 2nd edition, 1863).

Herbert, A., 1945, Swithland Slate headstones: Transactions of the Leicester Archeological Society, vol. 22, p. 211-240.

Hill, E., and Bonney, T.G., 1877, On the Pre-Carboniferous rocks of Charnwood Forest, part 1: Quarterly Journal of the Geological Society of London, vol. 33, p. 754-789. Part 2 in vol. 34, p. 199-239, 1878; part 3 in vol. 36, p. 337-350, 1880 and supplementary paper in vol. 47, p. 78-100, 1891.

Holl, H.B., 1865, On the geological structure of the Malvern Hills and adjacent districts. (Charnwood Forest, p. 72): Quarterly Journal of the Geological Society of London, vol. 21, p. 72-102.

Hudleston, W.H., 1875, Excursion to Charnwood Forest, Leicestershire: Proceedings of the Geologists' Association, London, v. 4, p. 307-317.

Hull, E., 1860, The geology of the Leicestershire coalfield and the country around Ashby-de-la-Zouch: Memoirs of the Geological Survey of England and Wales.

Jenkins, R.J.F. and Gehling, J.G., 1978, A review of the frond-like fossils of the Ediacara assemblage: Records of the South Australian Museum, v. 17, no. 23, p. 347-359.

Jukes, J.B., 1842, The Geology of Charnwood Forest, *in* T.R. Potter, Appendix to The History and Antiquities of Charnwood Forest (p. 1-31): Hamilton Adams, London. (Reprinted 1857).

Kay, M., 1969, Continental Drift in the North Atlantic Ocean, *in* North Atlantic Geology and Continental Drift – a symposium (p. 965-973): American Association of Petroleum Geologists, Memoir 12.

Kent, P.E., 1967, A contour-map of the sub-Carboniferous surface in the north-east Midlands: Proceedings of the Yorkshire Geological Society, vol. 36, p. 127-134.

King, R.J., 1967, The minerals of Leicestershire, part 1 — The Elements: Transactions of the Leicester Literary and Philosophical Society, v. 61, p. 55-64.

Lapworth, C., Watts, W.W. and Harrison, J., 1898, Sketch of the geology of the Birmingham District: Proceedings of the Geologists' Association, London, v. 15, p. 313-416.

Le Bas, M.J., 1968, The Caledonian igneous rocks *in* Sylvester-Bradley, P.C. and Ford, T.D., eds., The Geology of the East Midlands: Leicester, 400 p. (p. 41-58).

——————, 1972, Caledonian Igneous rocks beneath Central and Eastern England: Proceedings of the Yorkshire Geological Society, v. 39, p. 71-86.

Maw, G., 1968, The Trias of Charnwood Forest: Geological Magazine, v. 5, p. 197-198.

Meneisy, M.Y. and Miller, J.A., 1963, A geochronological study of the crystalline rocks of Charnwood Forest, England: Geological Magazine, v. 100, p. 507-523.

Miller, J.A. and Podmore, J.S., 1961, Age of the Mountsorrel Granite: Geological Magazine, v. 98, p. 86-88.

Misra, S.B., 1969, Late Precambrian (?) fossils from southeastern Newfoundland: Bulletin of the Geological Society of America, v. 80, p. 2133-2140.

Moseley, J.B., 1978, The geology of Charnwood Forest: PhD thesis, University of Leicester, (in preparation).

Mott, F.T., 1868, Charnwood Forest, its air, its scenery and its natural curiosities, 3rd edition: London, Kent, 144 p.

——————, 1878, A sketch of the history of the Leicester Literary and Philosophical Society: Midland Naturalist v. 1, p. 299-302.

Papezik, V.S., 1970, Petrochemistry of volcanic rocks of the Harbour Main Group, Avalon Peninsula, Newfoundland: Canadian Journal of Earth Sciences, v. 7, p. 1485-1498.

Phillips, W.D. and Luck Kent, 1824, Observations on the rocks of Mountsorrel and Charnwood Forest etc: Annals of Philosophy, v. 7, p. 1-19.

Plant, J., 1865, Are the slate rocks of Charnwood Forest Laurentian?: Geological Magazine, v. 2, p. 233.

Rast, N. and Crimes T.P., 1969, Caledonian orogenic episodes in the British Isles and Northwestern France and their tectonic and chronological interpretation: Tectonophysics, v. 7, p. 277-307.

Rast, N., B.H. O'Brien and R.J. Wardle, 1976, Relationships between Precambrian and Lower Palaeozoic rocks of the Avalon Platform in New Brunswick, the Northeast Appalachians and the British Isles. Tectonophysics. v. 30. pp. 315-338.

Sedgwick, A., 1834, [Abstract of a paper on Charnwood]: Philosophical Magazine, ser. 3, v. 4, p. 68-69.

Shotton, F.W., 1959, New petrological groups based on axes from the West Midlands: Proceedings of the Prehistoric Society, vol. 20, p. 135-143.

Snowball, G.T., 1952, The Leicestershire Intrusions: Bulletin. Leicester Museum and Art Gallery, 3rd series, no. 1, p. 2-7.

Taylor, J.H., 1958, Precambrian sedimentation in England and Wales: Eclogae Geologicae Helvetiae, v. 51, p. 1078-1092.

Thorpe, R.S., 1972, The geochemistry and correlation of the Warren House, the Uriconian and the Charnian volcanic rocks from the English Pre-Cambrian: Proceedings of the Geologists' Association, London, v. 83, p. 269-285.

——————, 1974, Aspects of magmatism and plate tectonics in the Precambrian of England and Wales: Geological Journal, v. 8, p. 115-136.

Watts, W.W., 1910, The geology of Charnwood Forest, *in* H.W. Monckton and R.S. Herries (eds.) Geology in the Field: the Jubilee Volume of the Geologists' Association, London (p. 770-785): Stanford, for the Geologists' Association, 916 p.

——————, 1947, Geology of the Ancient Rocks of Charnwood Forest: Leicester Literary and Philosophical Society, 160 p.

White, W., 1846, History, Gazetteer and Directory of Leicestershire, 1st edition: Sheffield; White, 704 p. (2nd edition including Rutland, 1863, 852 p.)

Wills, L.J. and Shotton, F.W., 1934, New sections showing the junction of the Cambrian and Precambrian at Nuneaton: Geological Magazine, v. 71, p. 512-521.

Wright, A.E., 1969, Precambrian rocks of England, Wales and Southeast Ireland, *in* Kay, Marshall (ed.), North Atlantic – Geology and Continental Drift (p. 93-109): American Association of Petroleum Geologists, Memoir no. 12.

History of Concepts in Precambrian Geology, edited by W.O. Kupsch and W.A.S. Sarjeant, Geological Association of Canada Special Paper 19, 1979

HISTORY OF THE CONCEPTS OF PRECAMBRIAN GEOLOGY AS APPLIED IN THE MYSORE AREA OF INDIA

S. Sambe Gowda and R. Srinivasan
Department of Geology, Bangalore University, Bangalore, India
Consulting Geologist, Geomysore Services, 12 Palace Road, Bangalore 560 052 India

---

ABSTRACT

The Mysore area is considered as one of the oldest geological provinces of the Indian sub-continent and has served as type area for the 3000 to 2100 m.y. old Precambrian. Attempts made to unravel the geology of this area have therefore profoundly influenced the thinking about the evolution of the Peninsular Shield. This paper gives a summary of the evolution of concepts of the Precambrian of Mysore State with regard to origin, nomenclature, classification, correlation, structure and micropalaeontology.

The geological investigations of Mysore date back to Newbold (1844). They can be considered as three phases of study: the Pre-Smeeth period (1844-1900), the Smeeth period (1900-1920) and the Post-Smeeth period (1920 to present).

In the Pre-Smeeth period, Newbold's "protogine schists" were named the "Dharwar System" by Foote (1886) and were considered as metasediments overlying the Archaean 'fundamental gneiss'. During the Smeeth period, the 'fundamental gneiss' was renamed the "Peninsular Gneiss" and considered as intrusive into the Dharwars. Both the gneisses and the Dharwars were included into the Archaean (Holland's "Vedic"). The existence of sediments in the Dharwars was not acceptable during this period.

In the Post-Smeeth period, definite evidence of sedimentation in the Dharwars was obtained. The geosynclinal concept (Holmes, 1955; Narayanaswami, 1964) and the continental accretion hypothesis (Pichamuthu, 1962) have been introduced. Evolving Sreenivasa, 1969) and the accumulating radiometric age data have forced a revision of the concepts of geosynclinal evolution, the evidence of life in these sediments (Gowda and earlier classifications, those formulated by Smeeth (1916), Rama Rao (1940) and Pichamuthu (1951) having been revised by Nautiyal (1966), Radhakrishna (1964, 1967) and Srinivasan and Sreenivas (1971). The inclusion of the Dharwars into the Palaeoproterozoic, rather than the Archaean, is another consequence of these new observations.

RESUME

La région de Mysore est considérée comme l'une des plus vieilles provinces géologiques du sub-continent indien et a servi de région type pour le Precambrien compris entre 3000 et 2100 millions d'années. Les tentatives qui furent faites pour démêler la géologie de cette région ont donc profondément influencé la pensée sur l'évolution du

Bouclier Péninsule. Cet article donne un résumé de l'évolution des concepts sur le Précambrien de l'Etat de Mysore en ce qui concerne l'origine, la nomenclature, la classification, la corrélation, la structure et la micropaléontologie.

Les recherches géologiques sur Mysore remontent à Newbold (1844). Elles peuvent être considérées comme comportant trois phases d'étude: la période Pré-Smeeth (1844-1900), la période Smeeth (1900-1920) et la période Post-Smeeth (de 1920 à la période actuelle).

Dans la période Pré-Smeeth, les "schistes protogines" de Newbold furent appelés par Foote (1886) le "Système Dharwar" et furent considérés comme des métasédiments recouvrant le "gneiss fondamental" archéen. Durant la période Smeeth, le "gneiss fondamental" fut renommé "gneiss Péninsulaire" et considéré intrusif dans le Dharwar. Les gneiss et le Dharwar furent inclus dan l'Archéen ("Vedic" de Holland). L'existence de sédiments dans les Dharwar ne fut pas acceptés au cours de cette période.

Dans la période Post-Smeeth, de nettes évidences de sédimentation dans les Dharwar furent démontrées. Le concept de géosynclinal (Holmes, 1955; Narayanaswami, 1964) et l'hypothèse de l'accrétion continentale (Pichamuthu, 1962) furent introduits. Le développement des concepts de l'évolution géosynclinale, l'évidence de la vie dans ces sédiments (Gowda et Sreenivasa, 1969) et l'accumulation des données d'âge radiométrique ont forcé une révision des premières classifications, celles formulées par Smeeth (1916), Rama Rao (1940) et Pichamuthu (1951) ayant été révisées par Nautiyal (1966), Radhakrishna (1964 et 1967) et Srinivasan et Sreenivas (1971). L'inclusion des Dharwar dans le Paléoprotérozoïque plutôt que dans l'Archéen, est une autre conséquence de ces nouvelles observations.

INTRODUCTION

Mysore State (Fig. 1) is the type area for the 3000 to 2100 m. y. old Precambrian rocks of India. The present paper attempts to give an account of the concepts applied over a period of a hundred and forty years to the nomenclature, origin, classification, correlation and micropalaeontology of the Precambrian rocks of this State. Smeeth, of the Mysore Geological Department, was mainly responsible for the detailed mapping of the State in 1916 and dominated the geological thinking of his time; therefore we propose to discuss the development of concepts under three headings – Pre-Smeeth Period (before 1900); Smeeth Period (1900-1920); and Post-Smeeth Period (1920 to date).

THE PRE-SMEETH PERIOD

Geological investigations in Mysore State were initiated by Christie (1836), who considered the rocks of the area as being the "transitional rocks of Werner". A comprehensive and systematic description of the geology of South India was given by Newbold (1846, 1848 and 1850). He described the Precambrian rocks under review as "protogine" or "hypogine schists and gneisses" intruded by various granites, considering them as "primitive rocks" because of their supposed great antiquity and their crystalline character. He gave a stratigraphic succession for the rocks of South India and also a cross-section covering the present Mysore State (Figs. 2 and 3). No rock was designated by him as Archaean because no such term existed then.

King (1872) divided the Precambrian rocks of the adjoining Madras State into "crystalline rocks" and "transition rocks", which were termed as the Lower and Upper Transitional System respectively. Foote (1886, p. 98) gave the name "Dharwar System" to the Lower Transitional rocks occurring near the type area of Dharwar, for the following reason:

> "The schistose rocks forming the band near Dharwar, and still more in the central part near Shimoga and Honnali (Mysore), show much less metamorphism than elsewhere generally; and it was here that I was for the first time forcibly impressed by the necessity of trying to establish a separation between these old schists and the granitoid gneiss system in South India. For this reason, and from the fact that this band shows the greatest development of the old schists, I have proposed to call the new system they form the Dharwar system . . . The study of the Sandur hills since then has satisfied me that the schists, or, as they should henceforth be called, the Dharwar rocks, rest on the granite gneiss with marked unconformity."

A map accompanying this article showed Dharwar schist bands separated from the granitoid gneisses (Fig. 4). Foote (1900) maintained that "Dharwar beds" overlaid "Archaean granitoid gneisses".

The Dharwar system was recognised by Newbold (1846) and Foote (1900) as being composed of metasediments and volcanics; the latter, according to Foote, appeared particularly at the base. Foote recognised ripple marks in the quartzites (1900, p. 31). Newbold (1846, p. 51), commenting on the scarcity of limestones in the hypogene series, considered that lime-secreting organisms must have then existed in far lower numbers than they did subsequently. However, he also thought (1850) p. 91) that many of them might have been obliterated due to the "heat metamor-

Figure 1. Index Map of Mysore State and Peninsular India.

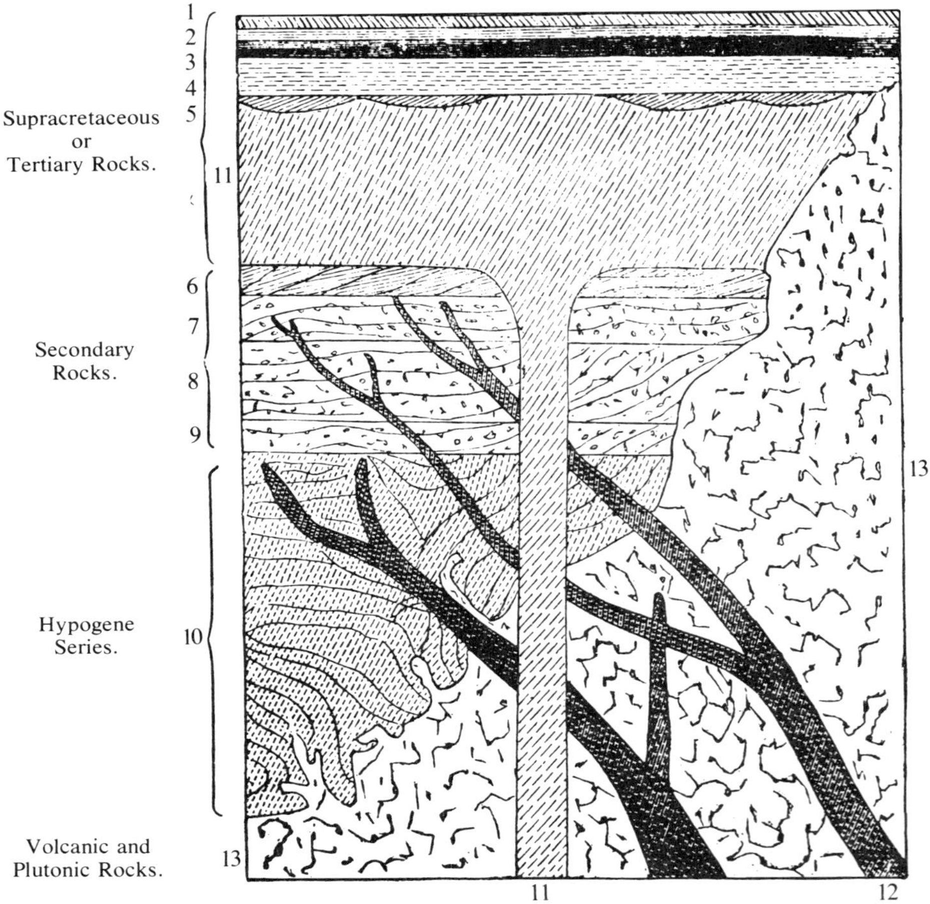

SUPRACRETACEOUS OR TERTIARY STRATA.

1—Sandstone of Coromandel and Paumbum, and Cape Comorin, imbedding existing marine shells.

2—Coromandel Black Clay underlying Madras, &c., and Regur (Pleiocene period).

3—Ancient Kunker, and Gravel imbedding Remains of Mastodon (Pleiocene period).

4—Silicified Wood Deposit of Pondicherry, and older Laterite (Miocene period).

5—Freshwater Limestone of Nirmul, Hydrabad, and Rajahmundry (Eocene period).

SECONDARY STRATA.

6—Limestone beds of Trichinopoly, Verdachillum, and Pondicherry (Neocomien or Lower Chalk).

7, 8, and 9—Diamond Sandstone Group—8 is the Limestone, and 9 the lower Sandstone. (Carboniferous or Devonian.)

10—HYPOGENE SERIES.

Clayslate	Gneiss
Quartzite	Protogine
Talcose	Eurite
Chloritic	Serpentine.
Actinolitic	
Hornblende	
Mica Schists	

VOLCANIC AND PLUTONIC ROCKS.

11—Newer or Overlying Trap.
12—Basaltic Greenstone.
13—Granite
Syenite
Syenitic Granite
Pegmatite
Porphyritic Granite
Protogine
Eurite
Serpentine
Diallage.

Figure 2. Newbold's Synoptical Table of the rocks of Southern India.

phism"; this suggests that he was, after all, not too sure of the unfossiliferous character of the hypogene strata!

THE SMEETH PERIOD

W.F. Smeeth, who became State Geologist in 1899, initiated detailed geological mapping in the State. In the meanwhile, Holland (1909) proposed a comprehensive classification of the geological column of India; he included the schists (Dharwars) and the gneisses into the Archaean and correlated the Dharwars with the Huronian of Canada. In the course of mapping, many workers recognised both igneous and sedimentary schists in Dharwars. Classification on the basis of unconformable horizons, represented by conglomerates and abrupt termination of dykes at stratigraphic contacts, was attempted (Iyengar, 1905, 1908). It is surprising to find that the use of dykes for the purpose of stratigraphic classification has never been used so effectively as during Post-Smeeth time.

Smeeth (1910), struck by the autoclastic nature of the conglomerates of the Kolar Gold Field, considered other conglomerates of Mysore also as autoclastic. He was reluctant to admit a sedimentary origin for any Dharwar rocks. He argued that the quartzites were crushed vein quartz, felsite, and quartz porphyry; that the limestones were calcified traps and metasomatically replaced gneisses; that the ferruginous quartzites were products of metamorphic breakdown of rocks rich in iron amphiboles; and that the kyanite-staurolite schists and sillimanite-cordierite schists were contact-metamorphosed mafic igneous rocks.

Under Smeeth's authoritarian rule, almost all his contemporaries except Sambasiva Iyer (1918) changed their views in favour of an igneous-metamorphic origin for the whole of the Dharwars; Smeeth (1924) accordingly claimed that he had raised a "hornet's nest of autoclastic conglomerates". It was a strange event indeed that such a situation should have arisen in an area where the neptunistic concept had been the first to be applied. This extremist view seems to be unique to the Precambrian of Mysore, for in no other Precambrian of the world has the existence of sediments been so totally denied.

It is also interesting to find that, during the Smeeth Period, the associated mafic igneous rocks were all considered to be lava flows (Sambasiva Iyer, 1899; Slater, 1906, 1908; Smeeth, 1899; Hatch, 1902; and Smith, 1904). Later, however, they all came to be regarded as intrusive (Wetherell, 1904; Jayaram, 1920, 1922; Smeeth, 1924).

The fact that the existence of sediments in the Dharwars was not acceptable to Smeeth prompted him to reject the classification formulated by Iyengar. On the basis of lithology and degree of metamorphism, he proposed a twofold classification for the Dharwars; thus the concept of increasing grade of metamorphism with age came to be applied in the Mysore Precambrian. (The classification proposed by Smeeth is given in Table I).

Detailed geological mapping in the State suggested that there were no rocks older than the Dharwars and that the associated granites and gneisses were intrusive into them. Evidence was discovered of assimilation contact metamorphism and of

Figure 3. Newbold's cross section of the rocks along N lat. 13° and passing through Mysore State.

the disturbance of structural trends at the contacts (Smeeth, 1916, Iyengar, 1920). Four epochs of igneous intrusions were recognised. The earliest of these, represented by the Champion Gneiss, was considered to have a genetic relation with the Champion lode of the Kolar Gold Field; hence the name. The second is the "fundamental gneiss" or "granitoid gneiss" of the earlier workers. As these rocks form a major part of Peninsular India, as it has been shown that they are intrusive into Dharwar schists, and as the fundamental gneiss was not intrusive elsewhere and had a different connotation outside Mysore, Smeeth (1916), at the request of Hayden (the then Director of the Geological Survey of India), gave the name "Peninsular Gneiss" to them. Maclaren (1906) and Middlemiss (1917) did not agree that these gneisses were younger, but Fermor (1909, 1919) endorsed the opinion of Smeeth.

The Peninsular gneisses were considered by Smeeth to be horizontal, sheet-like igneous intrusions. Since the gneisses came to be reckoned as intrusive by some and to represent the basement by others, this led to a controversy about the basement of the Dharwars. In order to explain the absence of a basement gneiss, Smeeth (1926) introduced the concept of refusion of the basement.

The third epoch of igneous intrusion is represented by charnockites, which were recognised in the State by Smeeth (1904). There was unanimity of opinion among Mysore geologists during the Smeeth period that those rocks were of igneous origin. The fourth and final epoch of plutonic igneous activity is represented by the Closepet Granites, which were considered as constituting a "central axial mountain system" by Iyengar (1920). Several periods of dyke activity were believed to have occurred later.

Some changes to the succession suggested by Smeeth (1916) became necessary within a few years. Iyengar (1920) proposed that the eruptive unconformity between Dharwars and intrusives should be placed between the Champion and Peninsular Gneisses, since the alteration products of the Champion Gneisses (to which the name "Champion Gneiss Series" had been given in the Smeeth period) constituted an integral part of the Dharwars. Rama Rao (1921), finding hornblendic rocks intrusive into chlorite and mica schists, could not reconcile his observations with the view that hornblendic rocks always belonged to the lower division of the Dharwars. Smeeth (1924) agreed with these new views, but by that time he had retired from service.

The existence of life in the Dharwars was considered plausible during the early years of the Smeeth period. Iyengar (1905) recognised a fossiliferous sandstone member in the G. R. Formations of the Chitaldrug belt and described petrified lamellibranchs. After Smeeth criticised the sedimentary origin of the G. R. Formations, however, Iyengar rejected his earlier views!

THE POST-SMEETH PERIOD

Following the retirement of Smeeth from the service of the Mysore Geological Department, the prevailing concept of his period was quickly abandoned. Rama Rao (1921) opined that the siliceous grits of the Holenarasipur schist belt, and also the associated kyanite-staurolite schists, were metasediments. Jayaram (1925) proposed a sedimentary origin for the cordierite-sillimanite bearing rocks of Sakarsanhalli. The graphitic schists and garnetiferous granulites were regarded as being metasediments and, in the latter, Rama Rao (1927) recognised relict cross-stratifica-

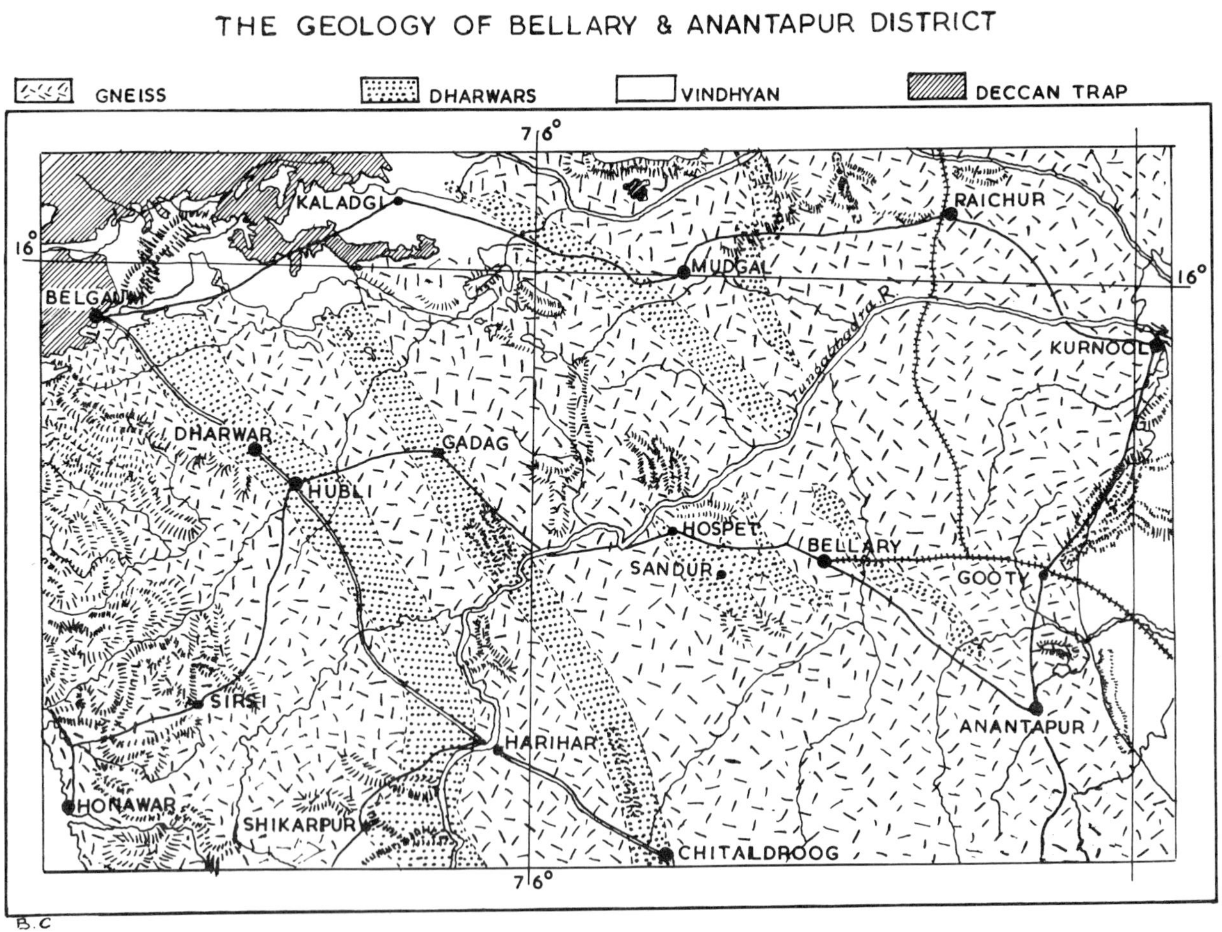

Figure 4. Foote's geological map of the Dharwars and other rocks of the area under discussion.

tion. On petrological and chemical grounds he considered (1934) that the cummingtonite limestones must be metasediments.

At about the same time Pichamuthu (1935), carrying out detailed investigations in the Bababudan schist belt, found graded bedding in conglomerates and grits, current bedding in quartzites and intraformational folds in ferruginous quartzites. Rama Rao (1935) found current bedding, ripple marks, sun cracks and rain-prints in the Dharwars of the Shimoga belt. Convinced by these confirmatory evidences, he as Director swung the official opinion of the Mysore Geological Department in favour of a sedimentary origin for certain components of the Dharwars (1936). (1936).

This new awareness of the sedimentary origin of the Dharwars led to interesting researches on the associated mafic igneous rocks, interpretation of structure and tectonics, understanding of metamorphism, delineation of sedimentary environments, discovery of life and revision of stratigraphy.

Mafic Igneous Rocks

The mafic igneous rocks were all considered to be intrusive during the Smeeth Period. The discovery of an abundance of pillow structures in them in almost all the schist belts (Raghunatha Rao, 1937; Pichamuthu, 1950; Rama Murthy, 1954; Narayanaswamy, 1955) led Pichamuthu (1957) to suggest that these were submarine lava flows of spilitic affinity. Utilising the well-established concepts of aerial extent, the presence or absence of pillow structures and interbedded sediments, and modern treatments of chemical data, Srinivasan and Sreenivas (1969b) recognised the existence of both subaerial flood basalts and submarine pillow lavas in the Dharwars. All of these are believed to be low-potash tholeiitic lavas (Srivinasan and Sreenivas, 1971).

Structure

Pichamuthu (1937) used for the first time the concept of orientation of current bedding in the Dharwars to work out the stratigraphic sequence. The use of sedimentary marker beds in interpreting the structure of the Dharwars led him to suggest that the Dharwars of the Chitaldrug and Shimoga belts constituted limbs of a north-plunging anticlinorium (Pichamuthu, 1951). However, Rama Rao (1962) opined that they are folded into a south-plunging synclinorium, while Narayanaswami (1970) has more recently advocated a north-plunging synclinorial structure. The closure of beds on the erosion surface has been taken by these authors as an evidence of the structure, based on the study of the bedding dips; this has indicated a domal nature for the area. It has been suggested that this dome is a Pre-Dharwarian feature, controlling the sedimentary basin configuration of the Chitaldrug and Shimoga belts during Dharwarian time (Srinivasan and Sreenivas, 1971).

Metamorphism

Pichamuthu and Rama Rao used the structures they were proposing to explain the increasing grade of metamorphism from north to south in Mysore State from green schist through amphibolite to the granulite facies; they attributed this to the progressive increase in depth or load. Thus the concepts of depth zonation in metamorphism of Grubenmann (1904) and the facies concepts of Eskola (1920) came to be applied in Mysore.

An important outcome of this interpretation was that the charnockites of Mysore were shown to be metamorphic, not igneous as had been held during the Smeeth period (Rama Rao, 1945; Pichamuthu, 1953). However, Pichamuthu (1953) considered that there are both metamorphic and rheomorphic charnockites. While Rama Rao (1945) believed that the Dharwars had been repeatedly metamorphosed to give rise to the charnockites, Pichamuthu (1960, 1961) introduced the concept of charnockitisation of Peninsular Gneisses. In contrast, Devaraju and Sadasiviah (1969) claimed that charnockites have retrograded to gneisses. Pichamuthu (1953) recognised two important periods of charnockite formation, one metamorphic and another rheomorphic; but Naidu (1967), on the basis of radiometric age data by then available for the charnockites, has suggested that there is nothing like a charnockite period.

Applying the concepts of metamorphic facies series and metamorphic belts of Miyashiro (1961), Srinivasan and Sreenivas (1971) have suggested that the charnockites of Mysore are low pressure metamorphic rocks. Srinivasan and Sreenivas (1971) have shown that intermediate to high pressure metamorphism prevailed within the Dharwar Schist belts, while negative pressure metamorphism characterises the gneissic country.

The development of concepts of metamorphism in the Dharwars had its implications for the other plutonic intrusives as well. The Champion Gneisses, considered as igneous intrusives into the Dharwars during Smeeth's time, were considered to be metamorphosed rhyolites and tuff by Burdon (1947). This view has been revived by Ziauddin (1971), the associated metaconglomerates being considered to be volcanic agglomerates. Nautiyal (1966) considered that the Champion Gneisses are metagreywackes and that the associated conglomerates are tilloids. Sreenivas and Srinivasan (1968a and b) argued that these latter are early-syntectonic mylonite gneisses and that the associated conglomerates are protomylonites and hence tectogenetic. Srinivas (1969), employing heavy mineral studies of zircons and modal analyses in the light of the concepts of Chayes (1956) and others, concluded that the Champion Gneisses are intrusives evolved through assimilation.

The Peninsular Gneisses, considered as horizontal sheet-like igneous intrusions during the Smeeth period, came to be recognised as products of granitization of the Dharwar schists (Rama Rao, 1939). The concept of migmatization was applied to them by Pichamuthu (1936, 1938), Babu (1955) and Srinivasan and Sreenivas (1969). By employing the ideas of degranitization at the granulite layer of the continental crust proposed by Beloussov (1966), the necessary fluids for regional migmatization could be derived (Srinivasan and Sreenivas, 1969). The concept of stratigraphy in the gneisses, based on evidence provided by cross-cutting relationship and the rule of enclosure, was introduced by Pichamuthu (1970).

The concepts of granitization and alkali metasomatism were extended to the origin of the Closepet Granites also (Radhakrishna, 1956; Suryanarayana, 1960, 1964; Divakara Rao, 1970). Reynolds' concept of basic front was used to explain the occurrence of high-magnesium cordierite-bearing rocks bordering the Closepet Granites (Radhakrishna, 1956).

Sedimentary Environments

Recognition of sediments in the Dharwars enabled an understanding of the paleogeography and resulted in a further modification of the tectonic picture. It was considered that the Dharwar sediments accumulated in isolated basins, in conformity with the concept of Archaean environments as proposed by Van Hise and others (Rama Rao, 1940, p. 48). The geosynclinal concept for the sedimentary environment was first introduced by Pichamuthu (1947), when he also introduced the idea of orogeny in the Dharwars.

Narayanaswami (1964), utilising as evidence the known association of active vulcanism with geosynclines, suggested that the Dharwar geosynclines were eugeosynclinal. Srinivasan and Sreenivas (1971), developing the geosynclinal concept, showed that they were of an ensialic intracratonic type that developed amidst mantle-deep fractures in the fashion of grabens. Radhakrishna (1964) estimated that over 50,000 ft. of sediments accumulated in these geosynclines. Srinivasan and Sreenivas (1968, 1972), on the basis of tectonolithological characters, established that the sediments belong to pregeosynclinal-shelf, geosynclinal-flysch and apogeosynclinal-molasse types, accumulated on platforms, in geosynclines and in intramontane basins respectively; these regimens are shown in Fig. 5.

Dharwar Life

The belated recognition of the existence of sediments in the Dharwars probably hindered the early discovery of Precambrian life in this terrain. Although Iyengar (1905) claimed to have noticed organic remains in the G. R. Formation, he withdrew his views under pressure from Smeeth. Only in 1945 were further fossils reported, Pichamuthu suggesting the existence of blue green algae in the Dharwars cherts. The occurrence of Cyanophycean algae and fungi with a very good histological organisation was subsequently recognised by Gowda (1970) from these same cherts. In addition, Gowda and Sreenivasa (1969) and Gowda and Kumar (1971) have discovered acritarchs in the type Dharwars and the G. R. Formations of the Precambrian complex of Mysore.

Revision of Stratigraphy

The lithological classification established during the Smeeth period was likewise subject to revision once sediments had been recognised in the Dharwars. Rama Rao (1936, 1940) and Pichamuthu (1947, 1951), respectively using the recognition of unconformable horizons and applying the concept of cyclicity in sedimentation, proposed a three-fold classification for the Dharwars. Subsequent to the introduction of the geosynclinal concept, Radhakrishna (1964, 1967) proposed instead a five-fold classification. Applying orogenic concepts, Nautiyal (1966) proposed a further reclassification. Using the concepts of sedimentation, tectonics and magmatic cycles in an evolving geosyncline, Srinivasan and Sreenivas (1968, 1972) proposed a fourfold classification for the Dharwars in line with concepts of the general trends of Precambrian crustal evolution. Iyengar and Jayaram (1970) and Iyengar (1971) likewise proposed reclassifications of the Dharwars on the basis of geosynclinal and orogenic concepts. All of these classifications except for the last two mentioned (whose complete details are not known to the authors) are given, along with Smeeth's classification, in accompanying Tables I to VI.

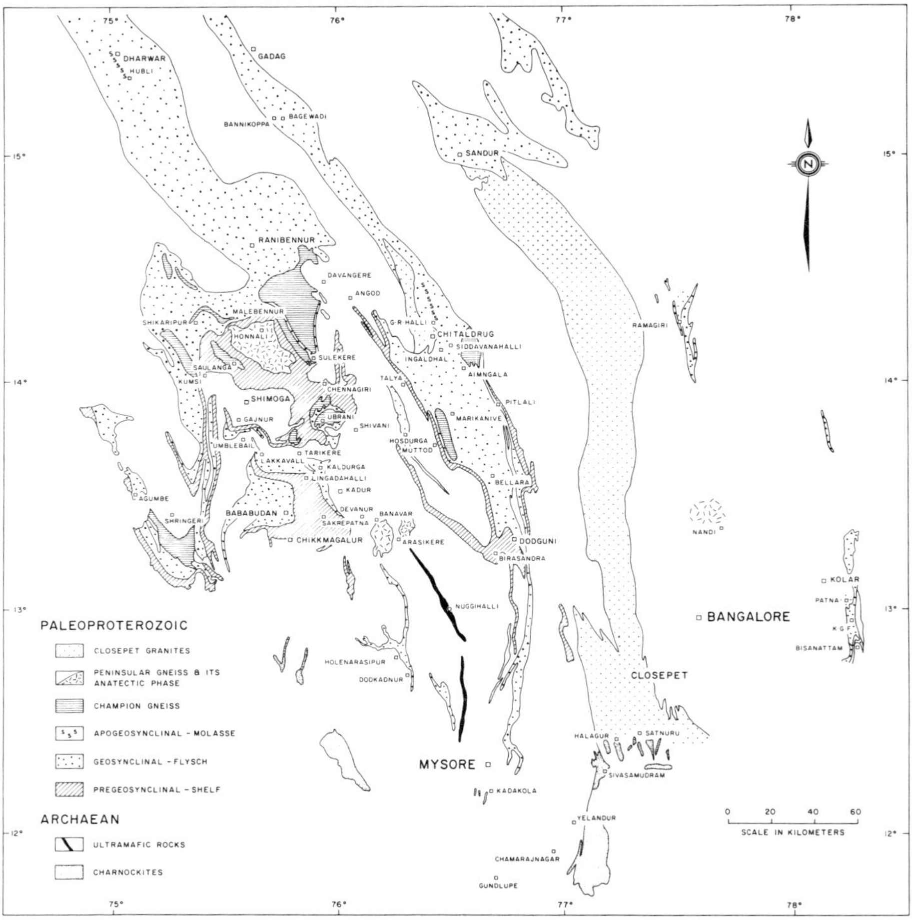

Figure 5. Geological map showing the distribution of the Precambrian rocks and their tectonic interpretation (after Srinivasan and Sreenivas, 1971).

Nautiyal (1966), in the light of the studies of Salop (1964) on Precambrian deposits, and Srinivasan and Sreenivas (1971, 1972), using as their basis the general trends of Precambrian crustal evolution as envisaged by Goodwin (1968) and Vinogradov and Tugarinov (1968), found it necessary to elevate the stratigraphical position of the Dharwars from Archaean to Proterozoic; the view is supported by the discovery of fossil life in them by Gowda (1970), Gowda and Sreenivasa (1969) and Gowda and Kumar (1971). Srinivasan and Sreenivas (1971, 1972) accepted the placement of the Archaean-Proterozoic boundary at about 2,600 to 2,700 m.y. ago, as is done in Canada.

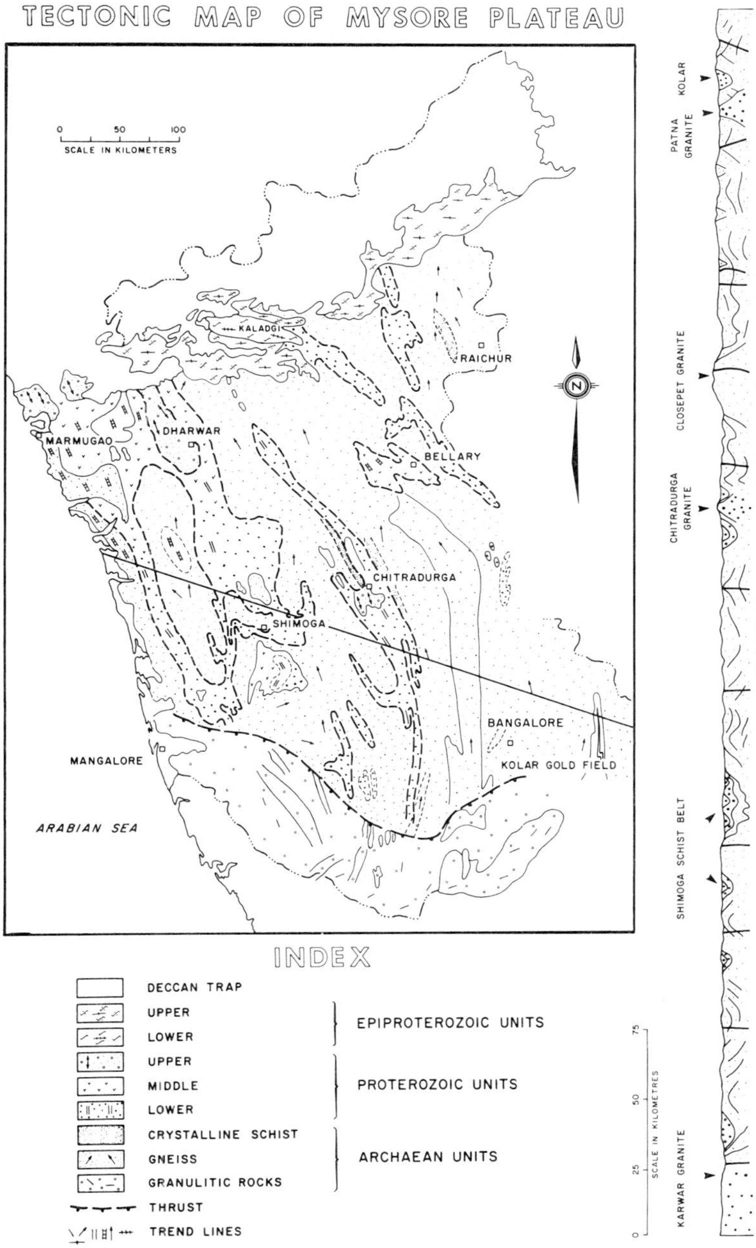

Figure 6. Nautiyal's Tectonic Map of Mysore Plateau and cross section.

TABLE I

		1	Recent soils	
	Possibly Tertiary	2	Laterite, horizontal sheet capping Archaeans	
	Precambrian	3	Basic dykes, chiefly dolerites	
GREAT EPARCHAEAN INTERVAL				
Archaean		4	Felsite and porphyry dykes	
Archaean		5	Closepet Granite and other massifs of corresponding age	
Archaean		6	Charnockite, norite and pyroxenite dykes	
Archaean		7	Charnockite massifs	
Archaean		8	Various hornblendic and pyroxene granulite dykes	
Archaean		9	Peninsular Gneiss:	Granite and gneissic complex.
Archaean		10	Champion Gneiss:	Granite porphyry, micaceous gneisses, felsites, porphyries containing opalescent quartz and frequently associated with autoclastic conglomerates.
Archaean	ERUPTIVE UNCONFORMITY			
Archaean	Dharwar System	11	Upper Chloritic Division (green stones and chlorite schists)	Amphibolites, peridotites etc., mostly intrusive. Conglomerates, autoclastic. Banded ferruginous quartzites, origin probably igneous. Quartzites and quartz schists mostly intrusive. Limestone, probably secondary.
Archaean	Dharwar System	12	Lower Hornblendic Division (epidiorites and hornblendic schists)	Mica schists, metamorphic and igneous. Intrusive masses of dioritic and diabasic character.
Archaean	Unknown			

THE SUCCESSION ACCORDING TO W.F. SMEETH (1916)

TABLE II

GNEISSIC SYSTEM		Felsites and porphyries. Closepet granites. Recrystallization and reconstitution of older rocks into complex type of Charnockite series. Norite dykes and hornblendic dykes. Peninsular gneiss.
DHARWAR SYSTEM	Upper Division	Some ferruginous cherty silts, clay and calcareous silts and clays, impure quartzites and conglomerates forming part of the G. R. Formation (local).
		Granitic and porphyry and some granitic rocks, fine and coarse.
		Basic and ultrabasic rocks.
	Middle Division	Ironstones, limestones, argillites, greywackes, quartzites, conglomerates and perhaps their intercalations of volcanic rocks. Rhyolites, felsites, quartz porphyries and other acid volcanic rocks with opalescent quartz and ashes, tuffs and other volcanic products, volcanic sediments.
	Lower Division	Basic volcanic lava flows, sills sheets and dykes.
ORIGINAL BASEMENT NOT RECOGNISED.		

THE SUCCESSION ACCORDING TO B. RAMA RAO (1962)

TABLE III

Proterozoic		Kaladgi	Quartz veins and younger basic dykes.
		SHEARED CONTACT	
	Upper Dharwars	Sandstones, greywackes, shales, calcareous phyllites of Dharwar, Chitradurga and North Goa.	Quartz veins.
		UNCONFORMITY	
	Middle Dharwars	Conglomerates, quartzites, limestones, dolomites, greywackes, pink shales, banded ferruginous quartzites.	Younger granites of Karwar and Goa, basic dykes and pillow lavas.
		DECOLLEMENT	
	Lower Dharwars	Acid, basic and ultrabasic intrusives, metagreywackes, schists, chloritic schists, metavolcanics, sheared tilloids, quartzites, interbedded greywackes of Chitradurga, Kolar, Shimoga etc.	Syntectonic Champion and post-tectonic diapir granites.
Archaean		THRUST	
		Narrow bands of sillimanite kyanite garnet schists and migmatites (Paragneisses), flaggy micaceous quartzites, calciphyres, amphibolites, epidiorites, ultrabasics of Karigatta and Kengalareas.	Closepet Granites
		THRUST	
		Gneisses with xenoliths of older metasediments.	Peninsular Gneissic complex.
		THRUST	
		Granulitic gneiss such as charnockites, pyroxene granulites, calc granulites, garnetiferous amphibolite.	
ORIGINAL CRUST NOT RECOGNISABLE			

THE SUCCESSION ACCORDING TO NAUTIYAL (1966)

TABLE IV

	Kaladgis
Supa Dandeli Series	Phyllites, slates, ankeritic limestones and minor beds of ferruginous quartzites.
Sirsi Ranibennur Series	Greywackes, slates and tuffs, minor beds of brecciated ferruginous quartzite.
Grey Trap Series	Epidiorite flows, pillow lavas, agglomerates and tuffs.
Dodguni Series	Phyllites, clay schists, dolomitic and high calcium limestones, banded ferruginous quartzites and iron ores.
Bababudan Series	Massive beds of iron ore, banded ferruginous and magnetite quartzites, argillites, quartzites and basic lava flows.
Older gneisses	Gneissic basement (a complex of granites and gneisses with lenses and patches of an older series of sediments and basic igneous rocks).

THE SUCCESSION ACCORDING TO B.P. RADHAKRISHNA (1967)

TABLE V

	Mafic dykes Felsites and porphyry dykes Charnockite Series Peninsular Gneissic Complex
Upper Dharwars	Cherty ferruginous and calcareous quartzites and conglomerates
Middle Dharwars	Granitic rocks and granite porphyry Mafic and ultramafic intrusives Ironstones, limestones, argillites, quartzites and conglomerates
Lower Dharwars	Champion gneiss Rhyolites, felsites and quartz porphyry Mafic volcanic flows and intrusions

THE SUCCESSION ACCORDING TO C.S. PICHAMUTHU (1967)

TABLE VI

			Supracrustal Sediments and Volcanics	Plutons
PALEOPROTEROZOIC = DHARWARS	Apogeosynclinal Molasse	G. R. Series	Ferruginous shales, silts and sandstones, ferruginous cherts and ankerite beds	Dolerite dykes
	Geosynclinal	Grey trap and Ranibennur series	Banded ferruginous quartzites (haematitic and pyritic) Metagreywackes and pillow lavas Polymictic conglomerates	Closepet granites Peninsular gneiss Champion gneiss
	Pregeosynclinal Shelf	Dodguni Series	Limestones and dolomites Metashales Orthoquartzites Oligomictic conglomerates	Ultramafic intrusions
		Bababudan Series	Magnetite quartzites Orthoquartzites Pyritiferous oligomictic conglomerates Subaerial basaltic lava flows	
EPARCHAEAN UNCONFORMITY				
ARCHAEAN			Sialic crust and charnockites	Chromite bearing Ultramafics

THE SUCCESSION ACCORDING TO R. SRINAVASAN AND B.L. SREENIVAS (1971)

Radiometric Dating, the Geosynclinal Concept and the Continental Accretion Hypothesis

An attempt to classify the Precambrian of Peninsular India into distinct regions and provinces was made by Fermor (1936), in his correlation of the "unfossiliferous" Archaean rock formations of India. Holmes (1949), who initiated radiogenic dating of the rocks of India, showed that the Precambrian of India is characterised by structural trends that resulted from several orogenic cycles (Holmes, 1955). These orogenic cycles were dated by Aswathanarayana (1956), Sarkar *et al.*, (1964) and Sarkar (1968). The Dharwar orogenic cycle – 2,300 m.y. ago – produced the NS to NNW structural trend of the Dharwars. Some very ancient dates, of the order of 2,900 to 3,295 m.y. ago, have been obtained from the Dharwar region. On the basis of these and of certain features considered characteristic of continental nuclei by Tuzo Wilson (1967), Pichamuthu (1962) postulated that the Dharwar province is a continental nucleus and that Peninsular India grew by accretion through several orogenic cycles – namely the Iron Ore, Eastern Ghat, Satpura-Aravalli and Delhi Cycles. This nucleus was suggested to be older than the Middle Dharwars and was called "Sugriva Land"; it was thought that the Middle Dharwar iron ores were deposited on the margins of this landmass (Sikka *et al.*, 1964).

However, that the Dharwar province is the oldest has been doubted by Narayanaswami (1971), who has shown that the Charnockite-Khondalite province is characterised by structural trends older than the Dharwars and that the Charnockite-Khondalite province must in consequence be considered older than the Dharwar. Furthermore, the radiometric age data given by Sarkar and his co-workers demonstrates that there are equally ancient rocks outside the Dharwar Province (Sarkar, 1968). This might suggest a possible application of the concept of several protocontinents to the Precambrian in India, of which the Dharwar province of Mysore may only be one.

REFERENCES

Aswathanarayana, U., 1956, Absolute ages of the Archaean orogenic cycles of India: American Journal of Science, v. 254, p. 19-31.

Babu, S.K., 1955, Migmatisation studies in the Katriguppe quarry, Bangalore District, Mysore State: Journal of Madras University, v. 25 B, no. 2, p. 129-146.

Beloussov, V.V., 1966, Modern conceptions about the structure and development of Earth's crust and upper Mantle of continents: Geological Society of India Journal, v. 7, p. 1-13.

Burdon, D.J., 1947; Petrology of the Kolar Gold Field: Transactions of the Kolar Gold Field Mining and Metal Society, Bull. 66, v. 12, p. 1-235.

Chayes, F., 1956, Petrographic and modal analysis: New York and London, J. Wiley, 113 p.

Christie, A.T., 1836, Madras Journal of Literature and Science (not seen), p. 457.

Devaraju, T.C. and Sadasiviah, M.S., 1969, Charnockites of Satnur Halaguru area: Indian Mineralogist, v. 10, p. 67-88.

Eskola, A., 1920, The mineral facies of rocks: Norsk Geologisk Tidsskrift, v. 6, p. 143-194.

Fermor, L.L., 1909, The manganese ore deposits of India: Memoirs of the Geological Survey of India, v. 37, pt. 2, p. 235-405.

————————, 1919, Some problems of ore genesis in the Archaean of India: Presidential address, Geology Section, 6th Indian Science Congress: Proceedings of the Asiatic Society of Bengal, v. 15, CIXX-CXCV.

———, 1936, An attempt at the correlation of the ancient schistose formations of Peninsular India: Memoirs of the Geological Survey of India, v. 70, pt. 1, 51 p.

Foote, R.B., 1886, Notes on the geology of part of Bellary and Anantapur districts: Records of the Geological Survey of India, v. 19, p. 97-111.

———, 1900, Geological notes on traverses through the Mysore State: Memoirs of the Geological Department, Mysore State, v. 1, 103 p.

Goodwin, A.M., 1968, Archaean protocontinental growth and the early crustal history of the Canadian shield: Proceedings of the International Geological Congress, Section Upper Mantle, (Geological Processes), 23rd session, p. 69-90.

Gowda, S.S., 1970, Fossil blue green algae and fungi from Archaean complex of Mysore, South India (Abst.): Proceedings of the International Symposium, Taxonomy and Biology of Blue-green Algae, Madras, India, p. 42-43.

Gowda, S.S. and Sreenivasa, T.N., 1969, Microfossils from the Archaean complex of Mysore: Geological Society of India Journal, v. 10, p. 201-208.

Gowda, S.S. and Kumar, S.S., 1971, Microfossils from the type rocks of the Dharwar Group: Proceedings of a Colloquium on Micropaleontology and Stratigraphy: Current Science, v. 40, p. 617-618.

Grubenmann, U., 1904, Krystallinen Schiefer: 2 vols. Borntraeger, Berlin, 280 p.

Hatch, F.H., 1902, The Kolar Gold Field; being a description of quartz mining and gold recovery as practiced in India: Geological Survey of India Memoirs, v. 33, p. 1-72.

Holland, T.H., 1909, Geology: Imperial Gazetteer of India, v. 1, p. 50-102.

Holmes, A., 1949 The age of uraninite and monazite from the post-Delhi pegmatites of Rajaputana: Geological Magazine, v. 86, p. 288-302.

———, 1955, Dating the Precambrians of Peninsular India and Ceylon: Proceedings of the Geological Association of Canada, v. 7, p. 81-106.

Iyengar, B.R.C. and Jayaram, B.N., 1970, Stratigraphy of the Dharwars: Journal of the Indian Geoscience Association, v. 12, p. 1-4.

Iyengar, P.S., 1905, Report on the survey work in Chitaldrug district: Records of the Geological Department, Mysore State, v. 6.

———, 1908, Geology of the Hassan hills and Kadur Districts, Mysore Province: Records of the Geological Department, Mysore State, v. 9, p. 59-84.

Iyengar, S.V.P., 1971, Structure, stratigraphy and correlation of the Dharwar super-group of rocks of Mysore and adjoining areas: Proceedings of a Seminar on Precambrian Geology of the Peninsular Shield, (Abst.), p. 52.

———, 1920, The acid rocks of Mysore: Geological Department Bulletin, Mysore State, no. 9, 31 p.

Jayaram, B., 1920, Traverse notes: Records of the Geological Department, Mysore State, v. 17, pt. 2, p. 86-90.

———, 1922, Report on the work done during the field session of 1917-18. Records of the Geological Department, Mysore State, v. 18, pt. 2, p. 66.

———, 1925, Progress report of the work done during field session of the year 1921: Records of the Geological Department, Mysore State, v. 22, p. 32-35.

King, W., 1872, On the Kadapah and Kurnool formations in Madras Presidency: Memoirs of the Geological Survey of India, v. 8, pt. 1, p. 1-346.

MacLaren, J.M., 1906, Notes on some auriferous tracts of Southern India: Records of the Geological Survey of India, v. 34, pt. 2, p. 96-132.

Middlemiss, C.S., 1917, Complexities of Archaean geology in India: Presidential address, Geology Section, 4th Indian Science Congress, Proceedings of the Asiatic Society of Bengal Journal, v. 13, p. CXCV-CCII.

Miyashiro, A., 1961, Evolution of metamorphic belts: Journal of Petrology, v. 2, p. 277-311.
Naidu, P.R.J., 1967, Charnockitic period: Indian Mineralogist, v. 8, p. 70-71.
Narayanaswami, S., 1955, Occurrence of pillow lavas in Kolar schist belt: Current Science, v. 24, p. 265-266.
———, 1964, Tectonic and lithologic provinces within Precambrian rocks of Peninsular India: Proceedings of a Symposium on Stratigraphy, age and correlation of the Archaean provinces of India, Abst., p. 32.
———, 1970, Tectonic setting and manifestation of the Upper Mantle in the Precambrian rocks of South India: Second Upper Mantle Project Symposium Proceedings, Hyderabad, p. 377-403.
———, 1971, Proposal for a charnockite-khondalite system in the Archaean shield of Peninsular India: Proceedings of a Seminar on the Precambrian Geology of the Peninsular Shield, (Abst.), p. 54 (Misc. Publ., No. 23, pt. 1, p. 1-14, Geological Survey of India, 1975).
Nautiyal, S.P., 1966, Precambrian of Mysore plateau: Presidential address, Geology and Geography Section, Indian Science Congress, 53rd Session, p. 1-14.
Newbold, T.J., 1846, Summary of the geology of Southern India: Journal of the Royal Asiatic Society, v. 8.
———, 1848, Summary of the geology of Southern India: Journal of the Royal Asiatic Society, v. 9, p. 1-42.
———, 1850, Summary of the geology of Southern India: Journal of the Royal Asiatic Society, v. 12, p. 78-96.
Pichamuthu, C.S., 1935a, The conglomerates and grits of Kaldurga, Kadur District, Mysore: Proceedings of the Indian Academy of Sciences, v. 2, p. 254-279.
———, 1935b, The quartzites of the Bababudan area, Kadur District, Mysore. Current Science, v. 4, p. 179-181.
———, 1935c, The iron formations and associated rocks of eastern Bababudans, Kadur District, Mysore: Half Yearly Journal of Mysore University, v. 8, no. 1, p. 1-48.
———, 1936, The gneisses and granites of Mount Joy, Bangalore: Half yearly Journal of Mysore University, v. 8, no. 2, p. 11-21.
———, 1937, On the use of "current bedding" in the interpretation of tectonics: Current Science, v. 6, p. 95-96.
———, 1938, On composite gneisses: Presidency College, Magazine, vol. 4.
———, 1945, Algal structures in the cherts from Dodguni (Abst.): Proceedings of the Indian Science Congress, pt. III, p. 60.
———, 1947, Some aspects of Dharwar geology with special reference to Mysore: Presidential address, Geology and Geography Section, 34th Indian Science Congress, 16 p.
———, 1950, Pillow structures in the lavas of Dharwar age in Chitaldrug District, Mysore State: Current Science, v. 19, p. 110-111.
———, 1951, Some observations on structure and classification of the Dharwars of Mysore State: Current Science, v. 20, p. 117-119.
———, 1953, The Charnockite problem: Mysore Geologists' Association, India.
———, 1957, Pillow lavas from Mysore State, India: Mysore Geologists' Association, Bulletin 15, 26 p.
———, 1960, Charnockite in the making: Nature, v. 188, p. 135-136.
———, 1961, Transformation of Peninsular gneiss into charnockites in Mysore State, India: Journal of the Geological Society of India, v. 2, p. 46-49.
———, 1962, Some observations on the structure, metamorphism and geological evolution of Peninsular India: Geological Society of India Journal, v. 3, p. 106-118.

———, 1967, The Precambrians of India *in* Rankama, K. (ed.), Geological Systems: The Precambrian, vol. 3, p. 3-96: New York, Wiley Interscience.
———, 1970, Some observations on the Peninsular gneiss complex: West Commemoration Volume, p. 82-97.
Radhakrishna, B.P., 1956, The Closepet granites of Mysore State, India: Mysore Geologists' Association, Bangalore, 110 p.
———, 1964, Some aspects of the stratigraphy and structure of the Dharwars of Mysore, South India (Abst.): Proceedings of the Symposium on Stratigraphy, age and correlation of the Archaean Provinces of India, p. 33-34.
———, 1967, Reconsideration of some problems in the Archaean complex of Mysore: Journal of the Geological Society of India, v. 8, p. 102-109.
Rama Murthy, V., 1954, A note on the occurrence of pillow structures in the Bellara traps of Dharwar age, Tumkur District, Mysore State (Abst.): Proceedings of the 41st Indian Science Congress, p. 292.
Rao, S. Divakara, 1970, Metasomatism and origin of granite: Proceedings of the Upper Mantle Symposium, Hyderabad, p. 46-63 (Pre-print volume).
Rao, B.N. Raghunatha, 1937, Pillow structures in the mica traps of Chitaldrug schist belt: Current Science, v. 6 p. 279-280.
Rao, B. Rama, 1921, Report on the revision survey work of parts of Holenarasipur Taluk, Hassan District, Mysore: Records of the Geological Department, Mysore State, v. 21, p. 146-203.
———, 1927, The Kodamites or cordierite micaceous gneisses near Channapatna area: Records of the Geological Department, Mysore State: v. 24, pt. 2, p. 134-152.
———, 1934, Cummingtonite in the limestones of Kudurekanive (Mysore State): Geological Department Bulletin, Mysore State, 36 p.
———, 1935, Evidences for the sedimentary origin of some of the constituents of the Dharwar schists: Records of the Geological Department, Mysore State, v. 34, p. 22-32.
———, 1936, Recent studies in the Archaean complex of Mysore: Presidential address, Geology and Geography Section, 23rd Session, Indian Science Congress, p. 215-244.
———, 1939, Remarks on the mode of occurrence and origin of Peninsular gneiss, Mysore: Records of the Geological Department, Mysore State, v. 38.
———, 1940, The Archaean complex of Mysore: Bulletin of the Geological Department, Mysore State, no. 17, 101 p.
———, 1945, The charnockite rocks of Mysore (Southern India): Bulletin of the Geological Department, Mysore State, no. 18, 199 p.
———, 1962, A handbook of the Geology of Mysore State, Southern India: Bangalore Printing and Publishing Co., 264 p.
Salop, L.I., 1964, Precambrian geochronology and some features of the early stages of the geological history of the Earth: International Geological Congress Proceedings, 22nd Session, Section 10, p. 131-149.
———, 1968, Precambrian of the U.S.S.R.: Proceedings of the International Geological Congress, Prague, Section Precambrian, p. 61-74.
Sambasiva Iyer, V., 1899, Report on the survey work in Chitaldroog and Tumkur Districts: Records of the Geological Department, Mysore State, v. 2.
———, 1918, On the distinct sedimentary origin of some quartzites of Mysore (Abst.): Proceedings of the 5th Indian Science Congress.
Sarkar, S.N., 1968, Precambrian Stratigraphy and Geochronology of Peninsular India: Dhanbad Publishers, India, 33 p.
Sarkar, S.N., Polkonov A.A., Gerling E.K., and Chukrov, F.V., 1964, Geochronology of

the Precambrians of Peninsular India: A synopsis: Science and Culture, v. 30, p. 527-537.
Sikka, D.B., Royce, J. and Nehru, C.E., 1964, Geosynclines in the Precambrian of India and their economic significance: Mineral Markets, December issue.
Slater, H.K., 1906, Report on the geological survey of portions of Shimoga, Tarikere and Kadur Taluks: Geological Department Records, Mysore State, v. 7, pt. 2, 20 p.
———, 1908, Report on the geological survey of the Kadur District: Geological Department Records, Mysore State, v. 9.
Smeeth, W., 1899, Notes on the geology of Kolar schists and surrounding rocks: Report of the Chief Inspector of Mines, Mysore State, Appendix, p. I-XIV.
———, 1904, General report of the work of the Department from July 1903 to June 1904: Records of the Geological Department, Mysore State, v. 5, pt. 1, 39 p.
———, 1910, General report of the work of the Department for the year 1909-10: Geological Department Records, Mysore State, v. 11, pt. 1, p. 9-65.
———, 1916, Outline of the geological history of Mysore State: Bulletin, Department of Mines and Geology, no. 6, p. 1-21.
———, 1924, Some views about the Archeans of Southern India. Presidential Address: Geology Section, Indian Science Congress: Records of the Geological Department, Mysore State, v. 23, p. 37-68.
Smith, A.M., 1904, Geology of the Kolar Gold Field: Transactions of the Institute of Mining and Metals, v. 13, p. 152-218.
Sreenivas, B.L. and Srinivasan, R., 1968a, Dharwar conglomerates of Mysore – A restudy: Geological Society of India Journal, v. 9, no. 2, p. 197-205.
———, 1968b, Geology of the Kolar Gold Field, Mysore State, India (Abst.): International Geological Congress Proceedings Prague, p. 125.
Srinivas, G., 1969, Quantitative petrography of the Champion gneisses (Kolar Gold Field): Proceedings of the 55th Indian Science Congress (Abst.) p. 209.
Srinivasan, R. and Sreenivas, B.L., 1968, Sedimentation and tectonics in the Dharwars (Archaeans) of Mysore State, India: Indian Mineralogist, v. 9, p. 47-58.
———, 1969a, Migmatization studies in the Peninsular gneisses of Bangalore: Indian Mineralogist, v. 10, p. 280-288.
———, 1969b, Flood basalts from the Dharwars of Mysore State, India: Proceedings of a Symposium on Deccan Traps and other flood eruptions, Bulletin Volcanologique, 1971, Tome. 35, Fasc. 4, p. 824-840.
———, 1971, Reappraisal of some aspects of Geology of the Dharwars of Mysore: Proceedings of a Seminar on Precambrian Geology of the Peninsular shield: Miscellaneous Publication No. 23, Geological Survey of India, 1975, p. 49-69.
———, 1972, Dharwar stratigraphy: Journal of the Geological Society of India, v. 13, no. 1, p. 75-85.
Suryanarayana, K.V., 1960, The Closepet granite and associated rocks: Indian Mineralogis,t v. 1, no. 1, p. 86-100.
———, 1964a, The Closepet granites and associated rocks II: Indian Mineralogist, v. 5, p. 60-68.
———, 1964b, The Closepet granites and associated rocks III: Indian Mineralogist, v. 5, p. 143-146.
Vinogradov, A.P. and Tugarinov, A.I., 1968, Geochronological scale of the Precambrian: Proceedings of the International Geological Congress, 23rd Session, Section Geochemistry, p. 205-222.
Wetherell, E.W., 1904, Geological report on the Chitaldrug District: Records of the Geological Department, Mysore State, vol. 5.

Wilson, J.T., 1967, Theories of building of Continents: *in* T.F. Gaskell (ed.), The Earth's mantle: Academic Press, London and New York, 509 p.
Ziauddin, Md., 1971, The nature of some of the conglomerates in Mysore State: Proceedings of a Seminar on Precambrian Geology of the Peninsular shield, Abst., p. 11-12.

History of Concepts in Precambrian Geology, edited by W.O. Kupsch and W.A.S. Sarjeant, Geological Association of Canada Special Paper 19, 1979

PRECAMBRIAN ORGANIC-WALLED MICROFOSSILS OF SWEDEN

Göran Kjellström
Sveriges Geologiska Undersökning
Box 670 S-751 28 Uppsala, Sweden

ABSTRACT

An account is given of early and recently published studies of organic-walled microorganisms from the Precambrian of Sweden. The diversity of opinions as regards the true biological relationships of these microfossils is outlined. Research in the stratigraphical application of Swedish Precambrian microfossils has just begun; the first tentative stratigraphical correlations have been made within the Precambrian.

RESUME

Un exposé est présenté des études publiées dans le passé et récemment de microorganismes à parois organiques du Précambrien de la Suède. La diversité d'opinions envers les vraies relations biologiques de ces microfossiles est présentée. La recherche dans l'application stratigraphique des microfossiles précambriens suédois à été récemment commencée; les premières tentatives de corrélations stratigraphiques ont été tirées dans le Precambrien.

HISTORY OF STUDIES

During the last few years, a certain revival of interest in organic-walled microorganisms of the Precambrian has taken place in Sweden. The nature of Precambrian microorganisms has long attracted the attention of Swedish paleontologists: the beginning of such studies can be traced back as early as the last two decades of the nineteenth century. However, many of the earlier contributions were based upon unconvincing, fragmentary structures which were described insufficiently and illustrated poorly or not at all. It is therefore understandable that many of these contributions have been viewed with scepticism. Today, improved paleontological techniques and well-preserved and well-documented material have aided much to elucidate the biology and biostratigraphy of these early microorganisms.

The first record of Precambrian organic-walled microfossils from Sweden appears to have been that of Nathorst (1879a, 1879b), who mentioned incidentally the presence of spherical, organic microstructures in the Visingsö Beds at Lake Vättern in the County of Östergötland, southern Sweden. The following year Linnarsson (1880), in his description of the Palaeozoic strata of the vicinity of the town of Motala in Östergötland, reported similar microstructures from the same beds. Later on, the finds of Nathorst and Linnarsson were confirmed with regards to their organic origin by

Holm (1885), who suggested that the round organic objects might either be the spat of brachiopod belonging to the genus *Discina* or might represent plant remains of unknown affinities. Nathorst, in a series of publications (1884, 1886, 1894), took issue. However, with the assumptions of Holm instead, Nathorst (1886) suggested a possible affinity to the conchostraceans (branchiopods) exemplified by the genus *Estheria*. This was in fact an unlikely affinity, since this genus is only known from fresh-water deposits of Devonian and from marine deposits of Middle Ordovician age (Dr. Jan Bergström, pers. commun. 1977).

The literature hitherto quoted did not contain accompanying illustrations or photos and it was not until the end of the nineteenth century, when Wiman (1895) carried out the first detailed investigation of the microorganisms of the Visingsö Beds, that fairly good microphotographs were given (Fig. 1). Wiman was the first to recognize the acid-resistant nature of the cell wall of the Precambrian microfossils. As a consequence of this chemical property, Wiman concluded that the microorganisms may represent eggs of trilobites.

After 1895, the interest in Precambrian organic-walled microorganisms of Sweden virtually ceased for nearly forty years. In the early 1930s, Ewetz (1932, 1933) focussed his interest on the microfossils of the Visingsö Beds. Previously he had studied the general geological setting of the Visingsö Beds (Ewetz, 1927, 1929); now he turned his attention to the microorganisms originating from the phosphate nodules in them. He was the first researcher to suggest that at least some of the brownish, spherical microorganisms were resting spores (Fig. 2). Some years later Brotzen (1941), in his investigation of the stratigraphy and tectonic history of the Visingsö Beds, compared the microorganisms illustrated by Wiman with *Chuaria circularis,* described from the Chuar Formation of the Grand Canyon by Walcott (1899), Walcott had at first considered *C. circularis* " . . . as concretions but, with the discovery of more perfect specimens, this view was abandoned and the objects were referred to as specimens of a discinioid shell" (Walcott, 1899). Brotzen followed the taxonomical proposals given by Walcott and erected a new species named after Wiman, *Chuaria wimani;* it was considered to be a chitinous foraminifer.

As stated above, Holm (1885) had much earlier presented his alternative hypothesis, namely that the organic-walled microorganisms might represent some plant remains, a theory that had been further advanced by Ewetz (1933). In 1960, the possible plant affinity of the Precambrian microfossils of the Visingsö Beds was discussed again, This time in a paper dealing with the Visingsö assemblages by the Russian paleontologist Timofeev. He regarded the microorganisms studied (13 species were recorded and described) as being trilete spores, indicating terrestrial or semi-terrestrial conditions. However, Timofeev's taxonomic deductions evoked strong criticism from Eisenack (1966), who rejected Timofeev's megaspore concept and followed the view expressed previously by Brotzen with regard to *Chuaria wimani*. The possible plant affinity later became once more a matter of dispute. In their comprehensive bibliography of the algal group Tasmanaceae, Muir and Sarjeant (1971) tentatively suggested that some of the microfossils mentioned by Holm (1885), Linnarsson (1880), Nathorst (1886), Timofeev (1960) and Wiman (1895) might belong to the Tasmanaceae, a group of fossils now known to represent the abandoned zoosporangia of Prasinophycean algae.

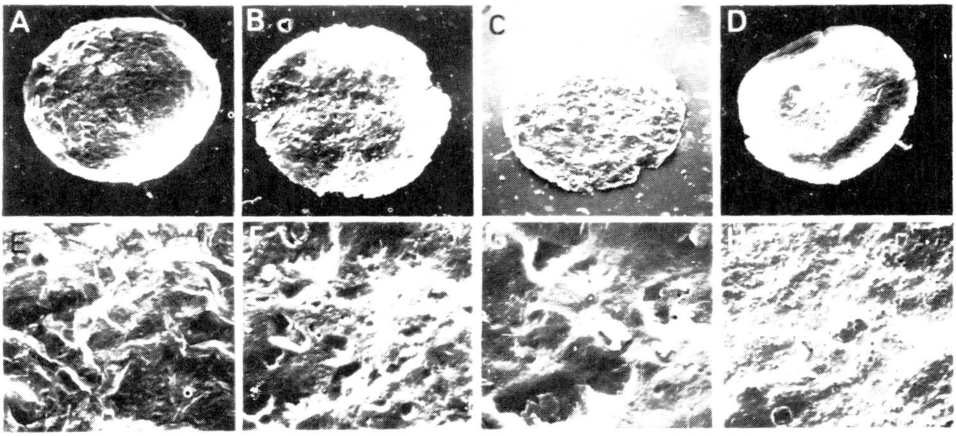

Figure 1. *Chuaria circularis* Walcott 1899. Stereoscan micrography. Specimens from the Visingsö Beds, southern Sweden. Photo: G. Vidal 1976. x 1000.

Up to this date, the research on Precambrian organic-walled microorganisms had been restricted to the Visingsö Beds. In the beginning of the 1970s, Vidal and Röshoff (1971) reported for the first time organic microfossils from other areas in Sweden and in rock sequences much older than the Visingsö Beds – the metasedimentary and metatuffitic rocks of the Vetlanda Series in the county of Smaland in southern Sweden. According to Vidal and Röshoff, the varied microfossil assemblage of the greater than 1740 year old rocks yielded four morphotypes of microorganisms: 1) spheroidal to ellipsoidal spore-like bodies, 20 to 100 microns in diameter, possessing a characteristic reticulate ornamentation. In some of the spheroidal bodies the individual components of the reticula are circular, and smooth areas can be observed alternating within the ornamentation. The size of the external wall structure varies inversely with the diameter of the bodies. Similar organization has, according to Vidal and Röshoff, been reported in *Huroniospora* from the Gunflint Formation of Canada (Cloud and Hagen, 1965); 2) spheroidal to sub-triangular bodies indistinctly ornamented but many displaying equatorial constrictions, a feature Vidal and Röshoff interpret as a stage of vegetative cell division comparable with similar ones reported from the Gunflint assemblages (Licari and Cloud, 1968); 3) translucent, amber to reddish-brown spheroidal reticulated bodies with elliptical cavities; 4) filaments of organic debris ranging from 20 to 200 microns in diameter.

Recently, attention has once again been focussed on the microorganisms of the Visingsö Beds (Vidal, 1974, 1976). By means of improved techniques, e.g., refined

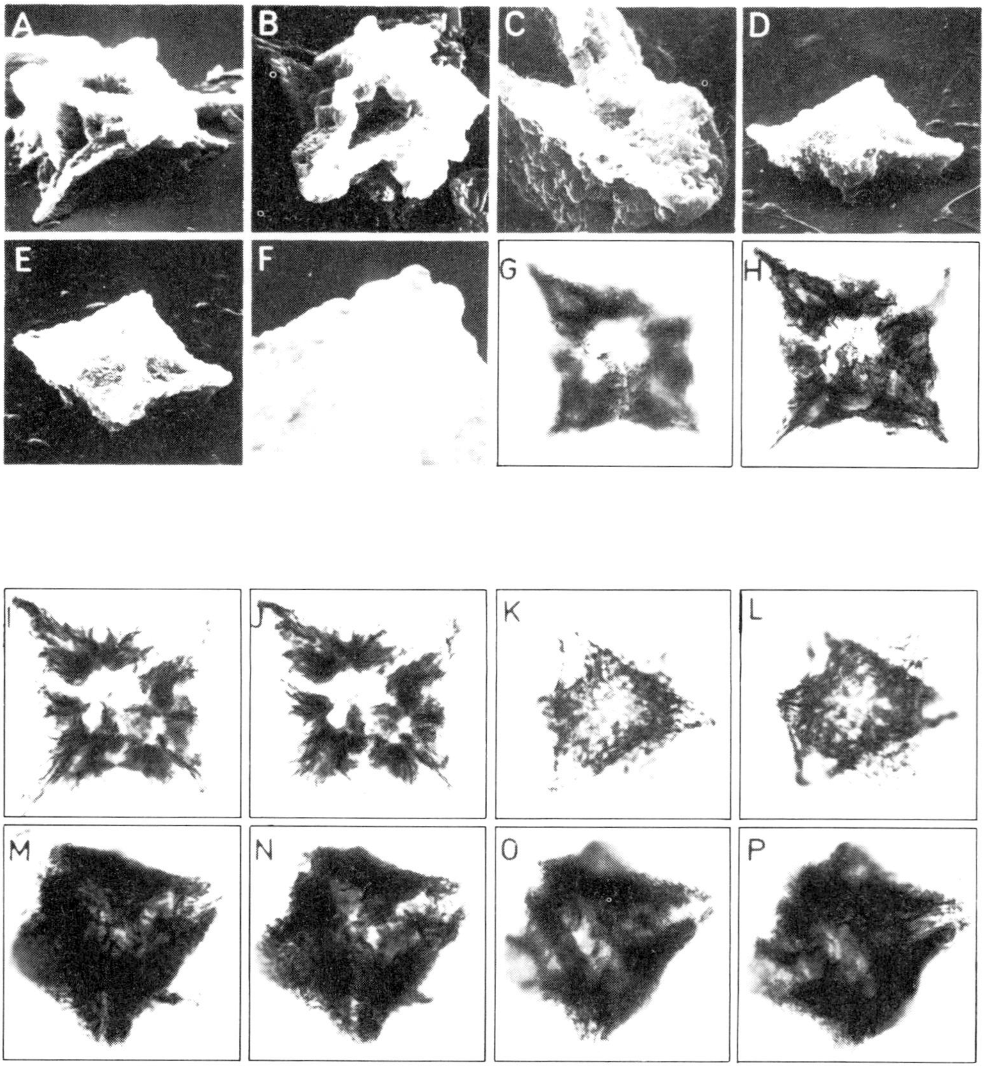

Figure 2. *Octoedryxium truncatum* (Rudavskaja). Stereoscan micrograph. Specimens from the Visingsö Beds, southern Sweden. Photo: G. Vidal 1976.

	2000	1500	1350	1100	650	560 m.y.					
	PROTEROZOIC					PALAEOZOIC			VISINGSÖ BEDS		
	LOWER & MIDDLE	UPPER RIPHEAN				LOWER CAMBRIAN					
		LOWER	MIDDLE	UPPER	VENDIAN	L	M	U	L	M	U
Bavlinella faveolata (Shepeleva)				?	●						●
Chuaria circularis Walcott				●	?				●	●	●
Favososphaeridium favosum Timofeev					●	●	●				●
Kildinella hyperboreica Timofeev				●	●				●	●	●
K. cf. *sinica* Timofeev		●	●	●	●				●	●	●
K. cf. *vesljanica* Timofeev				●	●				●	●	●
Octoedryxium truncatum (Rudavskaja)					●	●					●
Protosphaeridium cf. *flexuosum* Timofeev		?	●	●	●	●	●			●	●
P. laccatum (Timofeev) Vidal	●	●	●	●	●	●	●	●	●	●	●
P. cf. *papyraceum* Timofeev				●	●	●	●	●			●
P. cf. *scabridium* Timofeev			●	●	●					●	●
P. cf. *tuberculiferum* Timofeev		●	●	●	?						●
Pteinosphaeridium reticulatum n.sp.					○						●
Pterospermopsimorpha concentrica (Sin & Liu)					●						●
P.? densicoronata n.sp.				○	○					●	●
cf. *Stictosphaeridium* sp.				○	○					●	●
S. cf. *sinapticuliferum* Timofeev				●	●	●	●				●
S. verrucatum n.sp.					○						●
Synsphaeridium sp.				○	○					●	●
Trachysphaeridium apertum n.sp.				○						●	
T. laminaritum (Timofeev)				●	●				●	●	●
T. laufeldi n.sp.				○	○					●	●
T. levis (Lopukhin) Vidal			●	●	?				●	●	●
T. timofeevi n.sp.				○	○				●	●	●
Trematosphaeridium holtedahli Timofeev				●	●	?				●	●
Sphaeromorphs				○	○				●	●	●

Figure 3. Stratigraphical occurrence of Precambrian microorganisms form the Visingsö Beds, southern Sweden. (After Vidal, 1976). Black dots indicate occurrences. Open circles indicate possible stratigraphic range.

preparation methods and application of stereoscan micrography, Vidal (1976) has presented in monograph form a paleontological and biostratigraphical analysis of the Visingsö assemblages. This time they have been taxonomically incorporated into the Group Acritarcha Evitt 1963, a residue group of probably polyphyletic composition (acritarch = *akritos* + *arche*, uncertain + origin). Spheromorph as well as polygonomorph and acanthomorph species have been encountered (Figs. 1-2) and in total Vidal reports 21 species (five new species are erected). The acritarchs from the middle unit of the Visingsö beds are considered to indicate a Late Riphean age (Fig. 3) and the assemblages of the upper unit suggest a Vendian age. On the basis of the acritarchs, Vidal demonstrates that the rocks of the basal part of the middle unit of the Visingsö Beds were deposited in a prodelta zone with marine ingressions, whereas in the upper part of the middle unit and in the entire upper unit shallow marine conditions prevailed. The acritarch assemblage of the upper unit is similar to the acritarch assemblages from the Vendian Valdai series in Russia, the Upper Sinian in northern China, and Late Precambrian deposits in other areas, related to the Vendian glacial event. On this basis, Vidal concludes that the boulder beds occurring in the lower part of the middle unit of the Visingsö Beds may represent a glacial event older than the North European Varanger Glaciation.

If these inferences are correct, Precambrian organic-walled microorganisms must be considered most promising tools in biostratigraphy. In the opinion of the present writer, future research must, however, concentrate upon elucidating the polyphyletic status of these organisms, and so we are back once again where we started, still trying to resolve the biological origin of the Precambrian microorganisms of Sweden – a challenging problem indeed!

ACKNOWLEDGEMENTS

The writer is grateful to Dr. Gonzalo Vidal, who kindly supplied the illustrations.

REFERENCES

Brotzen, F., 1941, Några bidrag till Visingsöformationens stratigrafi och tektonik: Geologiska Föreningens i Stockholm Förhandlingar, v. 63, p. 245-261.

Cloud, P.E., Jr. and Hagen, H., 1965, Electron microscopy of the Gunflint microflora: Preliminary results: Proceedings of National Academy of Sciences, v. 54, p. 1-8.

Eisenack, A., 1966, Uber *Chuaria wimani* Brotzen: Neues Jahrbuch für Geologie und Paläontologie, Monatshefte, v. 1, p. 52-56.

Ewetz, C.E., 1927, Beiträge zur Kenntnis der Visingsö formation: Zeitschrift für Deutsche Geologische Gesellschaft, v. 79, p. 136-177.

——————, 1929, Weitere Beiträge zur Kenntnis der Visingsöformation: Jönköping, Sweden.

——————, 1932, Fossilrester i Visingsöformationen: Geologiska Föreningens i Stockholm Förhandlingar, v. 54, p. 310-314.

——————, 1933, Einige neue Fossilfunde in der Visingsöformation: Geologiska Föreningens i Stockholm Förhandlingar, v. 55, p. 506-518.

Holm, G., 1885, Om Vettern och Visingsöformationen: Bihang till Kungliga Svenska Vetenskapsakademiens Handlingar, v. 117, p. 1-49.

Licari, G.R. and Cloud, P.E., Jr., 1968, Reproductive structures and taxonomic affinities of some nannofossils from the Gunflint Iron Formation: Proceedings of National Academy of Sciences, v. 59, p. 1053-1060.

Linnarsson, G., 1880, De äldsta paleozoiska lagren i trakten kring Motala: Geologiska Föreningens i Stockholm Förhandlingar, v. 5, p. 23-30.

Muir, M.D. and Sarjeant, W.A.S., 1971, An annotated bibliography of the Tasmanaceae and of related living forms (Algae: Pracinophyceae): in C.I.M.P. Microfossiles organiques du Paleozoique, 3, Les Acritarches (Sécretaire de Rédaction: S. Jardiné); Editions du Centre National de la Recherche Scientifique, p. 1-117.

Nathorst, A.C., 1879a, En egeondomlig strukturvarietet af lerhaltig kalksten fran Grennatrakten: Geologiska Föreningens i Stockholm Förhandlingar, v. 4, p. 216.

——————, 1879b, Om de aldre sandstens – och skifferbildningarne vid Vettern: Geologiska Föreningens i Stockholm Förhandlingar, v. 4.

——————, 1884, Upplysningar till Geologisk Öfversigtskarta öfver Sverige. Södra bladet: Sveriges geologiska undersökning, v. 4, p. 14.

——————, 1886, Nagra ord om Visingsöserien: Geologiska Föreningens i Stockholm Förhandlingar, v. 8, p. 5-23.

——————, 1894, Jordens Historia: Beijers Bokförlagsaktiebolag, Stockholm, 1128 p.

Timofeev, B.V., 1960, Sur la caractéristique micropaléontogique de la formation de Visingsö: Geologiska Föreningens i Stockholm Förhandlingar, v. 82, p. 28-42.

Vidal, G., 1974, Late Precambrian microfossils from the basal sandstone unit of the Visingsö beds, South Sweden: Geologica et Palaeontologica, v. 8, p. 1-14.

——————, 1976, Late Precambrian microfossils from the Visingsö Beds in southern Sweden: Fossils and Strata, v. 9, 1-57.

Vidal, G. and Röshoff, K., 1971, Organic remains in metasedimentary and metatuffitic rocks of the Vetlanda Series, South Sweden: A preliminary report: Geologiska Föreningens i Stockholm Förhandlingar, v. 93, p. 773-778.

Walcott, Ch. D., 1899, Pre-Cambrian fossiliferous formations: Bulletin of the Geological Society of America, v. 10, p. 199-244.

Wiman, C., 1895, Paläontologische Notizen 1: ein Präkambrisches Fossil: Bulletin of the Geological Institutions of the University of Upsala, v. 2, p. 109-113.

History of Concepts in Precambrian Geology, edited by W.O. Kupsch and W.A.S. Sarjeant, Geological Association of Canada Special Paper 19, 1979

THE RUSSIAN SCHOOL OF GEOCHRONOLOGICAL RESEARCH ON THE PRECAMBRIAN

K.O. Kratz, V.A. Maslenikov, and N.N. Barkhatova
Institute of Precambrian Geology and Geochronology
Leningrad, U.S.S.R.

ABSTRACT

An account is presented of the Russian school of geochronological research in the Precambrian. V.I. Vernadsky was the earliest Russian scientist to recognize the significance of radioactivity and to begin the study of radioactive minerals. He outlined the problems that might be solved by their use in geochronology.

K.A. Nenadkevich (1926), using the uranium-lead method, determined the age of uraninite from White Sea pegmatites by classical analytical techniques, arriving at a date of 1850-2100 million years. This served as a basis for concepts of the duration of the Precambrian for 20 years.

In 1946, V.G. Khlopin and E.K. Gerling began to utilise the potassium-argon method, enabling age determinations for a much wider range of rock types. In the years 1950 to 1970, this field of research flourished and age determinations were made from rocks in the Baltic, Ukrainian and Aldan Shields, under the guidance of a Commission of the Academy of Sciences of the USSR. Problems considered were the Infracambrian Interval, the position of the base of the Proterozoic, and the influence of metamorphic effects on the ages obtained.

RESUME

On présente un compte-rendu sur l'école russe de recherche géochronologique sur le Précambrien. V.I. Vernadsky fut le premier scientifique russe à reconnaître la signification de la radioactivité et à commencer l'étude des minéraux radioactifs. Il ébaucha les problèmes pouvant être résolus par leur utilisation en géochronologie.

K.A. Nenadkevich (1926), utilisant la méthode uranium – plomb, détermina l'âge de l'uraninite des pegmatites de la Mer Blanche par les techniques analytiques classiques, aboutissant à un âge 1850-2100 millions d'années. Cela servit de base pendant 20 ans aux concepts sur la durée du Précambrien.

En 1946, V.G. Khlojin et E.K. Gerling commencèrent à utiliser la méthode potassium-argon, permettant des déterminations d'âge sur un plus grand intervalle de types de roches. Dans les années 1950 à 1970 ce champ de recherche se développa et des déterminations d'âge furent faites sur des roches des boucliers Baltique, Ukrainien et Aldan, sous la conduite d'une Commission de l'Académie des Sciences de l'URSS. Les problèmes considérés furent l'Intervalle Infracambrian, la position de la base du Protérozoïque, et l'influence des effets du métamorphisme surles âges obtenus.

INTRODUCTION

The founder of radiogeology, Academician V.I. Vernadsky, was the first Russian geologist to realize the importance of radioactivity as a natural process of the Earth. He organized the first systematic and detailed studies of radioactive minerals in Russia (Vernadsky, 1910, 1911). Foreseeing the great importance that knowledge of radioactivity would have for the progress of the earth sciences, Vernadsky formulated the principal tasks for radiochemical research in geology (Vernadsky, 1935, 1937). A prominent place among them belonged to the problems of absolute geochronology; to this Vernadsky paid special attention, repeatedly stressing its significance for the reconstruction of the history of geological processes, during the Precambrian in particular. In his estimation, the problems included: a) the absolute dating of geological events and subdivision of the geological history of the Precambrian; b) absolute age determinations for the oldest rocks of the Earth; c) the study of the behaviour of radioactive and radiogenic elements and the conditions of their preservation in different parts of the crust; and d) determination of the time of the earliest metamorphism in Precambrian rocks.

Among the first geochronological studies of outstanding importance in the realization of these tasks was an absolute age determination in 1926 of uraninite in pegmatites from the White Sea area by the uranium-lead method by K.A. Nenadkevich. Owing to the unavailability at that time of mass-spectrometers, the isotope composition of lead was determined by the ordinary classical analytical method used for the determination of the atomic weights of elements. The date of 1850-2100 m.y. obtained was, at that time, the oldest date on Archaean rocks; for two decades, it served as a basis for concepts on the duration of the Precambrian and for theoretical speculations on Precambrian geology. For instance, the Committee on Geological Time Determination (USA) accepted Nenadkevich's figure as the age of the crust (Knopf, 1931).

An important step for the subsequent development of scientific and applied geochronology was the introduction in 1946 of the potassium-argon method by V.G. Khlopin and E.K.Gerling. This made possible age determinations of widely different rock forming minerals and marked the beginning of a new stage in the advance of geochronology. Instead of the former single determinations of rare uranium-bearing accessories from a limited number of rocks (indeed, from granites and pegmatites only), thousands of dates could be obtained that would characterize rock complexes of the most varied compositions (Khlopin and Gerling, 1947; Gerling, 1967).

The 1950 to 1970 period saw a flourishing of geochronological research that included study of an extremely wide range of geological subjects in the Baltic, Ukrainian and Aldan shields, in the Urals, Siberia and the Far East, in the Caucasus, and in Central Asia. Extensive and varied data, obtained as a result of studies carried out by numerous scientific groups, enabled the followers of Vernadsky to make great strides in the solution of the problems he had raised, to elaborate his ideas and to create a distinctive Russian scientific school of geologists, geochemists and geochronologists represented by A.P. Vinogradov, A.A. Polkanov, E.K. Gerling, I.E. Starik, A.I. Tugarinov, L.N. Ovchinnikov, N.P. Semenenko, G.D. Afanassiev, L.V. Kimlev, N.I. Polevaia, M.A. Garris, G.V. Voitkevich, E.V. Sobotovich and others.

In 1950 a Commission for the absolute age determination of geological formations was organized by the Academy of Sciences of the USSR. The Commission became a scientific-methodological centre for the determination of trends in geochronological research and for the discussion and publication of data obtained in all the geochronological laboratories of the USSR [Proceedings (*Trudy Komissii*) for the years 1954-1970; Bulletins (*Biulleteni Komisii*) for 1955-1970].

THE INFRACAMBRIAN INTERVAL

In the Precambrian geochronology established by the Russian school during the 1950-1970 period, the first major discovery was the determination in the 1950s of the absolute age of Rapakivi Granites from the Baltic and Ukrainian Shields. Their geological features indicated that the formation of these granites marked the end of the geosynclinal period in the Proterozoic and the beginning of the Russian Platform regime, which in the fifties was considered to have originated in the Cambrian. Consequently, Rapakivi Granites directly preceding the Cambrian were expected to have an age of about 600-650 m.y.; instead, their age proved to be 1600-1650 m.y. An interval amounting to about one billion years thus became evident in the geological history of the Baltic and Ukrainian Shields, not filled by any geological formations then known. A geological question of foremost importance arose as to what events had taken place during this time interval.

This problem was solved within the next five years. Underlying the Cambrian, the so-called Gothian, Riphaean and Vendian sedimentary sequences and associated magmatic rocks came to be recognized, thus filling in the great span of time between the Rapakivi Granites and the beginning of the Palaeozoic (Polkanov, 1956; Polkanov and Gerling, 1961; Polevaia and Kazakov, 1961; Semenenko, 1965).

THE BASE OF THE PROTEROZOIC

The second discovery, first made on the Baltic Shield, referred to the lower age boundary of the Proterozoic. At the end of the 1950s, rock complexes which, by their geological relations with the underlying Archaean and by their distinct lithological and metamorphic features, belonged to the Lower Proterozoic were found to range in age up to 2700 m.y. At first this absolute age did not inspire the confidence of geologists, inasmuch as it happened to be older than the earliest date for the Archaean obtained at that time (1900-2000 m.y. for the White Sea area). Soon, however, ages of 3000 to 3600 m.y. were obtained for other Archaean complexes in the Baltic Shield. Just as in the first two cases (the age determinations of Rapakivi Granites and of the Lower Proterozoic complexes), these at first caused a sensation, since their acceptance demanded fundamental changes in the concepts then generally accepted concerning the duration of the Precambrian. For the same reason, many scientists were very sceptical; indeed, the data led to lively discussion and criticism.

As time went by, however, similar ages for Proterozoic and Archaean formations were found in other regions of the USSR. As a result, by 1964 a date of 2600-2800 m.y. became the official accepted lower age boundary for the Proterozoic and was introduced into the geochronological scale of absolute dates (Afanasiev, Bagdassarian *et al.*, 1964).

Also important for Archaean geochronology was the discovery at different points (Baltic and Aldan Shields, Urals) and by different methods (potassium-argon and uranium-lead) of ancient rock fragments of the Earth having an age from 4 to

4.3 billion years. These data indicated the possibility of reconstructing the history of the earliest periods in the formation of the crust by geological and geochronological methods (Gerling, Maslenikov *et al.*, 1963; Ovchinnikov and Dunaev, 1967; Rudnik and Sobotovich, 1969).

METAMORPHIC EFFECTS

Along with the ancient ages found for the Precambrian rocks, which so greatly expanded the chronological limits of the Precambrian, geochronological research also disclosed in Precambrian complexes a wide occurrence of "rejuvenation" phenomena of radiometric ages. Precambrian rocks, during their long existence since the time of formation, were subjected, as a rule, to the effects of later geological processes (e.g., tectonic and metamorphic). These effects caused a disequilibrium in the relations between radioactive and radiogenic elements which introduced distortion into the dates computed for the rocks. It has been found that "rejuvenation" (updating by thermal effects, especially frequent in K-Ar relations) causes discordance in the age values obtained for Precambrian complexes. However, despite the difficulties arising in these cases in determining the real time of formation for the rocks in question, the discordance of dates represents a source of very important information on the subsequent geochemical (-geological) history of the rock (Komlev, Savonenkov *et al.*, 1962; Maslenikov, Bondarenko *et al.*, 1963).

The numerous results of this research have changed the concepts about the early periods of the history of the earth's crust. It has come to be recognized that the duration of the Precambrian is much longer, and its geological history much more complicated, than had been assumed previously; this has led to a revision of concepts of Precambrian stratigraphy, magmatism, tectonics, and the structural development of the ancient parts of the Earth's crust and, as a consequence, to a revision of the nomenclature of Precambrian chronological subdivisions (Vinogradov, Tarasov *et al.*, 1959; Kratz, 1964; Vinogradov and Tugarinov, 1968; Kratz and Lobach-Zhuchenko, 1970; Tugarinov and Voitkevich, 1970, Salop, 1970). The establishment of a new concept of a lengthy and complicated geological history for the Precambrian, with due consideration of the entire complex of radiometric data, is still in progress.

The importance of the discoveries made, their reliability confirmed and checked by subsequent laboratory determinations in the USSR and abroad, are witness to the progressive approach of Soviet science in accepting the evolution of scientific thought as a basis for further creative development.

The present state of geochronological research is characterized by an increasing perfection and progressively wider use of isotope methods for the solution of a broader range of problems in geology and petrology, features of magmatic differentiation, sedimentation, metamorphic petrology, geochemistry of endogenic processes, etcetera.

REFERENCES

Affanassiev, G.D., Bagdassarian, G.P., *et al.*, 1964, Geokhronologicheskaia shkala u absoliutnom letoischislenii po dannym laboratori SSSR na april 1964 g. s uchetom zarubezhnykh dannykh (The geochronological scale and absolute age determinations according to the laboratories in the USSR up to April 1964, taking foreign data into account): *in* Absoliutny vozrast geologicheskikh formatzi: Mezhdunar. geol. kongress, XXII sessia, Doklady sovetskikh geology (Reports of Soviet Geologists): Problema 3, Moscow, Izdatel, "Nauka", p. 287-324.

Gerling, E.K., 1961, Sovremennoe sostoianie argenovogo metoda opredelenia vozrasta i ego primenenie v geologii (The present status of the argon method of determining age and its use in geology): Moscow and Leningrad, Academy of Sciences Publication, p. 131.

Gerling, E.K., Maslenikov, V.A. *et al.*, 1968, Drevneishie gornye porody Baltiiskogo schita (The oldest rocks of the Baltic Shield) *in* Geologia dokembria (Geology of the Precambrian): Mezhdunar geol. kongress. XXIII Sessia. Doklady sovetskikh geologov (Reports of Soviet geologists): Problema 4, Leningrad, Izdatel, "Nauka", p. 136-141.

Khlopin, V.G., and Gerling, E.K., 1947, Novy method opredelenia absoliutnogo geologicheskogo vozrasta mineralov (A new method of determining the absolute geological age of minerals): Doklady of the Academy of Sciences of the U.S.S.R. new series, v. VIII, no. 7, p. 1415-1417.

Knopf, A., 1931, The age of the Earth: Summary of principal results: National Research Council Bulletin, no. 80, p. 3-9.

Komlev, L.V., Savonenkov, V.G., Danilevich, S.I. *et al.*, 1962, O geologicheskom znachenii regional nykh protzessov omolozhenia v drevnikh formatziakh iugo-zapada Ukrainskogo kristallicheskogo schita (Geological significance of the regional processes of rejuvenation in the ancient formations in the southwest of the Ukrainian Cristalline Shield): Geokhimia, no. 3, p. 195-206.

Kratz, K.O., 1964, Akademik A. A. Polkanov i ocherednye problemy dokembria Baltiskogo schita (A.A. Polkanov and the immediate problems of the Precambrian of the Baltic Shield): *in* Geologia i geokhronologia dokembria (Geology and geochronology of the Precambrian): Trudy Laboratorii Geologii Dokembriya, fasc. 19, (Moscow and Leningrad, Izdatel, "Nauka", p. 17-23.

Kratz, K.O. and Lobach-Zhuchenko, S.B., 1970, Izotopnaia geokhronologia i glubinnoe stroenie semnoi kory (Isotopic geochronology and the deep-seated structure of the earth's crust): Geotektonika, no. 2, p. 74-79.

Ovchinnikov, L.N. and Dunnaev, V.A., 1967, Iztandit—drevneishaia gornaia poroda Urala (Izrandit—oldest rocks in the Urals): *in* Voprosy datirovki drevneishikh (katarkheiskikh) geologicheskikh obtazovanii i o novnykh porod. (Problems of dating the oldest (Catarchich) geological formations and base rocks): Trudy XIV Sessii Kom. po opredeleniu absoliutnogo vozrasta geologicheskikh formatzi, Moscow, Izdatel, "Nauka", p. 16-18.

Polevaia, N. I. and Kazakov, G.A., 1961, Vozrastnot raschleneni i korreliatzi drevnikh nemykh otlozheni po otnosheniu K-Ar v glaukonitakh (Age determination and correlation of ancient nonfossiliferous deposits according to K-A ratio in glauconites): *in* Voprosy gekhronologii i geologii (Problems of geochronology and geology): Trudy Laboratorii Geologii Dokembriya, fasc. 12, Moscow and Leningrad, Izdatel, "Nauka", p. 103-122.

Polkanov, A.A., 1956, Geologia hoglandia-iotnia iuzhnoi chasti Baltiskogo schita i problema dokembriskogo pereryva (The geology of Hoglandian-Jotnian in the southern part of the Baltic Shield and the problem of the Precambrian interval): Izvestiya Academy of Sciences, U.S.S.R. Geological Series, no. 1, p. 5-28.

Polkanov, A.A. and Gerling, E.K., 1961, Geokhronologia and geologicheskaia evolutsia Baltiskogo schita i ego skladchatogo obramlenia (Geochronology and geological evolution of the Baltic Shield and its folded framework) *in* Voprosy geokhronologii i geologii (Questions of geochronology and geology): Trudy Laboratorii Geologii Dokembbiya, fasc. 12, Moscow and Leningrad, Izdatel, "Nauka", p. 7-102.

Rudnik, V.A. and Sobotovich, E.V.O., 1969, O vozraste polimetamorficheskikh komplek-sov iengrskoi serii Aldanskogo schita (One the age of the polymetamorphic complexes of the Yengskriy series of the Aldan shield): Doklady of the Academy of Sciences of the U.S.S.R., v. 189, no. 4, p. 834-837.

Salop, L.I., 1970, Peresmotr geokhronologischeskoi shkaly dokembria (A re-examination of the geochronological scale of the Precambrian): Byulleten' Moskovskogo Obshchestra Ispytatelei Prirody, Otdel Geologicheskii, v. XV, no. 4-5, p. 115-31; 5-26.

Semenenko, N.P., 1965, Geokhronologisccheskeia karta ukrainskogo dokembria (A geo-chronological map of the Ukrainian Precambrian): *in* Geokhronologia dokembria Ukrainy (Geochronology of the Ukrainian Precambrian): Kiev, Naukova Dumka.

Tugarinov, A.I. and Voitkevich, G.V., 1966, Dokembriskaia geokhronologia materikov. (Precambrian geochronology of the continents): Moscow, Izdatel, "Nedra", p. 387.

Vernadsky, V.I., 1910, O neobkhodimosti issledovania radioaktivnykh mineralov Rossiiskoi imperii (On the need to investigate the radio-active minerals of the Russian Empire): Spornik, Akademii Nauk Publication, 54 p.

———, 1911, Zadacha dnia v oblasti radia (Rech chitannaia v publichnom zasedanii obschego sobrania 29 dekabria 1910g [A critical problem in the area of radium (Lecture presented at a public session of the General Meeting on 29th December 1910)]: Izvestiya Academy of Sciences of the U.S.S.R., 6th series, v. 5, no. 1, p. 61-72.

———, 1935, O nekotorykh ocheredhykh problemakh radiogeologii (Doklad v Otdelenii matem, i estestv. nauk. 20 dekabria 1934 g.) (Some immediate problems in radio-geology – Report to the Section on Mathematics and Natural Sciences, 20th December 1910): Izvestiya Academy of Sciences of the U.S.S.R., 7th series, no. 1, p. 1-18.

———, 1937, O znachenii radiogeologii dlia sovremennoi geologii (Opredelenie geologicheskogo vozrasta) (The significance of radio-geology for modern geology — determination of geological age): *in* Trudy 17 Sessii Mezhdunarodn. geol. Kongressa, SSSR, 1937, v. 1. Moscow, GONTI, p. 215-239.

Vinogradov, A.P., Tarasov, L.S. and Zykov, S.I., 1959, Izotopny sostav rudnykh svintzov Baltiiskogo schita (The isotopic composition of common leads of the Baltic Shield): Geokhimia, no. 7, p. 571-607.

Vinogradov, A.P. and Tugarinov, A.I., 1968, O geokhronologicheskoi shkale dokembria. (The geochronological scale of the Precambrian): *in* Problemy geoghimii i kosmologii (Problems of geochemistry and cosmology): Mezhdunar. geol. kongress, XXIII Sessie, Doklady sovetskikh geologiv, Problemy 6 i 13a. Leningrad, Izdatel, "Nauka", p. 47-60.

History of Concepts in Precambrian Geology, edited by W.O. Kupsch and
W.A.S. Sarjeant, Geological Association of Canada Special Paper 19, 1979

BOUNDARY OF THE CANADIAN SHIELD

W.O. Kupsch
Department of Geological Sciences, University of Saskatchewan
Saskatoon S7N 0W0

ABSTRACT

The shape of the Canadian Shield, that great region of Precambrian rocks forming the elliptical core of the North American continent, became fully known only in the mid-20th century.

The parts of the Shield's outer perimeter in Quebec, Ontario, and the United States, which were relatively easily accessible, are depicted with accuracy on geologic maps of the continent published in the middle of the 19th century. In Canada the work by Sir William Logan laid the foundation of our present knowledge.

By the 1850s several maps were published showing the western mainland boundary in subarctic and arctic regions based on information obtained from fur-traders and from Sir John Richardson during his travels with two Franklin Expeditions and his own Searching Expedition.

The extension into northern Canada and the more accurate retracing of the boundary during the latter half of the 19th century is an intrinsic part of the history of the Geological Survey of Canada during that time.

Technical advances in air transportation and photography in the 20th century made it possible to complete the mapping of the boundary of the Canadian Shield in the Arctic Archipelago.

The mapping of the boundary of the Canadian Shield thus spans slightly more than a hundred years in the history of geology.

RESUME

La forme du Bouclier Canadien, cette grande région de roches précambriennes constituant le noyau elliptique du continent nord-américain, commença seulement au milieu du 20ième siècle à être complètement connue.

Au Québec, en Ontario et aux Etats-Unis, les parties du périmètre extérieur du Bouclier qui étaient d'un accès relativement facile furent décrites avec exactitude sur les cartes geologiques du continent publiées vers le milieu du 19iéme siècle. Au Canada le travail de Sir William Logan posa les fondations de notre connaissance actuelle.

Plusieurs cartes des limites orientales du continent dans les régions subarctiques et arctiques furent publiées vers lés années 1850; ces limites étaient basées sur l'information obtenue des commerçants en fourrure et des voyages de Sir John Richardson au cours de deux expéditions Franklin et de sa propre Expédition d'Exploration.

L'extension vers le Canada septentrional et le nouveau tracé plus précis des limites au cours de la seconde moitié du 19ième siècle font intrinséquement partie de l'histoire du Service Géologique du Canada au cours de cette période.

Les progrès techniques dans les transports aériens et la photographie au cours du 20ième siècle permirent de compléter la cartographie de la limite du Bouclier Canadien dans l'Archipel Arctique.

La cartographie des limites du Bouclier Canadien s'étend ainsi sur près d'une centaine d'années dans l'histoire de la géologie.

DEFINITION OF BOUNDARY

The geographical boundary of the Canadian Shield is the limit-line of those exposed Precambrian rocks that form the core of the North American continent in Canada and the United States. It is the contact with overlying younger rocks of Phanerozoic, mainly Paleozoic, age. In some places the contact is obscured by Quaternary deposits, in others by water and Recent sediments. Where the contact is covered by water the exact position of the Shield's boundary may not be known even today. Surficial deposits on land, however, are nowhere so extensive or thick that they prevented mapping of the boundary with reasonable accuracy by field geologists without the help of subsurface information.

The location of the boundary as shown in Figure 1 is essentially the same as the one presented by Stockwell *et al.* (1970, p. 46) or, on a larger scale, on the most recent Geological Map of Canada (Douglas, 1969). Minor simplifications, particularly in the Eastern Arctic and in the southeastern part of the Shield, where isolated masses of Precambrian rocks were incorporated into the main mass, were necessitated by differences in scale and projection.

SHAPE OF THE CANADIAN SHIELD

The shape of the exposed Canadian (Precambrian) Shield is now well known (Fig. 1). It is that of an ellipse with its longest axis, which is at least 4,500 km in length, trending in a 160° azimuth direction. It has a short axis of about 3,300 km and its centre, in the northern part of Hudson Bay, is at approximately 85° W. longitude and 60° N. latitude. Its main mass occupies all of Labrador and parts of Quebec, Ontario, Manitoba, Saskatchewan, Alberta, and the Northwest Territories in Canada; and New York, Michigan, Wisconsin, and Minnesota in the United States. Parts detached from the main mass are called "inliers" (Stockwell *et al.*, 1970, p. 46). Only the larger ones need to be mentioned here: the Brock-Minto, Sutton, and Great Northern Inliers. The term "outlier" is commonly used for isolated outcrops of Phanerozoic rocks on the main mass of the Shield. These will not be considered further, although it should be mentioned that some authors regard the Hudson Bay Lowland and the Foxe Basin as large outliers.

The total area occupied by the Canadian Shield as shown in Figure 1, including its inliers and contiguous parts in the United States but not those areas under marine waters, is 4,587,000 km^2 (1,771,000 square miles), which makes it the largest region of exposed Precambrian rocks in the world.

BOUNDARY SEGMENTS

The boundary of the Canadian Shield will here be divided into segments, the names of which are based mainly on geological regions as recognized by Douglas (1970, p. 3). The segments – from the northeast in a clockwise direction – are the

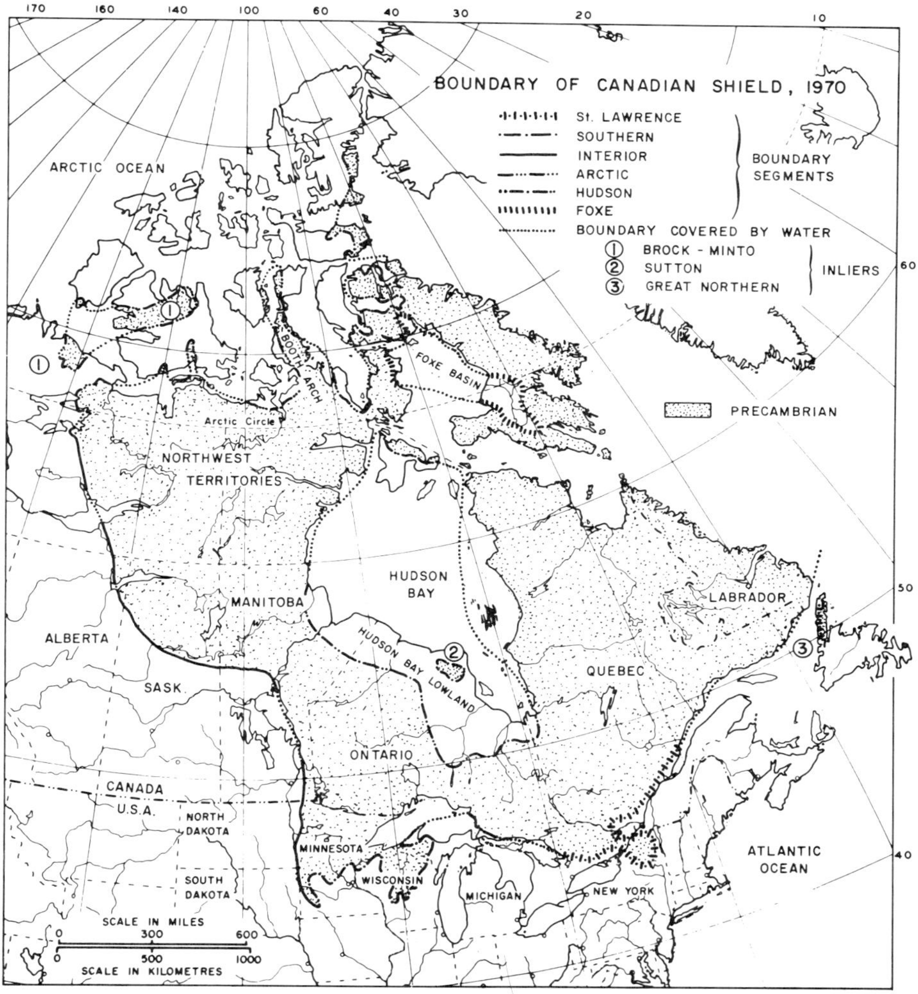

Figure 1. Boundary of Canadian Shield, 1970

St. Lawrence, Southern, Interior, and Arctic Boundary Segments. The Hudson and Foxe Boundary Segments are separate, interior segments (Fig. 1).

Of the various geological regions into which the Canadian Shield can be divided it suffices here to mention only the Boothia Arch, a north-northwesterly pointing finger-like extremity of the North American mainland.

STATE OF KNOWLEDGE BY 1857

Figure 2 is a hypothetical map showing the distribution of "crystalline schistous and granitoid rocks" (the Canadian Shield in modern parlance) as it may have been perceived in the year 1857 by a geologist who had to prepare such a map from other than already published maps covering parts but not the whole of the Shield region.

The St. Lawrence and Southern Boundary Segments are shown in detail on Rogers' (1855) impressive "Geological Map of the United States and British North America". In Lower and Upper Canada the requisite fieldwork had been done since the inception of the Geological Survey of Canada in 1842 by its first director, Sir William Logan, who began detailed mapping of the Shield with his survey of the Ottawa River in 1845. In the United States the older State Geological Surveys of

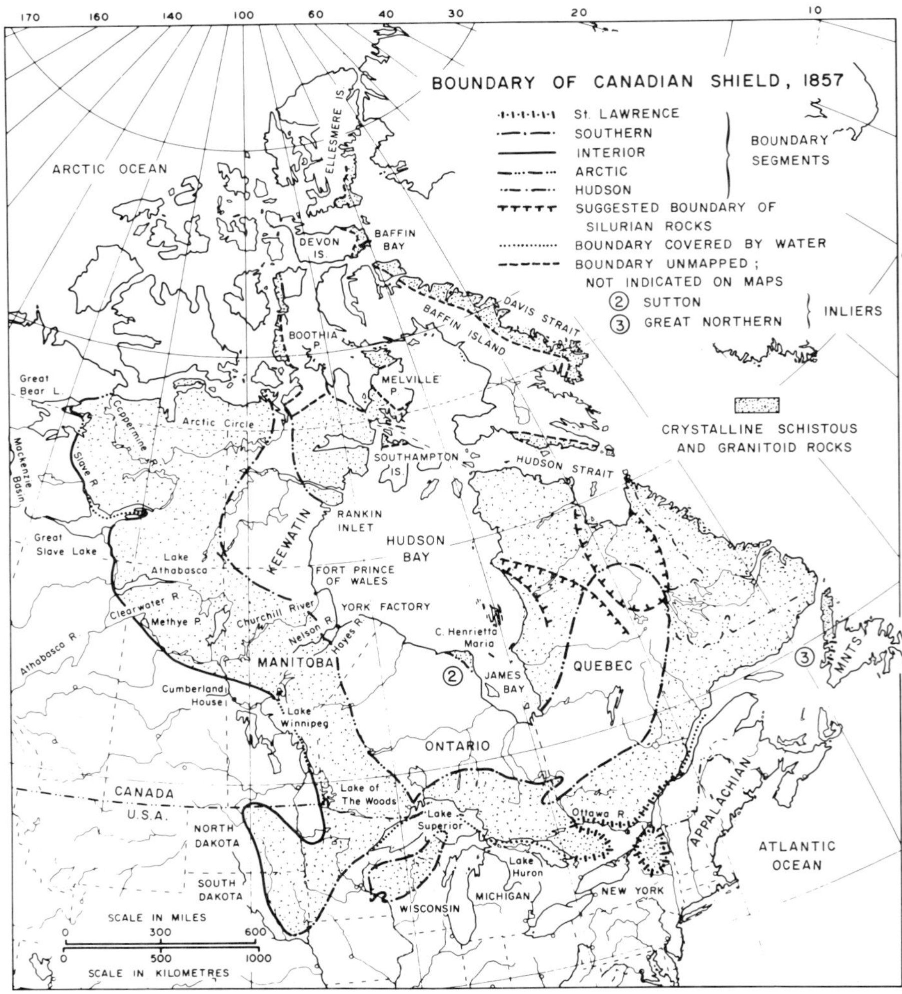

Figure 2. Boundary of Canadian Shield, 1857

New York and Michigan had contributed to the map "constructed [by Rogers] from the most recent documents and unpublished materials" which were duly acknowledged in the accompanying text. For Canada maps and reports by Bigsby, Logan, and Richardson were thus mentioned (Rogers, 1855, p. 32).

Although Rogers showed the position of both the St. Lawrence and Southern Boundary Segments for the most part in approximately the same position as on a modern geological map of similar scale, there are some notable and significant differences. Gneiss and granitic rocks are shown not only in the northern part of New York as a detached inlier but also in the core of the Appalachian Mountains in the New England States and Newfoundland. Although it means an anachronistic introduction of much later concepts regarding a cratonic Precambrian Shield *versus* the involvement of Precambrian rocks in a mobile belt, and the techniques of distinguishing granites of differing ages, Rogers' granites and gneisses are not shown in Figure 2 for the Appalachian Mountains whereas the Great Northern and New York inliers have been retained. Some justification can be given for this by pointing to the physiographic affinity of these two inliers to the main mass of the Canadian Shield rather than to the fold mountains to their east.

Opposite to the New York inlier on Rogers' map a much elongated, narrow indentation into the Shield, occupied by Phanerozoic strata, can be noted along the Ottawa River. It may have been depicted in this fashion to take account of an upstream outlier of Ordovician-Silurian rocks large enough to be shown clearly on the 1:5,000,000 Geological Map of Canada (Douglas, 1969).

The position of Lakes Huron and Superior on the edge of the "Metamorphic" rocks was recognized by Rogers as was the occurrence of these rocks in northern Michigan and Wisconsin. But whereas he showed an inlier in these states, modern maps depict a connection to the main mass of the Precambrian Shield in the west.

On mid-nineteenth century geological maps a large departure from the presently known position of the Southern and the extreme southern portion of the Interior Boundary Segments can be noticed in the southwestern corner of the Shield. Here the "Metamorphic" rocks of Rogers are shown to cover large parts of North and South Dakota and to extend into southwestern Manitoba (Fig. 2). Isbister (1855) and, following him, Devine (1857) carried the granites even farther onto the Western Plains.

The hypothetical map would, in all likelihood, have taken the Interior Boundary Segment beyond the area covered by Rogers' map from Richardson (1851), Isbister (1855), or Devine (1857). Of these, Richardson's map served as the prototype for the maps to follow. It is the oldest geological map covering all of present Canada with any degree of accuracy although only two fundamental geological units were recognized: 1) Metamorphic or Primitive Rocks, 2) Fossiliferous Rocks from the Silurian Rocks upwards. Although Isbister improved this classification with respect to the Phanerozoic rocks, his map is no improvement on Richardson's as far as the boundary of the Canadian Shield is concerned. Quite the contrary is actually true for the southern end of the Interior Boundary Segment where Isbister enlarged and stretched into a curious sickle-shape the extension of the Shield onto the Prairies from south of Lake of the Woods, an extension also mapped by Rogers (1855) but not by Richardson (1851).

In 1857 Thomas Devine compiled another map which, although it had the advantage of a larger scale (Table I), was, of course, not any more accurate or extensive than the maps of Richardson and Isbister on which it was based for those parts covering the "Indian Territories and Hudson's Bay". Isbister and Devine both acknowledged their debt to Richardson; and Devine his to Isbister, who was the obvious source for the even more wide-spread mapping by Devine of granites on the prairies.

That the Interior Boundary Segment was known to Sir John Richardson accurately enough to be shown on a map in the year 1851 can be understood only if one considers the fur-trade routes of the time. As Morse (1969, p. 27-32) points out, the Canadian Shield and its waterways dominated and moulded Canadian fur-trade history. The chain of large, connected lakes (Great Bear, Great Slave, Athabasca, Winnipeg, Lake of the Woods, Superior, and Huron) around its boundary all have one side, or end, in granite. There is for this reason a close correlation between the boundary of the Shield and the fur-trade canoe routes. Alexander Mackenzie (1801, p. 403), commenting on the presence of granite on the east side and of limestones on the west side of Lake Winnipeg, a relationship that is particularly striking at the narrows in the middle of the lake, stated that from " . . . the narrowest part of Lake Winipic . . . dark-grey granite . . . is to be found throughout the whole extent North of this country to the coast of Hudson's Bay and . . . from there onwards to the coast of Labrador; and it may be further observed that between these extensive ranges of granite and lime-stone are found all the great lakes of this country."

John Richardson, the first to record his and others' geological observations on a map, travelled the fur-trade "highway" from York Factory on Hudson Bay to the interior. That he in turn "stood on the shoulders of his predecessors" (in this case on those of the fur-traders and their Indian guides) has been well demonstrated by Warkentin (1967) who reviewed David Thompson's earlier contribution to the geology of western Canada. But because Thompson's *Geology*, a twelve-page description of the region from the Rocky Mountains to the eastern end of Lake Superior, was never printed until 1967 when it appeared in Warkentin's paper, it could not have served as a published source for anyone wanting to draw an outline map of the Canadian Shield in the mid-nineteenth century. Its importance is mainly that of a reminder to modern readers of the existence of hearsay sources known to have been used by Richardson in the preparation of his map.

The mapping of the boundary of the Shield in Arctic Canada, a land so unlike that of the Subarctic Boreal Forest in the Western Interior, did not grow out of the fur-trade but out of searches for a Northwest Passage, which in itself developed from fruitless efforts to find a shortcut to China. Again, it was Richardson (1851) who recorded his own work and that of others who undertook these maritime explorations. His own travels along the northern coast of the North American mainland brought him only as far east as the Coppermine River and the northwestern-most corner of the Shield. The period between the end of the Napoleonic wars and the disastrous end of Franklin's Third Expedition in 1848 has been called, with much justification, a Golden Age of exploration in the Canadian Arctic (Baird, 1949, pt. 2, p. 41). During that period, and through the observations of John and

TABLE I
COMPARISON OF SOME MAPS SHOWING THE WHOLE OR SUBSTANTIAL PARTS OF THE CANADIAN SHIELD, ARRANGED CHRONOLOGICALLY

REFERENCE	A. COVERAGE B. DIMENSIONS IN CM C. SCALE
Douglas, 1969	A. All Canada geologically coloured; geology in U.S.A. indicated but not coloured. B. 115 x 105 C. 1:5,000,000 (1 inch to 80 miles app.)
G.S.C., 1955	A. All Canada geologically coloured; geology in U.S.A. not indicated B. 73 x 71 C. 1:7,603,200 (1 inch to 120 miles)
Dawson, 1887	A. Canada north of 60° N lat. geologically coloured B. 46 x 22 C. 1:15,206,400 app. (1 inch to 240 miles app.)
Selwyn and Dawson, 1884	A. Canada south of 60° N lat. geologically coloured; geology in U.S.A. not indicated B. 204 x 60 C. 1:2,851,200 app. (1 inch to 45 miles app.)
Logan, 1864	A. Canada south of 52° N lat. and east of 100° W long. and adjacent U.S.A. geologically coloured B. 48 x 21 C. 1:7,920,000 (1 inch to 125 miles)
Devine, 1857	A. Canada and U.S.A. west of 80° W long. and north of 43° N lat. geologically coloured B. 153 x 144 C. 1:2,534,400 app. (1 inch to 40 miles app.)
Boué, 1855	A. All Canada and U.S.A. geologically coloured B. 60 x 49 C. 1:30,000,000 app. (1 inch to 500 miles app.)
Rogers, 1855	A. Canada south of 55° N lat. and all U.S.A. geologically coloured B. 60 x 49 C. 1:10,771,200 app. (1 inch to 170 miles app.)
Isbister, 1855	A. All Canada and U.S.A. geologically coloured B. 26 x 20 C. 1:22,176,000 app. (1 inch to 350 miles app.)
Richardson, 1851	A. All Canada and U.S.A. geologically coloured B. 48 x 38 C. 1:14,572,800 app. (1 inch to 230 miles app.)

James Ross, John Rae, and others, the existence of "Metamorphic or Primitive Rocks" became known not only from the mouth of the Coppermine River to the east, but also on Boothia and Melville Peninsulas and on northern Southampton Island in the Central Arctic.

"Crystalline rocks" are shown, but without definite boundaries and fading out into unexplored areas, on Isbister's map along the shores of Hudson Strait and the west shore of Davis Strait and Baffin Bay. Baffin Island, Devon Island, and southern Ellesmere Island, places of which only the sketchiest geographical outline could be drawn at the time, were thus known to have "Primitive" or Precambrian rocks along their shores; the interior parts of these islands had to await geological investigation during the second half of the nineteenth and the first half of the twentieth century. There is therefore only the barest indication of Foxe Basin on Richardson's, Isbister's, and Devine's maps, where Silurian rocks are, quite correctly, shown to occupy the northeastern coast of Melville Peninsula.

Of the six segments into which the limit of the Canadian Shield can be divided, the Foxe Boundary Segment was thus practically unknown. Of the Hudson Boundary Segment only the central sector was accurately located on geological maps of the 1850s. Again, this is understandable if one considers the canoe routes of the fur-traders. Their main route from Hudson Bay went from York Factory up the Hayes River to Lake Winnipeg and from there via Cumberland House to the Churchill River, across Portage La Loche or Methye Portage (the principal link between the Hudson Bay and Mackenzie River drainages), down the Clearwater River to the Athabasca River and Lake Athabsca and from there down the Slave River to Great Slave Lake into the Mackenzie Basin (Morse, 1968, map). It was this way, which crossed the Shield's boundary in several places, that Richardson travelled and which provided him with first-hand information on the location of the boundary between Paleozoic rocks and Precambrian granites, not only in the Hudson Bay Lowland, where he noted granites when 120 miles up the Hayes River from Fort York, but in the Western Interior as well. Information provided for him by others who had travelled the minor route from York Factory up the Nelson River and from Fort Prince of Wales up the Churchill River (where the contact is exceptionally well exposed) proved to be reliable, but to the north in Keewatin and to the south in Ontario and Quebec the Hudson Boundary Segment as shown on both Richardson's and Isbister's maps differs substantially in its location from that on modern maps. Fur-traders traversed neither the Keewatin nor the interior of Quebec and hence even the most elementary geological and physiographic information was unavailable. Before the exploratory surveys by officers of the Geological Survey during the second half of the nineteenth century these regions were usually presumed to be part of the Hudson Bay Lowland and its Phanerozoic rocks.

It is clear that a compiler of a geological map in 1857 would have faced his greatest difficulties with the Hudson Boundary Segment as he would have to choose from several alternatives presented on other maps. Except for the presence of "Primary stratifications . . ." in interior Keewatin and Quebec and for the lack of detail on account of its small scale, Ami Boué's (1855) map of the "Geological Structure of the Globe" has in time proved to be a better approximation of the geology around Hudson Bay than either Richardson's or Isbister's map. Firstly,

Boué shows the occurrence of "The Crystalline schistous formation comprehending all the Granitoid Rocks" from near Rankin Inlet north along the Keewatin Coast, and secondly he places the boundary in Ontario farther north and closer to its actual position. Because both Richardson and Isbister left the Keewatin coast blank on their maps, Boué's boundary has been copied on Figure 2, even though it is realized that if a compiler would have followed Devine (1857) the coast of Keewatin would have been shown to consist of Phanerozoic sedimentary rocks. The boundary of the Silurian as suggested by Isbister is, however, also shown on Figure 2.

In Figure 2 the southern or Ontario part of the Hudson Boundary Segment is located as on Isbister's map but in Quebec the definite, although inaccurate, limit of Boué has been preferred over the indefinite termination of the Shield as depicted by both Richardson and Isbister.

Although displaced toward the coast to Cape Henrietta Maria, at the northwestern entrance to James Bay, there is some indication on Isbister's map (but not on Richardson's nor on Devine's) that the occurrence of granitic rocks in what is now called the Sutton Inlier was suspected in 1855.

STATE OF KNOWLEDGE BY 1887

The task of drawing an outline map of the Canadian Shield in 1887 was much easier than 30 years earlier. Only three comprehensive works accompanied by geological maps needed to be consulted: Logan (1864), Selwyn and Dawson (1884), and Dawson (1887). It is these three maps that were used to draw figure 3.

The well-executed, colour-printed geological map of Logan depicted the St. Lawrence, Southern, and part (up to 52° N lat.) of the Interior Boundary Segments with a precision and clarity that has needed no essential revisions to this day. It improved Rogers' (1855) map in that the Precambrian in New York State became a part of the Shield, in that the limit of the Shield was shown more precisely in Michigan and Wisconsin, and in that the southern part of the Interior Boundary Segment was pulled back from the far western position which had indicated the presence of granites out on the Prairies. These improvements, all in the United States, were the result of a new compilation "under the authority of Prof. James Hall" as Logan states in the title of his map.

Logan divided the Azoic (later to become the Archaean) into the Laurentian, comprising light-coloured gneisses and granitic rocks, and the darker Huronion – highly deformed, schistose rocks representing metamorphosed sedimentary and volcanic material with metalliferous deposits of economic significance. He erroneously regarded the Laurentian as older than the Huronian. On Selwyn and Dawson's (1884) map the Laurentian and Huronian are also shown. Although Dawson (1887) makes the same differentiation in his text, on his accompanying map (which bears 1886 as the year of imprint) he shows only "Archaean".

Whereas Logan provides us with an insight into the state of knowledge in 1864 about the St. Lawrence and Southern Boundary Segments, Dawson does the same for the Arctic and Foxe Boundary Segments in 1886. By that time, as Dawson's map demonstrates, the Foxe Basin became a geological feature on the map of the Dominion, albeit in a more tentative outline than shown in Figure 3 where white, unmapped areas in the southwestern tip of Baffin Island and near Nettilling

Lake (Lake Kennedy on Dawson's map) have been shown as Precambrian right up to Dawson's broken line marking his assumed boundary with Phanerozoic rocks.

In Dawson's time the Arctic Boundary Segment was not yet too well known in northern Baffin Island where later detailed field studies were necessary to bring to light its intricate pattern. On eastern Devon and southeastern Ellesmere Island the Precambrian boundary stayed in about the same place as on Isbister's (1855) map, but Dawson also showed the occurrence of Archaean rocks in two isolated areas in the northernmost part of Ellesmere Island (then called Grinnell Land) based on information obtained by the British Arctic expedition under Sir George Nares (1875-1876). Although the Geological Map of Canada (Douglas, 1969) also shows the occurrence of Precambrian rocks in northeastern Ellesmere Island and thus verifies in detail the late nineteenth century observations, current interpretations assign these rocks not to the Shield but to the Franklinian Eugeosyncline.

The northwestern corner of the Shield west of the Coppermine River has an appearance on Dawson's map unlike that on the 1955 G.S.C. map, which again differs from the map published in 1969. Although remoteness is only one of the reasons that this area was until recently so poorly known, it is the likely explanation for the differences between the 1955 and 1969 G.S.C. maps. However, in order to understand Dawson's version one has to know that this corner of the Shield is not occupied by gneisses or schists but by Helikian sandstones which Dawson regarded as of Cambrian age and showed as such on his map. That the relationship of the only slightly metamorphosed and little deformed Helikian strata to the older Archaean and Aphebian rocks, on which they lie with marked unconformity, was not yet understood in the 1880s is also evident from Selwyn and Dawson's (1884) map where the Helikian Athabasca Formation is shown as "Sandstones Age Uncertain". In other places Helikian strata were shown as Cambrian or even Silurian. Where they are lying well inside the boundary of the Shield they did not hamper the determination of its shape but in the northwestern corner they caused considerable difficulties indeed.

Selwyn and Dawson's map rather nicely fills the gap between Logan's in the south and Dawson's in the north. It is especially important in its portrayal of the Hudson Boundary Segment south of 60° N latitude where the Lowland is shown to occupy the area between Churchill and the eastern end of James Bay in much the same outline as on current maps.

Figure 3 shows the Hudson Boundary Segment according to Selwyn and Dawson's map. It also shows, again as on that map, the central part of what is now northern Quebec as "Unexplored", although it was becoming obvious by 1887 that Archaean rocks were most likely to occur in this part of the country rather than an extensive outlier of Phanerozoic rocks separate from the Hudson Bay Lowland, the boundaries of which were now known.

Selwyn and Dawson regarded the mainland coast east and opposite of the Belcher Islands (which themselves were barely known at the time) as underlain by Cambrian strata. They are now classified as being of Helikian age.

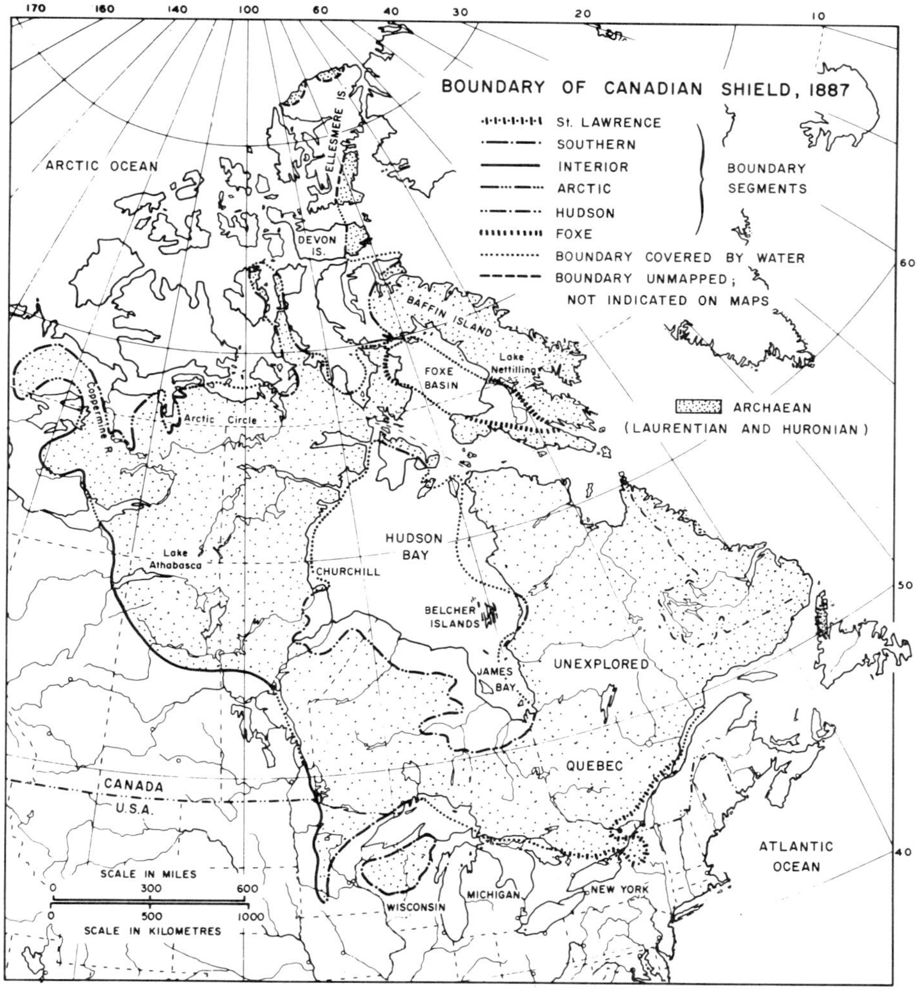

Figure 3. Boundary of Canadian Shield, 1887

LATE NINETEENTH AND FIRST HALF OF TWENTIETH CENTURY

During the last twenty years of the nineteenth century some remarkable exploratory traverses were made across the Shield by A.R.C. Selwyn, G.M. Dawson, McIntosh Bell, and many other officers of the Geological Survey of Canada. Besides geological exploration these men were expected tot describe and collect plants and animals, to record customs and language of the Indians and Eskimos, and to make their own topographical maps. The most demanding of such traverses were made by Robert Bell between Lake Superior and James Bay and between Lake Winnipeg and Hudson Bay; by J.B. Tyrrell in northern Saskatchewan and Manitoba, and in the Keewatin by canoe, by foot, or on snow-shoes; and by A.P. Low in the Labrador Peninsula. Through their efforts blank areas on the geological map of Canada became filled and the accuracy of the location of the Shield's boundary was checked. The much smaller extent of the Hudson Bay Lowlands, already shown on Selwyn and Dawson's (1884) map, was confirmed and thus large regions of Paleozoic rocks previously suspected to exist in the interior of Keewatin and Quebec permanently disappeared into oblivion.

The time after World War I brought aircraft and pilots to the "bush" of Canada. First the fixed-wing and then, after World War II, the helicopter changed the mode of travel of geologists from earth-bound to the almost limitless access enjoyed by those who fly. That a large area of Precambrian rocks, the Brock-Minto Inlier, appeared first with only tentative boundries on the 1955 Geological Map of Canada but later with a definite border on the newest map of 1969 is the direct result of better access by air rather than the availability of subsurface information regarding any cover by younger rocks that may obscure the boundary.

With the mapping of the Brock-Minto Inlier the accurate fixation of the boundary of the Canadian Shield was accomplished in the 1960s, more than 100 years after the publication of Richardson's map in 1851 which showed that outline for the first time.

ACKNOWLEDGEMENTS

The librarians of the Geology Library (Miss Anne Moran) and the Shortt Collection of Canadiana (Mrs. A.L. LaBrash), both part of the University of Saskatchewan's Library System, have been most helpful in finding relevant material. They are to be commended for their patience in assisting Mr. Dale Perry and Mrs. Valary Kennedy who did most of the initial searching and who drew prototypes of the maps.

That the University of Saskatchewan now has a copy of the magnificent 1856 edition of Johnston's *Physical Atlas of Natural Phenomena* in the Rare Book Room (as far as I know it is the only copy in Canada as McGill University's, although catalogued, cannot be located) is due to Professor George W. White, Department of Geology, University of Illinois, who, knowing through correspondence of my interest in Rogers' map in that atlas, called our librarian's attention to this item being for sale in Great Britain when I was away doing fieldwork in the Northwest Territories. We are indeed much obliged to Professor White for his alertness and thoughtfulness.

REFERENCES

Baird, P.D., 1949, Expeditions to the Canadian Arctic: The Beaver, pt. 1 (March, Outfit 279, p. 44-48); pt. 2 (June, Outfit 280, p. 44-47); pt. 3 (September, Outfit 280, p. 44-48).

Boué, Ami, 1855, The geological structure of the globe according to Ami Boué with corrections and additions to 1855 by A.K. Johnston, pl. 1 *in* Johnston, A.K., 1856 [sic!], New and enlarged edition: Physical atlas of natural phenomena: Edinburgh and London, W. and A.K. Johnston.

Dawson, G.M., 1887, Notes to accompany a geological map of the northern portion of the Dominion of Canada east of the Rocky Mountains: Report. Geological and Natural History Survey of Canada, (New series), v. 11 (1886), Report R, 62 p., geol. map [1886].

Devine, Thomas, 1857, Map of the north west part of Canada, Indian Territories and Hudson's Bay: Toronto, Derbishire and Desbarets, Printer to the Queen's Most Excellent Majesty.

Douglas, R.J.W., 1969, Geological map of Canada: Geological Survey of Canada, Map 1250A, scale 1:5,000,000.

————————, 1970, Introduction: *in* Douglas R.J.W. (Science Editor), Geology and Economic Minerals of Canada: Geological Survey of Canada, Economic Geology Report, no. 1, p. 2-6.

Geological Survey of Canada, 1955, Geological map of Canada: Map 1045A.

Isbister, A.K., 1855, On the geology of the Hudson's Bay Territories, and of portions of the Arctic and Northwestern regions of America: Quarterly Journal of the Geological Society of London, v. 11, p. 9-520, Geol. Map

Logan, W.E., 1864, Report on the geology of Canada: Montreal, Report: Geological Survey of Canada. Progress from its commencement to 1863, 1043 p., atlas [1865] of maps and sections [incl. Geological Map of Canada and the adjacent regions including parts of other British Provinces and of the United States . . .].

Mackenzie, [Sir] Alexander, 1801, Voyages from Montreal, on the River St. Laurence, through the continent of North America, to the Frozen and Pacific oceans, in the years 1789 and 1793 with a preliminary account of the rise, progress, and present state of the fur trade of that country: London, T. Cadell, jun. and W. Davies, etc., 412 p.

Morse, E.W., 1969, Fur trade canoe routes – then and now: Ottawa, Queen's Printer, 125 p.

Richardson, J., 1851, Arctic searching expedition; a journal of a boat voyage through Rupert's Land and the Arctic Sea in search of the discovery ships under command of Sir John Franklin, with an appendix on the physical geography of North America: London, Longman, Brown, Green, and Longmans, 2 vols., illustrated, including geological maps.

Rogers, H.D., 1855, Geological map of the United States and British North America, p. 29-32, pl. 8 (Geol. map): *in* Johnston, A.D., 1856 [sic!] New and enlarged edition: Physical atlas of natural phenomena: Edinburgh and London, W. and A.K. Johnston.

Selwyn, A.R.C. and Dawson, GM., 1884, Descriptive sketch of the physical geography and geology of the Dominion of Canada: Montreal, Geological and Natural History Survey of Canada, 55 p. (accompanied by Map of the Dominion of Canada geologically coloured from surveys made by the Geological Corps, 1842 to 1882).

Stockwell, C.H., *et al.*, 1970, Geology of the Canadian Shield: *in* Douglas, R.J.W. (Science Editor), Geology and Economic Minerals of Canada: Geological Survey of Canada, Economic Geology Report, no. 1, p. 44-150.

Warkentin, John, 1967, David Thompson's Geology – a Document: Journal of the West, v. 6, no. 3, p. 468-490.

History of Concepts in Precambrian Geology, edited by W.O. Kupsch and W.A.S. Sarjeant, Geological Association of Canada Special Paper 19, 1979

SOME GEOLOGICAL SIDE-EFFECTS OF THE SEARCH FOR RADIOACTIVE MINERALS IN THE CANADIAN SHIELD

A. H. Lang[1] and V. Ruzicka[2]
[1]Retired (formerly Geological Survey of Canada); [2]Geological Survey of Canada, 601 Booth St., Ottawa, Ont. K1A 0E8

ABSTRACT

Priorities for funds and personnel to provide field and laboratory work related to finding and evaluating radioactive deposits, mainly those of uranium, also led directly and indirectly to greatly increased knowledge of the general geology of the Canadian Shield. The paper outlines some of the main advances made in these ways, as well as some minor ones. The latter are included to make the coverage as complete as possible with regard to the geology of the Shield, rather than to discuss the results for uranium.

Main examples are: the acceleration of geological mapping and special studies of stratigraphy, structure and other aspects of several large parts of the Shield, including the Archaean-Proterozoic boundary; the acquisition of modern equipment capable of making large numbers of isotope age determinations, which was soon applied to general problems in the Shield, including revision of its regional provinces and sub-provinces and the subdivision of Proterozoic time; the obtaining of modern equipment for X-ray identification of minerals and for several kinds of rock and mineral analyses, which soon allowed more advanced and more extensive mineral, petrological and trace element studies of various parts of the Shield and a general study of metallogenic provinces, illustrated by metallogenic maps, which stemmed from a metallogenic map for uranium. The development of Geiger counters and other detectors is outlined because of their relation to the map for uranium and so to aeroradiometric surveys. By 1960, 26 producing uranium mines had been developed from an estimated 12,000 occurrences carrying more than 0.05% U_3O_8 equivalent. Current activities of the Geological Survey of Canada for radioactive minerals are summarized for whatever hints this may provide regarding further indirect effects.

Because this is essentially an historical paper it deals with steps in development of various topics rather than with details of these topics; reference to detailed and 'summary' publications are cited.

RESUME

Les priorités pour les fonds et le personnel afin de fournir l'ouvrage de terrain et de laboratoire ayant rapport à la découverte et l'évaluation de dépôts de minéraux radioactifs, surtout ceux de l'uranium, ont aussi conduit directement et indirectement à un accroîssement majeur des connaissances de la géologie générale du Bouclier canadien. Cette ouvrage-ci esquisse quelques uns des progrès majeurs effectués de cette façon, ainsi que quelques progrès mineurs. Ces derniers ont inclus afin de compléter l'ensemble de ces informations autant que possible en ce qui a trait à la géologie du Bouclier, plutôt que pour discuter les résultats en ce qui a trait à l'uranium.

Les exemples principaux sont: l'accélération de la cartographie géologique et les études spéciales de la stratigraphie, la structure et autres aspects de plusieurs larges régions du Bouclier, y inclu la limite archéenne-protérozoïque; l'acquisition d'équipement moderne capable de produire un grand nombre de déterminations d'age isotopique qui furent tôt appliquées aux problèmes généraux du Bouclier, y inclu la révision de ses provinces régionales et des sous-provinces et divisions du Protérozoïque, l'obtention d'équipement moderne pour l'identification radiographique de minéraux et pour plusieurs genres d'analyse de roches et minéraux qui ont permis des études de minéraux, de pétrologie et d'éléments en trace plus avancées et extensives dans plusieurs régions du Bouclier; et une étude générale des provinces métallogéniques illustrée par des ces cartes métallogéniques dérivées d'une carte métallogenique pour l'uranium. Les développements du compteur Geiger et d'autres détecteurs sont présentées à cause de leurs relation avec la carte pour l'uranium et, à sa suite, aux relevés aeroradiométriques. Par 1960, 26 mines produisant de l'uranium avaient été développées de 12,000 occurrences estimées portant plus de 0.05% de l'équivalent U_3O_8.

Les activités en cours de la Commission géologique du Canada portant sur les minéraux radioactifs sont résumés, afin de produire quelqu'indice envers d'autres effets secondaires possibles.

A cause du caractère essentiellement historique de cette présentation, elle se concerne avec les étages du développement de certains sujets plutôt qu'avec les détails; des références pour les détails et des publications "récapitulatives" sont citeés.

INTRODUCTION

Priorities for finding, evaluating and mining deposits of radioactive elements, especially from 1944 to 1957, led directly and indirectly to an unprecedented increase in knowledge of the general geology of the Canadian Shield. Much research begun in that way is still continuing. This paper outlines some of these results briefly and provides references to further details. Some of the references are themselves summary publications which contain many additional references.

This paper was not prepared for the symposium on which this Special Paper is based but was requested later. As the authors' time was limited by approaching field work and other duties it had to be brief and could not even list all the indirect contributions made by scientists of mining companies, Provincial and Federal organizations and universities. Only some of the side-effects that the authors recalled most readily are included. Inevitably, many of these concern work by the Geological Survey of Canada.

The history of discoveries and mining, and the data gained on the geology and mineralogy of radioactive deposits themselves, are mentioned only sufficiently to place the more general topics in context; many references to these topics are contained in a recently published selected bibliography (Garneau, 1976).

Canadian interest in radioactive minerals developed in five stages. From 1863 various reports of the Geological Survey of Canada mentioned and indexed occurrences of these minerals. When Provincial Mining Departments were established they followed this example. These and later Federal and Provincial reports formed a basis for investigations when commercial possibilities for radium deposits became significant in 1914, and for later investigations.

The second, or radium, stage began in 1914 when officers of the Geological Survey of Canada began studies of the more accessible known occurrences of radioactive minerals and the areas containing them. In 1921 H.V. Ellsworth, a skilled mineralogist and chemist appointed to that Survey two years previously, began a life-long study of so-called 'rare' element minerals (radioactive ones, niobium, tantalum, rare earths, etc.). His field work was confined almost entirely to occurrences in the Precambrian of southeastern Ontario on which he became an outstanding authority. The work of his earlier years culminated in a publication that became a classic (Ellsworth, 1932). In 1930 developments in aviation facilitated discovery and mining of a large deposit of pitchblende, close to the Arctic Circle. Difficulties of operation in wartime brought an end to the radium stage in 1940.

A ban on staking and mining for radioactive minerals, and other controls, began in 1943 a stage of secret Federal activities for uranium which lasted four years. During this stage the Eldorado company was bought by the Federal Government.

Easing of the ban on staking and mining late in 1947 quickly caused more persons to prospect for uranium than for any other metal and caused many companies to search for occurrences or to acquire and test discoveries made by independent prospectors. Top priorities were given to related governmental geological mapping and research. By 1958 uranium was, in dollar value, the leading metal produced in Canada. It came from 26 mines in five widely separated districts of the Shield.

The last, current stage began about 1958 when oversupply of uranium caused a decline in prospecting and mining for that metal. This decline has since been partly reversed. The emphasis of government geological mapping and research was shifted to metals that had to be neglected during the uranium 'boom'. Research on uranium and other radioactive deposits has, however, been continued. Relief from the extreme pressures of the boom stages and advances in data processing have not only permitted approaches regarding uranium that could not be attempted previously, but also permitted many laboratory facilities acquired for work on uranium to be used substantially for more general purposes.

GEOLOGICAL MAPPING AND STRUCTURAL AND STRATIGRAPHIC STUDIES

Discovery of the Eldorado radium deposit caused the Geological Survey of Canada to undertake geological mapping and studies there and in that region in 1931 and the next few seasons. This involved one of the first uses of aircraft for such work (Kidd, 1932). Also at this time studies, mainly of reconnaissance type, were carried out by the Geological Survey of Canada in the huge area between Great Bear and Great Slave Lakes and along the shores of the latter (Stockwell, 1933; Kidd, 1936).

When the Eldorado mine was reopened by the 'Crown' company the Geological Survey of Canada sent three parties in 1944 and the next few seasons to make detailed geological maps and studies of the mine area and of many square miles surrounding it. Being the first large detailed project in the northwestern part of the

Shield, it provided much information. The resulting maps were too large for publication when restrictions were removed, but some of the data were used in publication at a smaller size.

Also during the secret period a deposit of pitchblende that seemed of possible importance was found in the Athabasca region of northern Saskatchewan early in a joint prospecting and mapping project of the Eldorado company and the Geological Survey of Canada. A large region was mapped at 1 inch to ½ mile (1:31680). Because it soon became apparent that structural and stratigraphical controls of mineralization required more detailed mapping, L.P. Tremblay of the Geological Survey of Canada was assigned to begin this work in 1952. He continued it for five more seasons and C.K. Bell mapped a smaller adjoining area. Preliminary plotting was mainly at 1 inch to 550 feet, and publication at 1 inch to 800 feet (1:9600). The total area covered – about 170 square miles – still remains the largest single block of detailed geological mapping ever done in Canada. Special studies were also made on such aspects as structure, stratigraphy and provenance by Eldorado, the Geological Survey of Canada, the Saskatchewan Geological Survey, and certain private companies who acquired large holdings after the easing of controls (see, for example, Tremblay, 1972 and Fahrig, 1961). Four Ph.D. thesis projects were financed and supervised by Eldorado and two by the Geological Survey of Canada. Besides discovering additional radioactive occurrences, the mapping and studies in this region resulted in a vast amount of data on stratigraphy, structure, mineralogy and other aspects of geology, and in revision of stratigraphic units and names. Gulf Minerals Limited attempted the first seismic study for stratigraphic purposes. A seismic survey to provide data on the base of the Athabasca Formation, 100 miles south of Beaverlodge Area was made by the Geological Survey of Canada.

Farther east and south of Beaverlodge, mapping and studies were undertaken by geologists employed by the Province of Saskatchewan. Much of this work was done by Mawdsley (1950) and Mawdsley and Grout (1951) in areas where uraninite was found in numerous bodies of pegmatite, migmatite and related rocks. This resulted in new information, especially on rocks of the kinds mentioned.

The possibilities of uranium-bearing conglomerates in the Blind River of Ontario began to be realized in the early 1950s. These were near the base of the Huronian sedimentary sequence, which rests unconformably on Archaean rocks. As the conglomerates are in the classical area of the Huronian the relationships had previously been studied and mapped by various geologists. Growing importance of the discoveries caused much additional study and mapping and some revisions of units and names. As examples, the work of Robertson (1968), Roscoe (1957) and Frarey (in press) may be cited. Analysis of a large number of samples showed that in addition to uranium, the deposits contained significant amounts of thorium and rare earths. Studies of the lower Huronian and its possible equivalents were extended at intervals for hundreds of miles northeastward from Elliot Lake. Most geologists considered that the Elliot Lake ores were essentially of placer origin but were puzzled by the fact that uranium minerals are slowly soluble in waters of today, and that the deposits contain many grains of pyrite whereas modern placers generally contain much magnetite. Studies of these matters contributed to development of credible hypotheses regarding a lack of free oxygen in the Precambrian atmosphere, to studies of sediments containing mainly ferrous rather than ferric iron in other parts of the Shield, and to search for detrital deposits of possible interest for uranium.

Since the early 1930s a growing number of pegmatite and related deposits in the region east of Lake Huron were shown to contain radioactive minerals. Study of these during, before, and after, the uranium boom showed that simple granite pegmatites were unlikely to contain more than accessory amounts of radioactive minerals, but it was thought possible that other, more unusual deposits of the general pegmatitic or related classes might be found to carry more uranium. This hope was realized when two deposits were found in 1952 and 1953. One consisted of a zone of dykes of syenite and granite and the other of dykes of pegmatitic granite. As the sizes and uranium contents of these became evident much remapping and several special studies were done, notably by Satterly (1957) and Robinson and Hewitt (1958).

During the peak activity for uranium, the Geological Survey of Canada financed and supervised three Ph.D. thesis projects in Ontario. These were on such varied topics as the zonation of a large pegmatite deposit carrying radioactive minerals, the petrology and provenance of the Elliot Lake conglomerates, and a carbonatite deposit carrying niobium and uranium.

ISOTOPIC AGE DETERMINATIONS

The availability of large numbers of isotopic age determinations resulted in: (1) greatly improved knowledge of orogenies that affected the rocks of the Canadian Shield; (2) complete division of the Shield into tectonic provinces and subprovinces; and (3) time-stratigraphic revision of the Proterozoic.

Ellsworth had done valuable pioneer work on age determinations made by chemical analyses, and had contributed information and advice to an international Committee on the Measurement of Geological Time. About 1950, work on isotopic measurements, which offered advantages over chemical methods, was begun at a few Canadian universities. Collins *et al.* (1952) published results of 52 isotopic determinations made at the University of Toronto on samples obtained from various sources. These included twenty-one from recent field work by the Geological Survey of Canada and four from Ellsworth's earlier work.

The National Advisory Committee on Research in the Geological Sciences had recommended that the Geological Survey of Canada should be augmented by much increased, modern laboratory facilities. Funds, space and personnel were not made available until priorities for work on uranium made them imperative. Radiometric age determinations had been made at the University of Toronto and the Geological Survey was anxious to begin such work. In 1953 the Radioactive Resources Division was allotted $40,000 to obtain a mass spectrometer. An instrument designed especially for age determinations was built at McMaster University by R.K. Wanless under the direction of H.G. Thode. Wanless was then appointed to the Division to install and operate the equipment. The number of instruments was later increased to six, built under the supervision of Wanless, and equipment for radiocarbon datings was installed. When the requirements related to uranium permitted, the laboratory was transferred to an expanded Mineralogy Division. To date the laboratory has made more than 3000 potassium-argon determinations, more than half of them on samples from the Shield, and has made numerous determinations of other kinds.

These determinations, and those made at several universities, have contributed to many research projects related to the Shield. The most comprehensive study was carried out over several years by C.H. Stockwell (1970, 1972). He compiled available data on age determinations and added the results of determinations on a large number of samples that he collected himself from strategically located places in the Shield. Most datings were made on micas, by the potassium-argon method. Because the determinations indicated the times of metamorphism or latest crystallization rather than those of the original formation of rocks, Stockwell first applied them to attempts to define the later orogenies that affected the Shield. A histogram showed four pronounced groupings which were interpreted as representative of the last four major orogenies, the oldest being late in Archean time. The evidence of these supported Stockwell's idea on division of the Shield into provincs and subprovinces based on structural trends and kinds of folding. The four orogenies were regarded as separating five time-stratigraphic divisions, of Proterozoic time. Archean and Proterozoic were classed as Eons, the principal three divisions of the Proterozoic were called Eras, and two of the latter were designated as Sub-eras (Stockwell *et al.*, 1970). These regional and time-stratigraphic concepts have been generally adopted by Canadian geologists.

Goodwin (1968) based an interpretation of the evolution of the Canadian Shield partly on the available age determinations and the framework of orogenic and time-stratigraphic divisions proposed by Stockwell. Goodwin considered that Archean crustal growth was dominated by at least four protocontinents, each sialic and part of a thin crust resting on the mantle. In Proterozoic times thick bodies of sedimentary, volcanic and granitic rocks developed around the protocontinents.

Age determinations have aided in a recent study of the early Archean basement of the Shield (Baragar and McGlynn, 1976). Granitoid rocks from five separate localities were indicated to be more than 2,900 million years old. As these dates are markedly older than known determinations from belts of altered volcanic rocks in the Shield, it was concluded that the early crust was evidently formed of soda-rich granitoid rock.

ANALYSES OF MINERALS AND ROCKS

For many years the Geological Survey of Canada had only two mineralogists, who assisted the geologists occasionally with difficult identifications. It had one analytical chemist other than Ellsworth, who was engaged in his own research. The geologists were expected to do their own mineralogical, mineragraphic and petrographic work and any qualitative chemical tests they required; a few rock analyses could be obtained in rare instances.

Priorities for uranium, about 1951, allowed the Radioactive Resources Division to acquire equipment and staff for X-ray mineral identifications, X-ray fluorescence analyses, and greatly increased and modernized spectrographic and chemical analyses. These facilities contributed much to research on uranium deposits and were soon applied to deposits of other metals and to more basic studies of the Shield and other parts of Canada. The availability of equipment and methods for fast, cheap analyses also permitted other organizations to contribute to the marked increase in chemical knowledge of the Shield.

One of the first large-scale projects aided in this way was a study of geochemistry and isotopic datings in a wide belt some 800 miles long, between Elliot Lake, Ontario and Matagami in central Quebec (Roscoe, 1965). This work added significantly to evidence for the origin and emplacement of some of the ore deposits in the belt, which has accounted for the greater part of metal production from the Shield.

Only a few of the many studies of the general geology of the Shield that have depended wholly or partly on laboratories under discussion can be mentioned here. One of the broadest projects compiled by D.M. Shaw and associates at McMaster University, concluded that the average composition of the Shield approximates granodiorite and that the Shield contains more silicon and potassium and less magnesium and calcium than does the earth's crust in general (Shaw *et al.*, 1967).

This study was related partly to the expanded analytical facilities of the Geological Survey of Canada. The team obtained part of its samples from the Dominion Observatory which had collected samples during determinations of the gravitational field, and part from the Research Council of Alberta. The Survey contributed 8076 rock powders from samples collected during a comprehensive resource study of a large area in Western Ontario. After making analyses for 30 elements the team at McMaster University concluded:

1) that the Shield has regional compositional variations;
2) that the composition of major elements agrees well with other estimates for the Canadian and Fennoscandian Shields;
3) that the average composition approximates those of granodiorite, mica gneiss and greywacke;
4) that the trace-element composition agrees more closely with the composition of mica gneiss and greywacke; and that the Shield contains more silicon and potassium and less titanium, iron, manganese, magnesium and calcium than is indicated by recent estimates for the earth's crust as a whole.

A study of the geochemistry of volcanic rocks in the Yellowknife area of the Northwest Territories provided better understanding of some of the older volcanic rocks of the Shield (Baragar, 1966). Eade and Fahrig (1973) described variations in the abundances of certain trace elements over a large part of the Shield.

DEVELOPMENT OF PORTABLE RADIOACTIVITY DETECTORS

In 1934 physicists in the United States and at the University of British Columbia independently reported making heavy portable radioactivity detectors involving Geiger tubes (Curtiss, 1934; Shrum and Smith, 1934). The equipment of Shrum and Smith was used in locating radium 'needles' lost at a hospital. On reading of this equipment Ellsworth and two assistants made, in 1934-35, a much lighter counter weighing 21 pounds, including several heavy batteries. This was tested successfully at some known radioactive occurrences in Ontario. At the beginning of the secret search for uranium the Geological Survey of Canada asked the National Research Council to develop Ellsworth's counter further. Two of several models were adopted for testing, several hand-made copies of the two models being tested by parties working at Great Bear Lake. After imperfections were corrected more counters were made by the Research Council for use of Eldorado and Geological Survey parties; several were donated to Commonwealth countries. When these activities were no longer secret a Canadian manufacturer was permitted to make modifications of the two models. These were the first counters available to the Canadian public. By about that time counters had also been developed and manufactured in other countries. In 1949 professors at the University of Manitoba developed a practical scintillation counter, which had advantages for some purposes (Brownell, 1950).

METALLOGENIC MAPS AND PROVINCES

Three circumstances caused the reporting of thousands of occurrences of uranium or thorium in amounts of 0.05 per cent or more. One was the development of portable Geiger counters, with the aid of which a horde of prospectors and enthusiastic amateurs found numerous small occurrences, many of which would have been missed had counters not been available. Another factor was that a Regulation of the Atomic Energy Control Board required anyone who found or thought he had found radioactive material, amounting to 0.05 per cent or more, to give particulars to the Geological Survey of Canada which acted as agent for the Board in matters dealing with prospecting and mining. The other factor was free assays performed by the Survey. Notifications and assay results were treated confidentially if desired, but many were released later and others were spaced so close to released ones that they did not affect the patterns of discoveries. In these ways the locations and other data for about 12,000 occurrences at or above the legal limit for compulsory reporting, mainly in the Shield, became available, together with a much larger number below 0.05 per cent.

These occurrences seemed sufficient for making a trial metallogenic map showing not only producing mines but the areas containing prospects and all the occurrences at or above 0.05 per cent except a few confidential ones. Of course, maps had previously been published by various persons showing deposits of metals, other than uranium, in relation to geology, but this was the first Canadian metallogenic map specifically so called. To permit comparisons with the Geological Map of Canada the resulting map (Lang, 1958) was printed on transparent paper at the same scale. Maps for iron and beryllium were prepared by G.A. Gross and R.B. Rowe soon afterwards. These lacked the large coverage of minor occurrences available for uranium but considerable data were obtained from publications and files. Comments on these three maps, after publication, encouraged the preparation of maps for other metals by various officers. These were followed by two maps for combined metals, and a short preliminary summary on metallogenic provinces (Lang, 1961).

Roscoe (1966) extended the information of the early map for uranium by preparing a map and report including areas in which uranium had not been found but which offered theoretical possibilities. Little (1974) carried work of this kind a step further by issuing three maps, with an explanatory paper, which compared the uranium possibilities of the various geological or tectonic provinces of Canada and listed areas in the order of believed favourability for uranium discoveries of each producing type.

Whitmore *et al.* (1969) prepared a Mineral Deposits Map of Canada on which symbols of different colours, shapes and sizes indicate metals, geological classes and sizes of present and former producing mines. A comprehensive metallogenic map for all Canada, showing a large number of occurrences as well as present and past producing mines was compiled by G.B. Leech as a contribution to the Metallogenic Map of North America of the Commission for the Geological Map of the World. This has not yet been published but was displayed at the International Geological Congress in 1972.

Reviewers and other commenters (Zaslow, 1975, p. 432, 484) have indicated that the various maps mentioned above had aided in mineral exploration and furthered the knowledge of metallogenic provinces and methods of illustrating them. Certain other countries have adopted the methods of the early series, by including occurrences as well as producing mines and by printing on transparent paper. The maps are contributing toward the preparation of a metallogenic map of the world.

The Geological Survey of Canada has recently begun three advanced metallogenic projects in the southwestern and northwestern parts of the Shield and in the northern Cordillera. Studies related to a number of metals are planned to include the geology and genesis of deposits, detailed mapping, biostratigraphy, geochronology, geochemistry and data processing.

AERORADIOMETRIC SURVEYS

Progress in outlining, by means of airborne radioactivity detectors, areas favourable for prospecting for radioactive minerals has also had limited applications for aiding geological mapping within the Canadian Shield.

The Eldorado Company began experiments with large airborne Geiger Tubes in 1947. About a year later a joint program was begun by Eldorado, the National Research Council, the Geological Survey of Canada and the Royal Canadian Air Force, using scintillation counters and Survey personnel experienced in making and plotting aeromagnetic surveys. These trials were not as satisfactory as had been hoped because of difficulties in distinguishing between anomalies caused by uranium, thorium or potassium. Eldorado obtained some success with a scintillation counter in a low-flying helicopter used for prospecting and for tracing known occurrences (Gregory, 1955, 1960).

After gamma-ray spectrometers became available the Geological Survey of Canada began extensive airborne surveys of that kind in the late 1960s, resulting in maps showing areas of anomalies of different intensities and distinguishing between the three elements mentioned above. Darnley (1968) expressed the opinion that the methods showed promise, if combined with photogeology, as an aid to geological mapping of areas containing sufficient rock exposures, providing that some tests were made on the ground. Recent mapping by the Geological Survey of Canada in the Hearne Lake area, Northwest Territories, used these methods. The results, as yet unpublished, were promising except that much time had to be spent in skilful interpretation of air photos (A.G. Darnley, pers. commun.).

LIFE IN THE PRECAMBRIAN

Additional geological mapping and special studies along and near the East Arm of Great Slave Lake, Northwest Territories, resulted from radioactive discoveries there in the 1940s, though work on the discoveries themselves was suspended when the value of deposits in northern Saskatchewan became apparent. The initial discovery on the East Arm made by prospectors of the Eldorado Company, was in Proterozoic limestone containing abundant stromatolites. These had previously been reported from the area but geologists who examined the deposit obtained their first sight of evidence of Precambrian life in this way. Some of them speculated that the radioactive minerals may have been precipitated by the action of algae, but later work discounted this.

Recent studies of the uranium ores at Elliot Lake, Ontario showed considerable amounts of 'thucholite' a 'mineral' of variable composition including significant amounts of carbon (Ruzicka and Steacy, 1976). The composition and mode of occurrence is much like the 'carbon' in the Witwatersrand gold ores (Hallbauer, 1975). This has led to speculation by African and Canadian geologists regarding the possibility that it is related to some form of algae that existed in early Proterozoic time.

EXPERIENCE IN COMPREHENSIVE PROSPECTING AND EVALUATING PROGRAMS

The search for uranium and the 'screening' and testing of discoveries included some of the earliest modern multi-approach programs. Some of the experience acquired in this way was applied in later programs for other metals. Although some of the experience was published much was merely transferred by persons formerly engaged in uranium programs.

During the secret period, programs were evolved at Great Bear Lake and northern Saskatchewan by the Eldorado company, the Geological Survey of Canada and a firm of geological consultants. Geologists, mining engineers and prospectors experienced in searching for and testing deposits of gold and other metals were employed. Mainly in Saskatchewan, the prospectors and local residents with no experience in prospecting were taught to use Geiger counters. The experienced prospectors worked fairly independently in two-man teams. The inexperienced ones, who were attached to geological parties, left a paint-mark on a rock from which they obtained a significant count. A geologist or engineer appraised discoveries. Stripping of overburden, or trenching, was done where that seemed warranted, and the more promising 'showings' were tested by diamond drilling, first at fairly wide spacings and, if favourable results were obtained, at spacings sometimes as close as 25 feet. When programs by private companies became possible, fairly similar methods were adopted. Besides geological mapping and studies, governmental and private programs included geochemical prospecting of different types, systematic radiometric surveys on the ground and by aircraft, and at least one use of the seismic method.

RATIO OF OCCURRENCES TO PRODUCING MINES

Experienced Canadian 'mining men' have long known that for any metal, only a few mineral occurrences or prospects became viable mining operations. Such sayings as '100 occurrences for one prospect and 100 prospects for one mine' were accepted as conveying the proper concept, based on experience, but no statistics of occurrences were available to provide a more quantitative estimate until the reporting of radioactive occurrences was required.

This reporting permitted the senior author to estimate that by 1960 known radioactive occurrences carrying more than 0.05 per cent U_3O_8 equivalent known in Canada totalled 12,000. Almost all carried uranium rather than thorium and were in the Shield. The total could not be tallied exactly because several discoveries were restaked and reported again under different names; all such cases could not be eliminated but the total is believed to be reasonably accurate. Of these, 359 were tested by diamond drilling. Underground workings were driven to test 117 properties. This resulted in 19 producing uranium mines having their own treatment plants, and seven others from which ore was shipped to neighbouring mines.

It is not implied that this ratio of 26 producers to 12,000 occurrences is an absolute indication or that it would correspond exactly to other metals or to uranium over a different number of years or set of prospecting conditions. Use of radioactivity detectors probably accounted for more minor radioactive occurrences than would be the case for other metals. The comparisons do, however, appear to offer a more quantitative concept than was available before.

URANIUM RESOURCE STUDIES, 1975-77

Long-term energy studies conducted by various countries, by international organizations (e.g., International Atomic Energy Agency, Organization for Economic Co-operation and Development) and by industry showed in the late sixties and early seventies disproportions in the projected supply and demand balance of the nuclear raw material. This situation triggered a worldwide revival in uranium exploration, accentuated by increasing prices of uranium.

Canada, and particularly the Canadian Shield, became logical targets for uranium exploration. In an attempt to ensure sufficient reserves of nuclear fuel for domestic energy utilities the Minister of Energy, Mines and Resources announced on September 5, 1974 a new uranium policy for Canada. To implement this policy several nuclear energy programs were established, among them a) Assessment of Uranium Resources, by the Canada Centre for Mineral and Energy Technology (CANMET) and the Geological Survey of Canada and b) the Federal-Provincial Uranium Reconnaissance Program (E.M.R., 1977).

The Assessment of Resources includes quantitative estimation of resources and aids exploration organizations in development of geological guides for discovery of uranium resources (Ruzicka, 1976a). This task led to investigations of the nature and distribution of uranium deposits and their relationship to their geological environments, to provide better understandings of the metallogenic processes that govern the uranium potential of each region of Canada and to provide guidelines for discoveries of deposits (Ruzicka, 1977a). In this respect conceptual models simulating formation of fundamental types of uranium deposits were developed and areas considered favourable for occurrences of these types were delineated (Ruzicka, 1977b).

Studies on selected uranium-bearing areas in the Shield, namely Elliot Lake-Agnew Lake, Cobalt Embayment, Lake Nipissing, Bancroft, and Kenora in Ontario; the areas north of Lake Athabasca, Wollaston Lake Fold Belt and Carswell Structure in Saskatchewan; Amer Lake, Dubawnt Graben, Bear Structural Province, East Arm of Great Slave Lake in Northwest Territories; the Mont Laurier, Sakami Lake and

Johan Beetz in Quebec; and Makkovik-Seal Lake in Labrador led to interpretation of geological processes involved in uranium mineralization (Ruzicka, 1977c; Tremblay, 1978).

Studies on the conglomeratic ore from Elliot Lake, conducted at Queen's University (Theis, 1976) brought further support to the theory of sedimentary-syngenetic origin (paleoplacer) of the Elliot Lake deposits (Roscoe, 1969; Robertson, 1976).

Detailed mapping and subsequent mineralogic studies of known uranium mineralization of the Proterozoic Dubawnt succession in the Dubawnt Graben showed that the uranium-polymetallic mineralization within some of the formations may be attributed to volcanic and subsequent tectonic processes active in this area. Therefore study of the volcanic activity in similar geological environments in relationship to uranium mineralization is advocated (Blake *et al.*, 1977).

Mineralogical and isotopic studies on ore and country rocks from the Rabbit Lake deposit, Saskatchewan determined paragenetic relationships between uranium-bearing and non-radioactive minerals, and allowed interpretation of some ore-forming and alteration processes believed to have been important in the genesis of this deposit (Rimsaite, 1976, 1977a, 1977b). These studies led to development of new laboratory methods applicable in mineragraphic and metallogenic studies on uranium and other metallic elements.

Discoveries of uranium deposits in Saskatchewan in the late 1960s and in the 1970s, such as Rabbit Lake, Cluff Lake, Key Lake, led to investigation of geological processes that might participate in formation of these deposits. Detailed studies of the geology and genesis of these deposits (Knipping, 1974; Pagel, 1975; Sibbald, 1977; Dahlkamp and Tan, 1977; Tremblay, 1978; Hoeve, 1977; Ruzicka, 1976c; Beck, 1976; Harper, 1977; Langford, 1977) led to a broad discussion on their origin, whether they originated by ascending hydrothermal solutions or by descending meteoric waters. These studies and discussions contributed not only to a deeper knowledge of geology and metallogeny of existing uranium deposits and the surrounding areas, but also to formulation of criteria applicable to further exploration for uranium deposits. These studies undoubtedly helped in discoveries of new uranium deposits in Canada and elsewhere.

The Federal-Provincial Uranium Reconnaissance Program includes 1) airborne gamma-ray spectrometry surveys of potential uraniferous resource areas in Canada, covering approximately 150 thousand square miles annually, 2) national geochemical reconnaissance of selected potential uraniferous areas where geochemical methods are more effective than radiometric methods, and 3) ground follow-up and evaluation, including detailed geological, geophysical and geochemical surveys. Results of these surveys, which cover a substantial part of the Canadian Shield, are published mainly as open file reports. Application of these surveys necessitated development of new methods and equipment that were further modified by the industry.

The present uranium resources studies entail a broad spectrum of investigations such as lithologic, structural, stratigraphic, mineralogic, geochemical, metallogenic and geomathematical. These studies help to develop knowledge of the genesis of various types of uranium deposits and, in the process, add to the knowledge of the other geological disciplines.

CONCLUSIONS

This short outline is believed to show that basic knowledge of many aspects of the Canadian Shield was advanced greatly by 'side effects' of 'mission-oriented' projects of various kinds. Many of these required improvisations and, often, hasty planning. The results offer examples of how funding for a wise blend of individual initiative and comprehensive programs could advance knowledge of the Shield much further in the years to come.

REFERENCES

Baragar, W.R.A., 1966, Geochemistry of the Yellowknife volcanic rocks: Canadian Journal of Earth Sciences, v. 3, p. 9-30.

Baragar, W.R.A. and McGlynn, J.C., 1976, Early Archean Basement in the Canadian Shield: A Review of the Evidence: Geological Survey of Canada, Paper 76-14, 20 p.

Beck, L.S., 1976, History of uranium exploration in Northern Saskatchewan with special reference to changing ideas of metallogenesis; *in* Dunn, C.E., ed., Uranium in Saskatchewan: Saskatchewan Geological Society Special Publication No. 3, 396 p.

Blake, D., Miller, A., and Schau, M., 1977, Potassic volcanic rocks of the Paleohelikian Dubawnt Group and associated Cu, U and Se mineralization, Christopher Island Region, District of Keewatin, N.W.T.: Geological Association of Canada, 1977 Annual Meeting, Program Abstracts, v. 2, p. 8.

Brownell, C.M., 1950, Radiation Surveys with a Scintillation Counter: Economic Geology, v. 45, no. 2, p. 167-174.

Collins, C.B., Lang, A.H., Robinson, S.C., and Farquhar, R.M., 1952, Uranium Deposits in the Canadian Shield: Geological Association of Canada Proceedings, v. 5 p. 15-41.

Curtiss, L.F., 1934, Bureau of Standards Journal of Research, v. 12, p. 379-382 (This paper apparently was not titled).

Dahlkamp, F.J. and Tan, B., 1977, Geology and mineralogy of the Key Lake U-Ni deposits, northern Saskatchewan, Canada; Bulletin Institute Mining and Metallurgy, 1977, p. 145-157.

Darnley, A.G., 1968, Helicopter Tests with a Gamma-ray Spectrometer: Canadian Mining Journal, v. 89, no. 4, p. 104-106.

Energy, Mines and Resources Canada 1977, 1976 Assessment of Canada's Uranium Supply and Demand: Energy, Mines and Resources Canada Dept. EP 77-3, 19 p.

Eade, K.E. and Fahrig, W.F., 1973, Regional, Lithological, and Temporal Variation in the Abundances of Some Trace Elements in the Canadian Shield: Geological Survey of Canada, Paper 72-46.

Ellsworth, H.V., 1932, Rare-element Mineralogy of Canada: Geological Survey of Canada, Economic Geology Series No. 11, 272 p.

Fahrig, W.F., 1961, The Geology of the Athabasca Formation: Geological Survey of Canada, Bulletin 68, 41 p.

Frarey, M.J. in press, The Geology of the Huronian Belt between Sault Ste. Marie and Blind River, Ontario: Geological Survey of Canada, Memoir 383.

Garneau, D.M., 1976, Selected Bibliography on the Geology of Canadian Deposits and Occurrences of Uranium and Thorium: Geological Survey of Canada, Paper 75-45, 41 p.

Goodwin, A.M., 1968, Evolution of the Canadian Shield: Proceedings of the Geological Association of Canada, v. 19, p. 1-14.

————, 1970, Metallogenic Evolution of the Canadian Shield: *in* Geology and

Economic Minerals of Canada, Geological Survey of Canada, Economic Geology Series, Number 1, Fifth Edition, p. 156-164.

Gregory, A.F., 1955, Aerial detection of radioactive mineral deposits: Canadian Institute of Mining and Metallurgy, Transactions, v. 58, p. 261-267.

——— 1960, Geological Interpretation of Aeroradiometric Data: Geological Survey of Canada, Bulletin 66, 29 p.

Hallbauer, D.K., 1975, The Plant Origin of the Witwatersrand 'carbon' Minerals: Science Engineering, v. 7, no. 2, April 1975 p. 111-131.

Harper, C.T., 1977, Uranium Metallogenic Studies: III Cluff Lake: *in* Christopher, J.E. and Macdonald, R., eds., Summary of Investigations 1977: Saskatchewan Geological Survey, p. 136-146.

Hoeve, J., 1977, Uranium Metallogenic Studies: II Rabbit Lake, Mineralogy and Geochemistry: *in* Christopher, J.E. and Macdonald, R., eds., Summary of Investigations 1977: Saskatchewan Geological Survey, p. 124-135.

Kidd, D.F., 1932, Great Bear Lake-Coppermine River Area, Mackenzie District, N.W.T.: Geological Survey of Canada, Summary Report, 1931, Part C, p. 47-49.

———, 1936, Rae to Great Bear Lake, Mackenzie District, N.W.T.: Geological Survey of Canada, Memoir 187, 44 p.

Knipping, H., 1974, The concepts of supergene versus hypogene emplacement of Uranium at Rabbit Lake, Saskatchewan, Canada: *in* Formation of Uranium Ore Deposits ed. by International Atomic Energy Agency, IAEA-SM-186/38, p. 531-549.

Lang, A.H., 1953, The Development of Portable Geiger Counters: Canadian Mining Journal, v. 74, no. 4, p. 65-67.

———, 1958, Metallogenic Map, Uranium in Canada: Geological Survey of Canada, Map 1045A-M1.

———, 1961, A Preliminary Study of Canadian Metallogenic Provinces: Geological Survey of Canada, Paper 60-33, 54 p.

Langford, F.F., 1977, Surficial origin of North American pitchblende and related uranium deposits: American Association of Petroleum Geologists, Bulletin 61, p. 28-42.

Little, H.W., 1974, Uranium deposits in Canada – their exploration, reserves and potential: Canadian Institute of Mining and Metallurgy Bulletin, v. 67, no. 743 p. 155-163

Mawdsley, J.B., 1950, The Geology of Charlebois Lake Area: Saskatchewan Geological Survey, Precambrian Geological Series, Report Number 5, 30 p.

Mawdsley, J.B. and Grout, F.F., 1951, The Geology of the Stanley Map Area: Saskatchewan Geological Survey, Precambrian Geological Series, Report Number 4, 31 p.

Pagel, M., 1975, La diagenese des grès et les gisements d'uranium dans le basin Athabaska (Canada): *in* Rapport de l'activitées 1975: Centre de Recherches Petrographique et Geochimiques Nancy, France.

Rimsaite, J., 1976, Mineral assemblages at the Rabbit Lake uranium deposit, Saskatchewan: Geological Association of Canada, 1976 Annual Meeting, Program Abstracts, v. 1, p. 44.

———, 1977a, Mineral assemblages at the Rabbit Lake uranium deposit, Saskatchewan; *in* Report of Activities Part B, Geological Survey of Canada, Paper 77-1B, p. 235-246.

———, 1977b, Occurrences of rare secondary U- and Pb-bearing mineral aggregates in uranium deposits, northern Saskatchewan: *in* Report of Activities, Part C, Geological Survey of Canada, Paper 77-1C, p. 95-97.

Robertson, J.A., 1968, Uranium and Thorium Deposits of Northern Ontario: Ontario Department of Mines, Mineral Resources Circular 9.

———, 1976, The Blind River uranium deposits: the ores and their setting: Ontario Division of Mines, Misc. Paper 65, 45 p.

Robinson, S.C. and Hewitt, D.F., 1958, Uranium Deposits of Bancroft Region, Ontario: Second United Nations Conference on Peaceful Uses of Atomic Energy, Paper 224.

Roscoe, S.M., 1957, Geology and Uranium Deposits, Quirke Lake-Elliot Lake, Blind River Area, Ontario: Geological Survey of Canada, Paper 56-7 21 p.

——————, 1965, Geochemical and Isotopic Studies, Noranda and Mattagami Areas: Transactions, Canadian Institute of Mining and Metallurgy, v. 58, p. 965.

——————, 1966, Unexplored uranium and thorium resources of Canada: Geological Survey of Canada, Paper 66-12, 11 p.

——————, 1969, Huronian Rocks and Uraniferous Conglomerates: Geological Survey of Canada, Paper 68-40, 205 p.

Ruzicka, V., 1976a, Uranium Resources Evaluation Model as an Exploration Tool: *in* Exploration for Uranium Ore Deposits ed. by International Atomic Energy Agency, Vienna, 1976, p. 673-682.

——————, 1976b, Evaluation of uranium resources in the Elliot Lake - Blind River area Ontario; *in* Report of Activities, Part B, Geological Survey of Canada, Paper 77-1B, p. 127-129.

——————, 1976c, Some metallogenic features of the 'D' uranium deposit at Cluff Lake, Saskatchewan; *in* Report of Activities Part C, Geological Survey of Canada Paper 75-1C, p. 279-283.

——————, 1977a, Estimation of Undiscovered Uranium Resources in Canada: Energy, Mines and Resources Canada, Report ER 77-1, p. 37-45.

——————, 1977b, Conceptual models for uranium deposits and areas favourable for uranium mineralization in Canada: Report of Activities, Part A Geological Survey of Canada, Paper 77-1A p. 17-25.

——————, 1977c, Assessment of selected uranium occurrences and areas favourable for uranium mineralization in Canada; Report of Activities, Part A, Geological Survey of Canada, Paper 77-1A, p. 27-29.

Ruzicka, V. and Steacy, H.R., 1976, Some sedimentary features of conglomerate uranium ore from Elliot Lake, Ontario: Report of Activities, Part A, Geological Survey of Canada, Paper 76-1A, p. 343-346.

Satterly, J., 1957, Radioactive Mineral Occurrences in the Bancroft Area: Ontario Department of Mines, Annual Report, 1956, v. 65 part 6, 181 p.

Shaw, D.M., Reilly, G.A., Muysson, J.R., Pattendeu, G.E., and Campbell, F.E., 1967, Chemical Composition of the Canadian Precambrian Shield: Canadian Journal of Earth Sciences v. 4, p. 829-853.

Shrum, G.M., and Smith, R., 1934, A Portable Geiger-Muller Tube Counter as a Detector for Radioactive Ores: Canadian Journal of Research, v. 11, p. 652-657.

Sibbald, T.I.I., 1977, Uranium Metallogenic Studies 1 Rabbit Lake, Geology: *in* Christopher, J.E. and Macdonald, R., eds., Summary of Investigations 1977: Saskatchewan Geological Survey, p. 111-123.

Stockwell, C.H., 1933, Great Slave Lake-Coppermine River Area, Northwest Territories: Geological Survey of Canada, Summary Report 1932, part C, p. 37-63.

——————, 1972, Revised Precambrian Time Scale for the Canadian Shield: Geological Survey of Canada, Paper 72-52, 4 p.

Stockwell, C.H. et al., 1970, Geology of the Canadian Shield: *in* Geology and Economic Minerals of Canada: Geological Survey of Canada, Economic Geology Series Number 1, 5th Edition, p. 44-54.

Theis, N., 1976, Uranium-bearing and associated minerals in their geochemical and sedimentological context, Elliot Lake, Ontario: Unpublished Ph.D. Thesis, Queen's University.

Tremblay, L.P., 1972 Geology of the Beaverlodge Mining Area, Saskatchewan: Geological Survey of Canada, Memoir 367, 468 p.

————, 1978, Uranium subprovinces and types of uranium deposits in Precambrian rocks in Saskatchewan: Geological Survey of Canada, Current Research, Paper 78-1A, p. 437-439.

Whitmore, D.R.E., Johnston, A.G., MacLean, B., and Smith, A.Y., 1969, Mineral Deposits Map of Canada: Geological Survey of Canada, Map 1250A.

Zaslow, M., 1975, Reading the Rocks: MacMillan of Canada, Department of Energy, Mines and Resources, and Information Canada, 599 p.

History of Concepts in Precambrian Geology, edited by W.O. Kupsch and W.A.S. Sarjeant, Geological Association of Canada Special Paper 19, 1979

HISTORY OF THE STUDY OF THE SWEDISH PRECAMBRIAN

Per H. Lundegårdh
Assistant Director,
Sveriges Geologiska Undersökning, Box 670, S-75128, Uppsala, Sweden

ABSTRACT

Geological mapping of the Swedish Precambrian began in 1838, but the first important studies were those of A. Erdmann, whose prototype map (1856) served as a model for later studies. The iron ore-bearing region of sourth central Sweden was mapped by A.E. Törnebohm between 1876 and 1881. Subsequently J.J. Sederholm formulated a subdivision of Swedish and Finnish rocks on decreasing tectonization: this was modified by W. Ramsay and especially by Walter Wahl, who recognized three Precambrian orogenies; Wahl also recognized different groups of granites. His work was followed up by H.G. Backlund (1936, 1937) and by N.H. Magnusson (1963), whose revised concepts of basement evolution included five major orogenic phases.

The introduction of radiometric dating gave precision to these conclusions and the extension of mapping, by Lundegårdh and others, clarified the picture of the tectonic evolution of the Svecofennides. Work in western Sweden demonstrated the existence of a final, Sveconorwegian, tectogenesis.

In northern Sweden, investigations were begun by H. Lundbohm in Norrbotten County. Subsequently, Magnusson recognized that the Precambrian rocks were in part of eugeosynclinal character. More recent mapping by G. Kautsky and others has begun the elucidation of the tectonic history of this region, indicating correlations with the Precambrian rocks farther south.

RESUME

La cartographie géologique du Precambrien suédois date de 1838, mais les premières études importantes furent celles de A. Erdmann, dont la carte prototype (1856) servit de modèle pour les études qui suivirent. Entre 1876 et 1881, A.E. Törnebohm dressa une carte de la région contenant des minerais de fer dans le sud central de la Suède. Ensuite, L.L. Sederholm formula une division des roches suédoises et finlandaises basée sur la diminution de la tectonisation; ceci fut modifié par W. Ramsay et surtout par Walter Wahl, qui reconnurent trois orogénèses précambriennes; Wahl reconnut aussi des groupes différents de granites. Son oeuvre fut suivie par celle de H.G. Backlund (1936, 1937) et celle de N.H. Magnusson dont les concepts révisés de l'évolution du sous-sol (1963) inclurent cinq phases majeurs d'orogénèse.

L'introduction de la datation radiométrique ajouta de la précision à ces conclusions et l'extension de la cartographie par Lundegårdh et autres clarifia l'image de l'évolution tectonique des régions suédo-finlandaises. L'étude de l'ouest de la Suède démontra l'existence d'une tectogénèse suédo-norvégienne finale.

Dans le nord de la Suède, l'investigation fut commencée par H. Lundbohm dans le Comté de Norrbotten. Ensuite, Magnusson reconnut que les roches précambriennes avaient d'une part, un caractère eugeosynclinal. La cartographie encore plus récente par G. Kautsky et autres a initié la clarification de l'histoire tectonique de cette région, indiquant des corrélations de ces roches précambriennes avec celles plus loin au sud.

INTRODUCTION

Geological mapping of the Swedish Precambrian bedrock goes back to 1838, when Jacob af Forselles started his geognostical investigation of southern and central Sweden, extending as far north as the 64th latitude. The map was printed in a great number of sheets. However, it is not a geological map in the modern sense and thus has only historical interest.

The first map made using principles still valid appeared in 1856. This was produced by Axel Erdmann, the founder and first Director (from 1858) of the Geological Survey of Sweden, who advocated the view that geological maps ought to reflect not only the distribution of rocks and soils but also the character of the landscape. Before starting a regular mapping programme in 1858, he prepared a prototype. This was printed in 1856, at the scale of 1:100,000 and shows the geology around the river Fyris in central Uppland, with Uppsala city in the south. The map includes all visible outcrops; these are marked with different colours to display the various rock types encountered during the field work (gneiss, halleflinta, mica schist, crystalline limestone, granite, pegmatite, syenite, quartzite, diorite, hornblendite, and uralite porphyrite). Areas covered with soils were painted differently; the colours here reflect the superficial character of each special kind of soil. Six profiles accompany the map.

Previously Axel Erdmann had published his renowned manual (1855) on the rocks of the earth's crust, especially those of Sweden.

During the next 100 years the mapping of the Geological Survey followed the principles established by Axel Erdmann. The ordinary scale, however, was 1:50,000 (publication series Aa). Later, the subdivision of the rocks mapped became more refined, especially after the fundamental work of Alfred E. Törnebohm (Fig. 1). Director of the Geological Survey 1897-1906. Between 1876 and 1881 Törnebohm published a map in nine sheets, at a scale of 1:250,000, covering the whole iron ore-bearing region of south central Sweden. This should be considered a milestone in Swedish petrological mapping. Not only the map itself, but also the accompanying descriptions, have retained their value to the present day. Törnebohm was also the first to interpret the tectonic evolution of the Scandinavian Caledonian mountain range in a plausible way, including recognition of the great overthrusts; furthermore, he investigated the Dal Formation and the bedrock of Värmland county in western Sweden. In 1885 the first edition of his popular work on the elementary principles of the geology of Sweden appeared.

In 1893 the famous Finnish geologist J. Johannes Sederholm, describing the bedrock of southern Finland, pointed out that the ages of the rocks are fundamental parameters when preparing petrological maps. Earlier maps and papers concerning Precambrian geology in Sweden and Finland had been mainly based on the petrography and tectonics of the various rocks studied, without much regard to the rock ages. In a long series of papers, among which should be mentioned his account of the geology of Fennoscandia (1932), Sederholm discussed numerous aspects of the Finnish and

Figure 1.
A.E. Törnebohm, 1938-1911 (photograph from A. Högbom, 1912, Geological Magazine, v. 49, no. 2, p. 49-53.

Swedish bedrock. Among other things, he introduced geological cycles as principal periods of rock evolution in the Precambrian of Fennoscandia. He also divided the Swedish (and Finnish) granites into four groups characterized by decreasing tectonization, viz.

I. Older gneiss-granites.
II. Younger gneiss-granites and the Jörn granite of northern Sweden.
III. a. Småland (and Värmland) granites.
b. Bohus and Blekinge (Spinkamåla, Karlshamn, Eringsboda) granites.
IV. Rapakivi granites (Strömsbro, Dala, Härdal, Ragunda, and Nordingrå granites).

Another remarkable concept in Swedish Precambrian geology was also formulated by a Finnish geologist, Wilhelm Ramsay, in his text book on the fundamental principles of geology (1909). Ramsay therein defined the Precambrian basement of southern central Sweden and southern Finland as being the roots of an old mountain range, which he called the Svecofennian fold belt. Sederholm's Svionian rocks, of volcanic and sedimentary origin, were considered to be the oldest ones belonging to this belt.

Meanwhile, various Swedish petrologists contributed largely to the knowledge of the Swedish Precambrian, among others Otto Nordenskjöld (1894), Per J. Holmquist (1906), Per Geijer (1916, 1917, 1923), Nils Sundius (1922), and Nils H. Magnusson (1925).

Sundius' Grythyttan paper (1922) is really a most interesting one, because of his detailed stratigraphic division and advanced microscopic examination of the rocks of that region. The stratigraphic column of the ancient Grythyttan syncline commences with extremely sodic and potassic halleflintas, in part showing delicate pyroclastic textures (ash particles etc.). The upper part of the column is dominated by black clay slates displaying shearing schistosity. The youngest sedimentary rock is a polymict conglomerate, which contains pebbles of all rocks of the region except the granitoids and has been deposited disconformably on the older rocks.

WAHL'S DIVISION OF THE PRECAMBRIAN

In 1933 Nils H. Magnusson (Fig. 2, Director of the Geological Survey 1951 - 1958), influenced by the work of Sederholm, made his first attempt to define geological cycles in Swedish Precambrian rocks. This work on the age division of the central Swedish Precambrian was not completed, however, before the appearance of a remarkable man on the geological stage. In 1936 the Finnish chemist Walter Wahl gave a lecture and wrote a paper dealing with the granite groups and periods of folding in the bedrock of Sweden and Finland (1936). Wahl distinguished between three Precambrian orogenies or periods of folding, namely:

1. (oldest) Svecofennian
2. Gothian
3. Karelian

Wahl found empirically that each orogeny ought to include three groups of granite, viz.

a. Primorogenic (synorogenic) granites (foliated granitoids or gneiss-granites).
b. Serorogenic granites (granites associated with pegmatites, migmatites, and veined gneisses).
c. Postorogenic granites.

Wahl's statements have proved to be principally correct and have had an inspiring influence on subsequent work.

Immediately afterwards, in 1936, Nils H. Magnusson published another paper in which he presented once again his views on geological cycles in the Swedish Precambrian. For some unascertainable reason he also exchanged Wahl's Svecofennian for Sederholm's Svionian, which he widened to comprise a complete cycle.

Soon another contribution to the orogenetic discussion appeared. This was a study by Helge G. Backlund on aspects of granite groups, orogenies and cyles in Fennoscandia published in 1936, which adjusted Wahl's scheme thus:

1. (oldest) The Svecofennides, general strike E - W.
2. The Norwegosamides, general strike WNW - ESE.
3. The Gothokarelides, general strike N - S to NNW - SSE.

In 1937, Backlund proposed an alternative to this division:

1. (oldest) The Norwegosamides, 2,100 Ma±?
2. The Svekofennides, 1,200 Ma±?
3. The Gothokarelides, 800 Ma±?

This was before anyone had made reliable age determinations on the radioactive matter in rocks.

Backlund also interpreted the granites resulting from the orogenic evolutions as granitization products more or less *in situ*. Cooperating with Eugen Wegmann in Switzerland and Arthur Holmes in Great Britain, he proceeded to develop granitization concepts and ultimately accepted obviously magmatic rocks, such as the Finnish rapakivi, as products of granitization *in situ*. Backlund summarized his hypotheses regarding the Fennoscandian Precambrian in two papers (1941, 1943).

MAGNUSSON'S REVISED CONCEPTS OF BASEMENT EVOLUTION

Magnusson later revised his concepts of the evolution of the Swedish crystalline basement, especially in his part of the textbook on the geology of Sweden; this was published in four editions between 1936 and 1963 and written together with two Quaternary geologists, at first Erik Granlund and later Gosta Lundqvist, and one paleontologist, Gerhard Regnéll. His final scheme appeared in an atlas of Sweden, map-sheet 5-6, 1970, and was based upon Russian K/Ar age measurements (from the Leningrad laboratory):

1. (oldest) Presvekofennokarelian.
2. Svekofennokarelian (1,650 - 2,100 Ma).
3. Gothian and Subjotnian (1,200 - 1,650 Ma).
4. Dalslandian and Jotnian (850 - 1,200 Ma).
5. Eocambrian (600 - 850 Ma).

When radiometric investigations had put back the formation of the Karelian rocks to the Svekofennian evolution, the term Svekofennokarelian was constructed in 1968, at an international geological mapping conference in Freiberg, DDR (East Germany). The Finnish geologist Ahti Simonen abbreviated this term to Svekokarelian, which is now in current use.

A scheme adapted to the modern Swedish age measurements by Eric Welin and co-workers (see below) was constructed by the present writer in the fourth edition of his part of a textbook on rocks and soils in Sweden (this work appeared in five editions between 1964 and 1978 and was written together with Jan Lundqvist, the Quaternary geologist, and Maurits Lindström, the paleontologist):

Figure 2.
Nils Harald Magnusson, 1890-1976 (photograph from S. Gavelin, 1977, Geologiska Föreningens i Stockholm Förhandlingar, v. 99, pt. 3, p. 254-263.

Period	System	Orogeny	Age
	(from youngest to oldest)		
Algonkian	Eocambrian	Caledonian (Early phase)	600 Ma
			900 Ma
Dalslandian	Jotnian	Dalslandian (Svekonorwegian)	1 450 Ma
Gothian or Postsvekokarelian	sub-Jotnian		1 750 Ma
Karelian Svekofennian (Svekokarelian)		Svekokarelian	2 000 Ma

The terms Jotnian and Subjotnian (sub-Jotnian) go back to the Finnish Sederholm nomenclature, as revised by Arvid G. Högbom in 1910. Högbom distinguished between:

Upper Precambrian	Epijotnian Jotnian Subjotnian
Middle Precambrian	Epijatulian Jatulian Subjatulian
Lower Precambrian	Archean

In Sweden, the Gävle-Dala sandstones (with conglomerates) and the Åsby-Särna dolerites were classed as Jotnian, while the Dala porphyries and Dala granites were referred to the Subjotnian (Hjelmqvist, 1964; Gorbatschev, 1967; Lundegårdh, 1967). In his earlier schemes, Magnusson included Jotnian and Subjotnian in the Algonkian, which he paralleled with the Eocambrian in its widest sense. The rocks belonging to these systems thus ought, he considered, to be very late Precambrian. As was often the case during the geological investigations of yesteryear, the reasons were ridiculous: the rocks looked young and most of the supracrustal rocks had retained their original bedding and layering. But, as Thomas Lundqvist laconically remarked, the porphyries by and by became so compact that they ought to have an age of, at least, 1,000 Ma. (compare his work on the Precambrian geology of the Los-Hamra region, central Sweden, 1968).

INVESTIGATIONS OF ABSOLUTE AGE

Thanks to radiometric investigations, first by Frans E. Wickman and co-workers (1963) and later by Eric Welin and co-workers (Welin *et al.* 1966, Welin and Lundqvist, 1966, 1970), the age of the Dala porphyries and granites could ultimately be put back to 1,699 ± 32 Ma. Soon afterwards, Göran Åberg, working in Welin's laboratory, determined the age of some Subjotnian Smaland porphyries as 1,695 ± 20 Ma (1972). The Postsvecofennian (Postsvekokarelian), or Gothian, Smaland-Värmland granitoids have been investigated by Welin and co-workers and have yielded Rb/Sr ages between 1,745 and 1,690 Ma (Welin *et al.*, 1966, 1977).

In 1960 Magnusson published a K/Ar age of 1,185 Ma for the Jotnian sandstone. Finnish determinations have given still older ages, reaching 1,400 Ma. According to P.J. Patchett (1977), the Åsby dolerite, which is intrusive in the Dala sandstone, has an age about 1,250 Ma. A related dolerite at Nordingrå in eastern Sweden has yielded the same age (Welin and Lundqvist, 1975).

Many dolerites earlier considered as equivalents of the Asby type, have recently proved to be older. The east-west dolerite dikes in Södermanland County (Krokström, 1937; Gorbatschev, 1961), for example, have yielded Rb/Sr ages about 1,560 Ma, according to Patchett (1978). The hyperites of southern Sweden (boudinated dikes of black dolerite free from, or very poor in, olivine) seem to have similar Rb/Sr ages, 1,600 ± 230 Ma (Klingspor, 1976) whereas the foliated hyperites of Värmland County (black olivine dolerite in part changed to garnet amphibolite) are older (Lundegårdh, 1977).

Frans E. Wickman was a pioneer in determining absolute ages of Swedish rocks. He started his work in the 1940s, and in 1954 published a paper on the ages of radioactive minerals in various Swedish pegmatites, in cooperation with Alexander Parwel. The measurements on uraninite were considered by them to be the most reliable ones. A sample from pegmatite in the easternmost part of the central Swedish Svekofennides gave 1,700 Ma, and a sample from Dalslandian pegmatite 960 Ma. These values correspond quite well to those later obtained by Eric Welin and co-workers: 1,810 and about 900 Ma.

TECTONIC EVOLUTION OF THE SVECOFENNIDES

Though the principal elements of the Svecofennides have been known for a long time, the signs of the tectonic evolution were not fully deciphered until the late sixties. (The principal elements are early metavolcanics and metasediments, prim- or synorogenic gabbro, diorite, and granitoids, intraorogenic dike rocks, migmatites, serand postorogenic granites and pematites, postorogenic metavolvanics.) When mapping the bedrock of Gävleborg county in eastern central Sweden, I was able to distinguish between an early Svecofennian folding about horizontal axes striking N - NNW and a later Svecofennian folding about axes plunging moderately towards the east (Lundegardh, 1967). The early folding was shown to be older than the primorogenic granitoids, whereas the later folding is younger than these and has acted as a cross-folding. Soon afterwards, Göran Stålhös demonstrated a similar tectonic evolution in the Stockholm district (1969).

Independently, Hans J. Koark (1973) studied the tectonics of some central Swedish ore districts and observed the same tectonic elements. However, the cross-folds there have axes plunging more or less vertically.

In my Gävleborg monograph (1967) I included a series of drawings of the Svecofennian tectogenesis in central Sweden, based on the isostatic principles (compare also my 1971 paper). Strong anatexis in the basal parts of the Svecofennides would have resulted from the evolution proposed, making possible not only migmatization and granitization but also the formation and eventual eruption of voluminous syn-, ser- and postorogenic Svecofennian granitoid magmas. During his investigation of the Los-Hamra region, Thomas Lundqvist (1968) simultaneously arrived at a similar conclusion regarding the development of anatectic magmas in the Svecofennides.

The tectonic evolution of the other parts of the Swedish crystalline basement has also been clarified. Walter Larsson and the present writer established the presence of N - S and E - W foldings in southwestern Sweden during their mapping for the Geological Survey. (See, among others, Larsson, 1956, and Lundegårdh, 1958).

In the late sixties Roland Gorbatschev and Lennart Samuelsson began thorough investigations in western Sweden. Gorbatschev has concentrated on the Vänersborg - Åmål region and has been able to make distinctions between several Gothian granites (Hästlefjord, Ursand etc.) (1969). In his study of Precambrian geology in western Sweden (1971), he has also discovered a supracrustal and hypabyssal unit which is definitely older than the Dal formation, viz. the Ellenö complex including both volcanic and sedimentary rocks. Gorbatschev (1969) refers all his Gothian granites to the Åmål megaunit and separates them from the older Småland-Värmland granitoids.

Samuelsson works in and around the Göteborg (Gothenburg) district where, among other things, he has solved the principal problems concerning the N - S and E - W foldings. Reference should be made to three Geological Survey publications – the petrological map Göteborg SO, (1977); Göteborg NO, (1979) and Kungsbacka NO (1978).

The final tectogenesis of western Sweden is the Sveconorwegian, a term introduced in 1960 by Nils H. Magnusson. This tectogenesis comprises folds around axes trending N - S, together with thrust faults. It is especially well developed within the Dal Formation (Walter Larsson, 1956) but has also reworked a great part of south-western Sweden and caused a regional regeneration (Magnusson, 1960). It has been referred to the Dalslandian period. Two older foldings have been established, viz. an early F_1 with horizontal axes striking N - S to NNW - SSE, and a middle F_2 with horizontal axes striking E - W to ENE - WSW. F_2 has reworked F_1 so as to produce complicated tectonic patterns (see Larsson, 1966, and Lundegårdh, 1977). Age determinations by means of various methods point at 1,110 to 925 Ma as the age of the Sveconorwegian tectogenesis.

STUDIES IN NORTHERN SWEDEN

In northern Sweden, serious efforts have long been made to disentangle the petrological and tectonic evolution. The conception of an eugeosyncline stretching WNW - ESE immediately to the north of Gävleborg County originates from Sederholm's Bothnian Formation of greywackes, schists, and intermedic to basic volcanics in Finland displayed by the Tampere schist belt. In Sweden, Magnusson called these rocks the Härnö Formation and was the first to define them (in his account of the Geology of Sweden, 1936 to 1963) as eugeosynclinal deposits. In 1959, Gunnar Kautsky spoke of a great sea bordering on a shore in northern Västerbotten County marked by conglomerates. He also called attention to Josef Eklund's Arvidsjaur anticlinorium in the north, which is referred to in a 1923 paper (Eklund, 1923) on the geology of the Skellefteå field and which is a counterpart to my Alfta anticlinorium to the south of the eugeosyncline. In 1960, I described the southern marginal deposits of the eugeosyncline as the miogeosynclinal rocks of eastern central Sweden. In my Gävleborg monograph (Lundegårdh, 1967), I sketched the approximate boundaries of the eugeosyncline and the limits of its tectonization.

The geology of Västerbotten County has also been investigated by, among others, Sven Gavelin (1955), Olof H. Ödman (1941) and Erland Grip (1935, 1942 and 1973).

The renowned practical geologist and ore field manager, Hjalmar Lundbohm, initiated extensive petrological research in Norrbotten County during the early nineteen hundreds. The Kiruna district was the topic of monographs by Per Geijer (1910) and Nils Sundius (1915). Geijer afterwards widened his scope and published a great number of papers concerning the geology of Norrbotten County. In 1957 Olof H. Ödman published a map of the Precambrian bedrock of the county at a scale of 1:400,000. In the early sixties the Geological Survey of Sweden started detailed mapping, on the scale of 1:50,000, of the crystalline basement of the country. Leader on the project was Gunnar Kautsky. Already a great number of petrological and geophysical map sheets have been printed. Simultaneously, Rudyard Frietsch has published several papers on the Precambrian geology of the county. Among these should be mentioned his examination of trace elements in magnetite and haematite, mainly from northern Sweden (1970), and of Precambrian iron ores in Norrbotten County (1973). Frietsch here advocates a magmatic origin for the greater apatite-rich iron ores of Kiruna type, in agreement with earlier suggestions by Per Geijer but contrary to the sedimentary origin proposed by Sture Landergren in 1948 and Tibor Parak in 1975.

Stratigraphic division of the Precambrian of Norrbotten County remained problematic for a long time. A reliable scheme could not be constructed until Eric Welin started his program of absolute age determinations (1970, 1971). Earlier most Precambrian rocks of Norrbotten County had been referred to the Karelian cycle (Magnusson, 1936), whereas it is now quite clear that a great part of the bedrock of Norrbotten County belongs to the Svecofennides, or Svekokarelides according to Ahti Simonen. The migmatizing variety of Lina granite has Rb/St ages between 1,790 ± 60 and 1,820 ± 50 Ma, and the Postsvekofennian anatectic rocks of the county correspond well with those in other parts of Sweden. They comprise Subjotnian porphyries as well as the Sorsele granitoids (1,635 ± 65 Ma) and the younger Lina granite (1,565 ± 35 Ma).

The early Svecofennian volcanics and sedimentary rocks have been deposited on a Presvekofennian complex, including the Råstojaure granite with an age amounting to 2,700 Ma. This is the oldest rock radiometrically dated in Sweden, but probably other rocks of equal or similar ages will be found in the future. Investigations in Hälsingland, central Sweden, have already indicated the presence of such a rock, a granite older than the synorogenic Svecofennian granitoids; and in the neighbourhood of Uppsala an early Svecofennian volcanic formation contains granite xenoliths (Lundegårdh, 1956).

Much more could of course have been reported from the great field of investigations in the Swedish Precambrian. Southern interior Sweden has, for instance, been neglected, owing to the fact that a period of reinterpretation has started there. Only two monographs will be mentioned, namely those by Sven Hjelmqvist (1934) and Ulf Wiklander (1974).

REFERENCES

Aberg, G., 1972, An Rb/Sr age of Småland porphyries: Geologiska Föreningens i Stockholm Förhandlinger, v. 94, p. 377-379.

Backlund, H.G., 1936, Till frågan om granitgrupper, bergskedjeveckningar och cykelindelning i Fennoskandia [Aspects of granite groups, orogenies and cycles in Fennoscandia]: Geologiska Föreningens i Stockholm Förhandlingar, v. 58, p. 349-356.

———, 1937, Die Umgrenzung der Svekofenniden: Bulletin of the Geological Institute of the University of Uppsala, v. 27, p. 219-269.

———, 1941, Die ältesten Baueinheiten von Fennoskandia: Mitteilungen der Naturforschenden Gesellschaft Schaffhausen, v. XVII.

———, 1943, Einblicke in das geologische Geschehen des Präkambriums: Geologische Rundschau, v. 34, pts. 2-6, p. 79-149.

Eklund, J., 1923, Skelleftefältets geologi [Geology of the Skellefteå field]: Geologiska Föreningens i Stockholm Förhandlingar, v. 45, H. 1-2 p. 218-223.
Erdmann, A., 1855, Vägledning till bergarternas kännedom [Manual on the rocks of the earth's crust]: Stockholm, Isacc Marcus.
——————, 1856, Geologisk karta öfver Fyrisåns dalbäcken [Geological map of the river Fyris valley]: Gen. Stab. Lith. Inr., Stockholm.
Forselles, J. af, 1838-1855, Geognostisk karta öfver Sverige. Södra delen till 64:e breddgraden [Geognostic map of southern and central Sweden]: Brukssocieteten, Stockholm.
Frietsch, R., 1970, Trace elements in magnetite and hematite mainly from northern Sweden: Sveriges Geologiska Undersökning, v. 64, no. 3, ser. C, no. 646, 136 p.
——————, 1973, Järnmalm i Norrbottens urberg [Iron ores in the Precambrian bedrock of Norrbottens County] *in* Grip, E, and Frietsch, R.: Malm i Sverige pt 2: Uppsala, Almqvist and Wiksell.
Gavelin, S., 1955, Urbergsomradet inom Västerbottens län [The Precambrian bedrock of Vasterbötten County] in S. Gavelin & O. Kulling: Beskrivning till berggrundskarta över Vesterböttens län: Sveriges Geologiska Undersökning, ser. Ca, no. 37, p. 5-99.
Geijer, P., 1910, Igneous rocks and iron ores of Kiirunavaara, Luossavaara and Tuolluvaara: Scientific and practical researches in Lapland arranged by Luossavaara-Kiirunavaara Ltd; Geology of the Kiruna District, v. 4.
——————, 1916, On the intrusion mechanism of the Archean granite of central Sweden: Bulletin of the Geological Institute of the University of Uppsala, vol. 15, p. 47-60.
——————, 1917, Falutraktens berggrund och malmfyndigheter [Rocks and ores of the Falun district]: Sveriges Geologiska Undersökning, ser. C, no. 275, p. 1-316.
——————, 1923, Riddarhytte malmfält [The Riddarhyttan iron ore field]: Kungl. Kommerskollegii beskrivningar av malmfyndigheter, no. 1.
Gorbatschev, R., 1961, Dolerites of the Eskilstuna regions; Sveriges Geologiska Undersökning, ser. C, no. 580. Arsbok 55, no. 4, p. 1-39.
——————, 1967, Petrology of the Jotnian rocks in the Gavle area, east central Sweden: Sveriges Geologiska Undersökning, ser. C, no. 621. Arsbok 60, no. 6, p. 1-50.
——————, 1971, Aspects and problems of Precambrian geology in western Sweden: Sveriges Geologiska Undersökning, ser. C, no. 650. Arsbok 64, no. 7, p. 1-63, 19 figs.
——————, 1969, Fundamental subdivision of Precambrian granitoids in the Amal megaunit and the evolution of the southwestern Baltic Shield: Geologiska Föreningens i Stockholm Förhandlingar, v. 91.
Grip, E., 1935, Die Arvidsjaurporphyre: Bulletin of the Geological Institution of the University of Uppsala, v. 25.
——————, 1942, Die Tektonic und Stratigraphie der zentralen und östlichen Teile des Skelleftefeldes: Bulletin of the Geological Institute of the University of Uppsala, v. 30, no. 3, p. 67-90.
——————, 1973, Skelleftefältets sulfidmalmer [Sulfide ores of the Skelleftea field], *in* Grip, E. and Frietsch, R.: Malm i Sverige pt 2: Almqvist and Wiksell, Uppsala, 295 p.
Hjelmqvist, S., 1934, Zur Geologie des Südschwedischen Grundgebirges: Meddelanden fran Lunds Geologisk-Mineralogiska Institution, no. 58.
——————, 1964, Berggrundskarta over Kopparbergs län [Map of the pre-Quaternary rocks of the Kopparberg County, central Sweden]: Sveriges Geologiska Undersökning, ser. C, no. 40.
Högbom, A.G., 1910, Precambrian geology of Sweden: Bulletin of the Geological Institution of the University of Uppsala, v. 10, p. 1-80.
Holmquist P.J., 1960, Studien uber die Granite von Schweden: Bulletin of the Geological Institute of the University of Uppsala, v. 7, no. 4, p. 77-269.
Kautsky, G., 1959, Studien zur Paläogeographie des Archaikums im nördlichen Schweden: Geologiska Föreningens i Stockholm Förhandlingar, v. 81, pt 4 no. 499, p. 559-571.

Klingspor, I., 1976, Radiometric age determinations of basalts, dolerites and related syenite in Skane, southern Sweden: Geologiska Föreningens i Stockholm Förhandlingar, v. 98, p. 195-215.

Koark, H.J., 1973, Zur Entstehung des tektonischen Stengelbaus an Präkambrischen Eisen- und Sulfiderzkörpern der Zentralschwedischen Leptitserie: Mineralium Deposita, v. 8, p. 19-34.

Krokström, T., 1937, The Hallefors dolerite dike and some problems of basaltic rocks: Bulletin of the Geological Institute of the University of Uppsala, v. 26, p. 113-263.

Landergren, S., 1948, On the geochemistry of Swedish iron ores and associated rocks: Sveriges Geologiska Undersökning, ser. C, no. 496. Arsbok 42, no. 5, p. 1-185.

Larsson, W., 1956, Berggrunden [The bedrock], *in* W. Larsson & Sandegren: Beskrivning till kartbladet Vårvik: Sveriges Geologiska Undersökning, ser. Aa, No. 187, 166 p.

———————, 1966, Berggrunden [The bedrock] *in* Caldenius-Larsson-Mohrén-Linnman-Tullstrom: Beskrivning till kartbladet Halmstad: Sveriges Geologiska Undersökning, ser. Aa, no. 1984, 138 p.

Lundegårdh, P.H., 1957, Göteborgstraktens berggrund [Petrology of the Gothenburg-Kungäv region]: Sveriges Geologiska Undersökning, ser. C, no. 553. Arsbok 51, no. 4, p. 1-82.

———————, 1958, Petrology of the Uppsala region, eastern Sweden: Sveriges Geologiska Undersökning, ser. C, no. 544, p. 1-74.

———————, 1960, The miogeosynclinal rocks of eastern central Sweden: Sveriges Geologiska Undersökning, ser. C, no. 570, p. 1-68.

———————, 1967, Berggrunden i Gävleborgs län [Precambrian geology of the Gavleborg County]: Sveriges Geologiska Undersökning, ser. Ba, no. 22, 303 p.

———————, 1971, Neue Gesichtspunkte zum Schwedischen Präkambrium: Geologische Rundschau, v. 60, no. 4, p. 1392-1405.

———————, 1974, Jordklotet [The Earth] and Det svenska urberget [The Precambrian bedrock of Sweden], *in* Lundegårdh, P.H., Lundqvist, J. and Lindström, M.: Berg och jord i Sverige, fourth edition: Uppsala, Almqvist and Wiksell. (Fifth edition 1978).

———————, 1977, The Gräsmark Formation: Sveriges Geologiska Undersökning, ser C, no. 732.

Lundqvist, T., 1968, Precambrian geology of the Los-Hamra region: Sveriges Geologiska Undersökning, ser. Ba, no. 23, p. 1-255.

Magnusson, N.H., 1925, Persbergs malmtrakt [The Persberg ore district]: Kungl. Kommerskollegii beskrivningar av malmfyndigheter, no. 2

———————, 1933, Det mellansvenska urbergets åldersschema [Age division of the central Swedish Precambrian]: Geologiska Föreningens i Stockholm Förhandlingar, v. 55, p. 612-662.

———————, 1936, Om cykelindelningen i det svenska urberget [On the division of the Swedish Precambrian into orogenic cycles]: Geologiska Föreningens i Stockholm Förhandlingar, v. 58, p. 102-108.

———————, 1956, Berggrunden [The bedrock] *in* N.H. Magnusson and E. Granlund; Sveriges geologi: P.A. Norstedt Söner (Svenska Bokförlaget) Stockholm, 557, p., [p. 1-289].

———————, 1960, Age determinations of Swedish Precambrian rocks: Geologiska Föreningens i Stockholm Förhandlingar, v. 82, p. 407-432, 4 figs.

———————, 1963, Urberget [The Precambrian bedrock] *in* Magnusson, N.H., Lundqvist, G., and Regnéll, G.: Sveriges geologi, fourth edition: Svenska Bokförlaget (L) Norstedt, Stockholm.

———————, 1970, Stratigrafi och tektonik [Stratigraphy and tectonics]: Atlas över Sverige, map-sheets 5-6; Kartografiska institutet, Stockholm.

Nordenskjöld, O., 1894, Ueber Archaeische Ergussgesteine aus Småland: Bulletin of the Geological Institution of the University of Uppsala, v.1, p. 133-255.

Odman, O.H., 1941, Geology and ores of the Boliden deposit: Sveriges Geologiska Undersökning, ser. C, No. 438, Arsbok 35, no. 1, 190 p.

————, 1957, Berggrundskarta över urberget i Norrbotten län [Map of the Precambrian bedrock of Norrbotten County]: Sveriges Geologiska Undersökning, ser. Ca, no. 41, p. 1-151.
Parák, T., 1975, The origin of the Kiruna ores: Sveriges Geologiska Undersökning, ser. C, no. 709.
Patchett, P.J., 1978, Rb/Sr ages of Precambrian dolerites and syenites in southern and central Sweden: Sveriges Geologiska Undersökning, ser. C, no. 797.
Ramsay, W., 1909, Geologins grunder [Fundamental principles of geology]. Third edition: Holger Schildt, Helsinki, 1931.
Samuelsson, L., 1977, Berggrundskartan Göteborg NO [The petrological map Göteborg NO]: Sveriges Geologiska Undersökning, ser. Af, no. 117.
Sederholm, J.J. 1893, Om Bärggrunden i Södra Finland [On the bedrock of southern Finland]: Fennia, v. 8.
————, 1932, On the geology of Fennoscandia: Bulletin de la Commission Géologique de la Finlande, v. 16, no. 98, p. 1-30.
Stalhos, G., 1969, Stockholmstraktens berggrund [Precambrian geology of the Stockholm district]: Sveriges Geologiska Undersökning, ser. Ba, no. 24, p. 1-190.
Sundius, N., 1915, Beiträge zur Geologie des südlichen Teils des Kirunagebiets: [Scientific and practical researches in Lapland arranged by Luossavaara-Kiirunavaara Ltd.]: Geology of the Kiruna District, v. 4.
————, 1922, Grythyttefältets geologi [Geology of the Grythyttan field]: Sveriges Geologiska Undersökning, ser. C, no. 312. Arsbok 16, no. 2, p. 1-354.
Törnebohm, A.E., 1876-1881, Geologisk karta öfver Mellersta Sveriges bergslag [Geological map of the ore-bearing region of southern central Sweden]. With nine descriptions: Stockholm, Jernkontoret.
————, 1885, Grunddragen af Sveriges geologi [Elementary principles of the geology of Sweden], second edition, 1894: Stockholm, P.A. Norstedt and Soner.
Wahl, W., 1936, Om granitgrupperna och bergskedjeveckningarna i Sverige och Finland [On the granite groups and orogenies in Sweden and Finland]: Geologiska Föreningens i Stockholm Förhandlingar, v. 58, p. 123-124.
Welin, E., Blomqvist, G., and Parwel, A., 1966, Rb/Sr whole rock age data on some Swedish Precambrian rocks: Geologiska Föreningens in Stockholm Förhandlingar, v. 88, p. 19-28.
Welin, E., Christiansson, K., and Nilsson, O., 1971, Rb/Sr radiometric ages of extrusive and intrusive rocks in northern Sweden, I: Sveriges Geologiska Undersökning, ser. C, no. 666, p. 1-38.
Welin, E., Gorbatschev, R., and Lundegårdh, P.H., 1977, Rb/Sr dating of rocks in the Varmland granite group, Sweden: Geologiska Föreningens i Stockholm Förhandlingar, v. 99.
Welin, E., and Lundqvist, Th., 1970, New Rb/Sr age data for the sub-Jotnian volcanics (Dala porphyries) in the Los-Hamra region, Central Sweden: Geologiska Föreningens i Stockholm Förhandlingar, v. 92, p. 35-39.
Welin, E., 1971, Den svekofenniska orogena zonen i norra Sverige [The Svecofennian orogenic zone in Northern Sweden]: Geologiska Föreningens i Stockholm Förhandlingar, v. 92, p. 433-451.
Welin, E., and Lundqvist, Th., 1975, K-Ar ages of Jotnian dolerites in Västernorrland County, central Sweden: Geologiska Föreningens i Stockholm Förhandlingar, v. 97, p. 83-88.
Wickman, F.E., and Parwel, A., 1954, Några preliminära resultat av ålderbestämningar på svenska pegmatitmineral [Some preliminary results of age determinations on Swedish pegmatite minerals]: Geologiska Föreningens i Stockholm Förhandlingar, v. 76, p. 353-354.
Wickman, F.E., Blomqvist, G., Geijer, P., Parwel, A., von Ubisch, H. and Welin, E., 1963, Isotopic constitution of ore lead in Sweden: Arkiv for Mineralogi och Geologi, v.3, p. 193-257.
Wilklander, U., 1974, Precambrian petrology, geochemistry and age relations of northeastern Blekinge, southern Sweden: Sveriges Geologiska Undersökning, ser. C, no. 704, Arsbok 68, no. 11, p. 1-142.

History of Concepts in Precambrian Geology, edited by W.O. Kupsch and W.A.S. Sarjeant, Geological Association of Canada Special Paper 19, 1979

WERNER'S CONCEPT OF THE BASEMENT COMPLEX

Alexander M. Ospovat
Department of History, Oklahoma State University
Stillwater, Oklahoma 74074

ABSTRACT

Much of our present-day misunderstanding of the history of geology stems from attempts to deal with the past from the vantage point of the present and to use modern terms when we are dealing with old concepts. Werner did not have a concept of the basement complex. The term was unknown to him, and the present-day definition of it is thoroughly out of keeping with his theories.

Werner's primitive rocks are the only part of his system that can be considered analogous to the modern concept of the basement complex. To understand this analogy it is necessary to understand Werner's classification of rocks, the basis of his classification, the relationship of objective and subjective categories and of time units and rock units, as well as his general concept of the history of the earth's crust.

Werner's concept of the primitive period and his interpretation of that period are historical. The order of succession is important, but not to the exclusion of any deviation. What Werner attempted to do was to lay the "foundation for a complete canvass of the universal formation" of rock masses. Gradual changes in the formation and development of rocks are recognized, but not to the exclusion of occasional revolutionary changes.

Werner's concept of the primitive period was extremely fruitful. It inspired much work by both his supporters and his opponents.

RESUME

Beaucoup de nos actuelles conceptions erronées de l'histoire de la géologie proviennent de nos tentatives de considérer le passé à partir de la position avantageuse, qu'offre le présent, et d'utiliser des termes modernes alors que nous avons affaire à de vieux concepts. Werner n'avait aucun concept du complexe de base. Le terme lui était inconnu et son actuelle définition est en complet désaccord avec ses théories.

Les roches primitives de Werner constituent la seule part de son système pouvant être considérée comme analogue au concept moderne du complexe de base. Pour comprendre cette analogie il est nécessaire de comprendre la classification des roches de Werner, la base de sa classification, la parenté de catégories objectives et subjectives et d'unités de temps et de roches, aussi bien que son concept général de l'histoire de la croûte terrestre.

Le concept de Werner de la période primitive et son interprétation de cette période sont historiques. L'ordre de succession est important mais n'exclut pas toute déviation. Ce que Werner tentait de faire était de poser "les fondations d'un canevas complet de la formation universelle" des masses rocheuses. Il reconnaissait des changements graduels dans la formation et le développement des roches, mais n'excluait pas des changements occasionnels révolutionnaires.

Le concept de la période primitive de Werner fut extrêmement fructueux. Il inspira beaucoup de travaux à la fois à ses partisans et ses adversaires.

INTRODUCTION

When we survey the development of the various sciences, it becomes clear that, for a long period in our history, the geological sciences attracted less attention than other sciences, the physical, the chemical, and the biological. From time immemorial there has been an interest in the development of the earth and the earth's crust as well as a strong interest in minerals, but the actual study of geology, compared to the studies of other sciences, is young. When we survey the development of the history of science, we find that the historiography of geology is also not so well established as that of other fields. A perusal of the literature of the history of science, the Critical Bibliographies of *Isis,* and the various current journals dealing with the history of science will show that, compared with the work in the histories of other sciences, work in the history of geology is scarce. When Dr. Robert P. Multhauf assumed the editorship of *Isis* a few years ago, he remarked that there are very few articles in *Isis* which deal with the geological sciences. More important than the scarcity of work in the history of geology, however, is the fact that what most people know about geologists of the past and their ideas comes from histories of geology which were written by persons who did not have a fundamental understanding of historical theory and method but instead had a presentist point of view; that is, they wrote with the idea in mind that any past geological theory which is by and large acceptable today was right and good and furthered the advance of science and that all those conceptual schemes which are no longer acceptable were wrong and bad and hindered the advance of science.

Recently there has been an increasing awareness that the familiar histories of geology are not only inadequate, but in many instances also inaccurate. They do not describe the ideas of geologists of the past and the controversies that accompanied the progress of geology quite as they should be described. As a result, we are now in what I consider to be a revisionary, even iconoclastic period in the study of the history of the geological sciences. And this revisionism and iconoclasm require as accurate an understanding as possible of the geological concepts of the past and of the times in which they were advanced. What we are doing now would better have been done long ago, before general accounts of geology were written, that is, examining the sources and evaluating them in historical context.

The study of the history of geology differs from the studies of histories of other sciences in that we still do not have really definitive works on even the most illustrious geologists of the past. Various scholars are in the process of preparing such works, but at this time there are no definitive studies of even Lyell, Hutton, or Werner, to mention only the three best known and perhaps most important individuals in the history of geology. Instead we are burdened with misconceptions about the work and contributions of geologists of the past and many misleading labels and anachronisms. For instance, the labelling of geologists as Wernerian or Huttonian, vulcanist or neptunist, catastrophist or uniformitarian has, it seems to me, beclouded the study of 18th and 19th century geology rather than clarifying it. I cannot claim innocence here. In the past I have been as guilty as others in this respect. However, my own work on Werner has revealed that many geologists who have been labelled Wernerian followed Werner only in part, if at all, and that, contrary to what has been widely believed, none of his most illustrious and influential students ever followed him blindly. Work which Professor Kenneth Taylor is currently undertaking shows that Desmarest and even Dolomieux were not so vulcanist as they have been pictured (Taylor, 1969) and work that my son Dov has been doing reveals that even Hutton himself was not so Huttonian as has generally been believed (Ospovat, 1971). Labels can, of course, be useful when they are applied accurately, and even when they are used loosely, I suppose that they do not do much harm as long as they are not taken very seriously. But anachronisms have no place in any kind of history, especially in the history of science.

WERNER AND THE BASEMENT COMPLEX

My paper is entitled Werner's Concept of the Basement Complex because that is what I was asked to write on; but the simple truth is that I cannot discuss Werner's concept of the basement complex because Werner did not have a concept of the basement complex. The term was completely unknown to him, and as it is understood and defined today, it cannot possibly be used within the context of Werner's theories nor within the historical context of his time.

The *Glossary of Geology and Related Sciences* defines "basement complex" as "a series of rocks generally with complex structure beneath the dominantly sedimentary rocks. In many places they are igneous and metamorphic rocks of either Early or Late Precambrian, but in some places may be much younger, as Paleozoic, Mesozoic, or even Cenozoic." "Basement rock (complex)" is defined in two ways: "1) A name commonly applied to metamorphic or igneous rocks underlying the sedimentary sequence; 2) metamorphic and igneous Precambrian rocks." And "basement" is also defined two ways: "1) Complex, generally igneous and metamorphic rocks, overlain unconformably by sedimentary strata; 2) Crustal layer beneath a sedimentary layer and above the Mohorovičić discontinuity" (Glossary of Geology and Related Sciences, 1960, p. 72). Not one of these definitions makes any sense when applied to Werner's system. Not only were most of the terms used in these definitions, such as Precambrian, Paleozoic, Mesozoic, metamorphic, and Mohorovičić discontinuity, not known in Werner's day, but they are derived from modern geological concepts and reflect modern geological thinking.

However, the basic difficulty with using any of the definitions of *basement complex* in a discussion of Werner's ideas about the fundamental rocks is that they are incompatible with his geological system, which is an attempt to establish a neptunistic history of the earth's crust, in which virtually all the rocks are of aqueous origin. The processes of metamorphosis of rocks were not known in Werner's day, and the whole concept of metamorphism of rocks is incompatible with his neptunistic theory of the history of the earth's crust, in which the differences among the various rock masses and the various rocks are explained as the result of gradual changes in the conditions and modes of formation and in the contents of the waters from which the rock-making materials were derived.

Werner is rightly known as the father of historical geology. As a student of history, he understood very well that the exposition of the history of the earth, like any other kind of history, must be more than a mere recitation of events in chronological order, that it must attempt to show the interdependence of events and advance explanations for their causes. Because he included a search for causes in his history of the earth, he chose the term *geognosy* as the proper designation for the study, explaining that geognosy is "the abstract, systematic knowledge of the solid earth." (Lecture notes taken by Adolph L. Schippan, 1814-1815, in Werner's course on geognosy: Werner MSS).

THE WERNERIAN CLASSIFICATION OF ROCKS

Werner's classification of rocks was basically chronostratigraphic. As he himself put it:

> "I had to be guided completely in the classificatory presentation or tabulation of these masses by the discoverable time sequence of the particular formations if I wanted to remain true to my plan to sketch through this classification a foundation for a complete canvass of the universal formation of these masses."

(Unpublished reply by Werner to a review of his *Kurze Klassifikation* which appeared in the journal *Allgemeine Deutsche Bibliothek,* 1778: Werner MSS).

He used various means by which to discover this time sequence. One of these was the order of succession of rocks. Werner thought that this was a very important clue to the history of the earth's crust, especially since he believed that virtually all rocks are precipitates or sediments from an aqueous solution. However, he was fully aware of the fact that the order of succession cannot always be read directly from the facts as they are observed, but must be interpreted. And the study of stratigraphic relations, of the structure of rock masses, was in Werner's mind one of the most important clues to the understanding of the history of the earth's crust.

The compositional differences among the various rocks also served Werner in his classification of rocks and his interpretation of the earth's history. He also used textural differences in support of his chronostratigraphic classification. Thus he thought that rocks of a granular texture are the result of chemical precipitation, that their constituent parts are imperfect crystals of contemporaneous origin, and that they were formed primarily in the primitive period. Rocks of a slaty texture, he thought, were also the result of chemical precipitation and formed primarily in the primitive period. As another basis for his classification Werner used the presence or absence of fossils. Finally, he also used what he called formation suites. They are in a sense a control, a cross reference for his interpretations. The formation suites cut through the various periods of formation and tie together the order of succession, stratigraphic relations and structure of rock masses, and compositional and textural features of rocks.

> "Werner believed that nature repeats itself and that similar rocks were formed during different periods in the formation of the earth's crust. He explained that these rocks have common characteristics but that they also have characteristics which are peculiar to each of them, being the result of the circumstances under which they were formed and the period of their formation. Together they form

a series or suite of formations. The members of such a suite that are widely separated in time show such great differences that only with difficulty can they be recognized as belonging to the same suite. On the other hand, those members of a suite that were formed during periods that are not widely separated are so similar that it is difficult to distinguish among them. The outstanding example of Werner's formation suites is the series of limestone formations, which includes the primitive limestones, floetz limestones, chalk, and calc-tufa. Altogether Werner recognized seven formation suites: slate, limestone, trap, porphyry, gypsum, coal, and serpentine." (Werner, 1971).

Time of origin and mode of formation are basically subjective features of rocks, whereas stratigraphic relations, mineral content, and fossil content are basically objective features. While it is important to distinguish clearly between the two categories, it is also important to realize that an understanding of the earth's history demands a combination of the two. Interpretations, conclusions drawn on the basis of comparisons of factual data were, as far as Werner was concerned, the most essential part of geognosy.

There was no confusion in Werner's mind when he used petrography and lithology and fossils as bases for his chronostratigraphic classification of rocks. He was searching for a unified geological system, believing that all aspects of rocks – stratigraphic, structural, compositional, textural, etc. – could give testimony to their mode and time of formation.

Clearly then, when we discuss Werner's theory concerning the foundation rocks, we are dealing basically with a problem in stratigraphy, especially stratigraphic classification and nomenclature. Specifically, we are concerned with one part of Werner's geological system, that part which deals with the earliest history of the earth's crust, with what Werner called the primitive period and what he called the primitive rocks. According to his theory, the primitive rocks were crystalline precipitates from the universal ocean which adhered to an originally uneven surface. In general the siliceous, argillaceous, and talcose minerals predominate in them. Only the newer rocks of the primitive class contain carbon, and a gradual increase in carbon can be noticed as one proceeds to the even newer rocks (Werner, 1971, p. 105).

WERNER'S SPECULATIONS ON THE STRUCTURE OF THE EARTH

Other geologists, notably Buffon, had postulated an originally uneven surface and offered explanations of the unevenness. Werner speculated little about the core of the earth. He did believe, however, that the entire earth, including the core, had come from a fluid, and he did, to use his own word, "conjecture" about the origin of the earth's core. Referring to what he called "the order of succession of the five main classes of mechanically simple natural bodies," he wrote:

"We find that according to the laws of gravity, ether, or the purest form of fire, occupies the uppermost place on the planet Earth; following it is air; between air and the earth is water, under which comes the earth and stone mass of the solid earth. There still remains the fifth main class of mechanically simple natural bodies, the metals. However, these we find, as far as we know, scattered among the different earths and stones. But could it not be possible that the principal depository of them is located in the very interior of the earth, in other words beneath the earth and stone mass known to us, and that the depositories which occur in the earth and stone mass are only offsprings of this principal depository?" (Notes on Geognosie: Werner MSS).

This conjecture is interesting in that it seems to contradict Werner's general theory that virtually all the constituents of the earth's crust settled from above. However, it could also explain the unevenness of the earth's core. Since the different metals have different weights, they would not have settled at the same rate; and since, according to Werner's theory, the contents of the universal ocean were not everywhere the same, these metals would not necessarily have settled in the same places. Thus the uneven surface could have been produced. However, Werner does not mention this possibility. In general he scorned what he thought of as idle speculation, and in fact he concluded his own speculation about the earth's core as follows: "But we want to leave the dark abyss of the earth, where we can only wander about along the giddy paths of conjecture and return to that upper region of the solid earth, where the light of experience of searching minds illuminates the path." (Notes on Geognosie: Werner MSS).

Werner believed that granite is, in general, the foundation rock, or, as he put it in his *Kurze Klassifikation,* "Some granite seems to form the foundation rock." (Werner, 1786, p. 272-297, p. 277). He pointed out that granite is found in the highest as well as the lowest places on earth. This phenomenon is explained by the theory of the original unevenness of the earth's core and by the postulated diminution of the waters of the universal ocean. At the beginning of the primitive period, the water of the universal ocean stood deep and calm above the tops of even the highest mountains that we know today. But as the primitive rocks were being precipitated, the waters of the ocean were receding. Thus, since at later times the waters of the ocean no longer stood above the highest mountains, the granite which had been precipitated on the tops of those mountains would have no other rocks above it, whereas the granite which had been precipitated in the lowest portions of the original surface would have many layers of other rocks above it. In fact, this original unevenness, differential settling, and, later, erosion and destructive agents of a mechanical and chemical nature, in Werner's theory, explain most of the configuration of the earth's crust.

Formation of the Primitive Rocks

As mentioned above, Werner's classification was historical. Thus when we discuss the primitive rocks, we must also consider a primitive period of formation. Therefore, it is easy to see that a concept of the basement complex which allows rocks of different periods to form part of the foundation layer is incompatible with Werner's theory. According to that theory, all the primitive rocks were formed during the primitive period, although some were formed later in the period than others and could therefore be called "younger." And since Werner postulated some changes in the character of the universal ocean during the primitive period, there were also some differences in the character of the rocks formed at different times during the period, the very oldest ones, those formed when the waters were very deep and calm being highly crystalline; those formed at a later time when the ocean was stormy showing some signs of the mechanical action of water; and still later ones, formed when the ocean was once more becoming calm, once more showing a greater degree of crystallization. In general, Werner characterized the primitive rocks as showing all the characteristics of aqueous formation, as being chemical precipitates throughout, and as being entirely free of fossils.

In his classification of 1802, which appeared in Heinrich Wilhelm Meuder's translation of Robert Jameson's *Outline of the Mineralogy of the Scottish Isles*, Werner lists the primitive rocks as granite, gneiss, mica-schist, clay slate (true slate), primitive limestone, porphyry, syenite, primitive trap, serpentinite, quartzite, topaz rock, flinty slate (chert), and primitive gypsum (Meuder, 1802, p. III). Since he made very few changes in his system after the appearance of this classification, it can be assumed to represent substantially his final conclusions about the primitive rocks. In his *Kurze Klassifikation*, he had listed basalt as a primitive rock, but apparently his famed visit to the Scheibenberger Hill, which convinced him beyond doubt that basalt is of aqueous origin and laid the ground for the great basalt controversy, also convinced him that basalt belonged to the floetz rather than the primitive period. It should also be pointed out that Werner's definition of the trap rocks was not the same as the current definition. Werner considered all rocks in which hornblende is the predominant or only constituent to be trap rocks, and he classified all such rocks which in the transition and floetz formations grade into ironstone, basalt, and floetz greenstone as primitive trap (Werner, 1971, p. 106-107).

In commenting on the classification of 1802, Werner said that the rocks were listed according to their relative age, in his own words "in so far as this can be expressed to some extent in this manner" (Meuder, 1802, p. III), one above the other in the following order: granite, gneiss, mica schist, and clay slate. He thought that, in general, the other rocks mentioned do not lie above these but occur in conformable beds in gneiss, mica, schist, or clay slate, or even in all three. However, he recognized two porphyry formations, the later one being widely extended, though without much continuity, and unconformably overlying granite, gneiss, mica schist and clay slate. He also thought that there are at least two serpentinite formations, the newer ones unconformably overlying the older formations. Thus the geologic column according to this classification would consist of the following formations in this order: *granite*, including porphyritic granite, *gneiss*, including primitive limestone, the older porphyry, quartzite, primitive trap, and older serpentinite; *mica schist*, including primitive limestone, the older porphyry, primitive trap, primitive gypsum, the older serpentinite, and quartzite; *clay slate*, including various kinds of slate, primitive trap, chert, the older porphyry, the older serpentinite, quartzite, and primitive limestone; and the newer serpentinite and the newer porphyry, including syenite. However, Werner said nothing about the age of the newer serpentinite and the newer porphyry relative to each other (Werner, 1971, p. 106). This is, of course, an idealized concept. Since Werner believed that the contents of the universal ocean had varied from time to time and from place to place, there could be places where one or more of the formations were absent, and thus the order of the various formations would not always be the same. This could be true of the basic four, as well as of the later formations. In fact, in some of his "Notes on Geognosie," Werner pointed out that it would be difficult to prove that granite is everywhere the foundation rock, but that on the basis of geological observations no other foundation rock had ever been reported (Notes on Geognosie: Werner MSS).

The Transition Rocks

Athough Werner's primitive rocks are the only part of his system that can be considered as analogous to the modern concept of the basement complex, one can scarcely leave this subject without at least a brief mention of the transition rocks. Werner had characterized the primitive rocks as being entirely free of fossils. Since this was an integral part of his definition, rocks that contained fossils could not be primitive, and primitive rocks could not contain fossils. But just as modern geologists are now re-evaluating and redefining the Precambrian rocks, on much the same basis, Werner was forced to reevaluate and redefine some of his primitive rocks. Some time after the publication of his *Kurze Klassifikation* in 1786, apparently in the early 1790s, Werner added the transition period to his theory and the transition rocks to his classification. Observations that had been made during that time had revealed that some of the rocks that he had originally classified as primitive contained fossils. Of the transition period, he wrote, "It was during the transition period that the first, chaotic state gradually changed into a state capable of supporting life, and it is in the rocks of this period that we discover the earliest organic remains." (Ospovat, 1960, p. 147). The transition rocks make up only three formations: transition limestone, transition trap, and graywacke.

CONCLUSIONS

As indicated earlier, a comparison of Werner's primitive rocks with the modern concept of the basement complex really is not possible, and even if it were, it would be irrelevant. It is, however, quite proper to ask whether in the light of the knowledge of his day, Werner's theory saved the phenomena, as the Greeks would have said, whether the bases of his classification were sound and his classification historical, and whether his theory was fruitful.

The question of crystallization, which plays an important role in the formation of the primitive rocks, was not resolved for a long time after Werner's death. The theory generally accepted during his life, however, supported the neptunistic interpretation, although Playfair was arguing that granite is "crystallized from a state of simple fluidity, such as, of all known causes, heat alone is able to produce (Playfair, 1802, p. 87). As late as the 1840s, with the aid of chemistry, works were still being written supporting the neptunistic as well as the plutonic theory of the formation of rocks. Microscopic investigation of rocks was not known until about the middle of the 19th century. And the primitive rocks are characterized by the absence not the presence of fossils. Basically, therefore, geologists of the 18th and early 19th centuries had to base their interpretations on the study of the relationships of the various rock masses that make up the complex of primitive rocks and their compositional and textural features. Geology is primarily the history of the earth's crust and therefore has much in common with history. Since the geologist cannot recreate past geological events, he tries to explain how certain rocks came to be and what changes they have undergone on the basis of a theory agreed upon by at least a substantial segment of the geological profession.

Werner's interpretation of the primitive period and his order of succession of the rocks of that period is solidly based on his own investigations of the rocks of central Germany. On the basis of his own investigations and the work of other geologists in other parts of the world, he extended his interpretation of the early part of the history of the earth's crust to the rest of the globe. He read widely and was well acquainted with the geological literature of his day, and he found confirmation of his theory in much of it. An interesting example is Saussure's *Voyages dans les Alpes*, in which Werner and James Hutton both found support for their respective theories. Hutton, however, had to reinterpret Saussure's observations, for Saussure agreed substantially with Werner.

Much has been made of the fact that Werner did not travel widely, and the weaknesses of his system have been attributed to his limited observations. But to my knowledge it has never been established that there is any correlation between the extent of a geologist's personal observations and his creativity, especially his ability to bring forth hypotheses and theories of wide applicability. Hutton, if anything, traveled even less than Werner. He himself admitted that his observations were limited to Scotland, and at the end of a lengthy quotation from Saussure's work, he wrote:

> "It is peculiarly satisfactory to me, and I hope to my readers, to have the observations of so able a philosopher and so diligent a naturalist to offer in confirmation of a theory which had been formed from appearances of the same kind in a country so far distant from those of our author now described, as are the Alps of Savoy from those of Scotland." (Hutton, 195, p. 48).

As far as relationships that Werner established between time units and rock units are concerned, there is nothing wrong with this in principle, considering Werner's basic assumption about the mode and time of formation and that he believed that there are relatively few basic rock-forming substances which are recognizable in the rocks of various periods. Even in modern geology the stratigraphic categories of time and rock units cannot be completely separated from each other. As W.W. Ruby wrote "I can see no basis for any real conflict regarding the relative importance of time-lines and lithologic units. Either one of these two types of information is meaningless unless it is supplemented by the other." (American Commission on Stratigraphic Nomenclature, 1948).

Above all, Werner's concept of the primitive period and his interpretation of that period are historical. The order of succession is important, but not to the exclusion of any deviation. Gradual changes in the formation and development of rocks are recognized, but not to the exclusion of occasional revolutionary changes.

Finally, Werner's concept of the primitive period was extremely fruitful. The geological work that was spurred on by his theory substantiated many of his ideas, but it also contradicted others. As far as the origin and nature of the primitive period was concerned, there is still much that needs to be done. The concentration on the study of the transition and floetz rocks – that is, the fossil-containing rocks – has for a long time caused a neglect of the fossil-free rocks. If 19th century geologists had not reacted so violently to Werner's teachings, our knowledge of the earliest rocks might be greater than it is today.

ACKNOWLEDGEMENTS

I wish to acknowledge the support that I received from the Oklahoma State University which enabled me to do the research upon which this paper is based. I would also like to acknowledge the co-operation of the Bergakademie Freiberg in giving me unrestricted access to the Werner manuscripts in its possession.

Concerning the present-day location of Werner manuscripts, most of them are in the Bergakademie Freiberg Library. These include Werner's own notes, as well as notes taken by his students during his lectures, letters, and numerous other items. The Freiburg Bureau of Mines also has a large collection of Werner manuscripts, relative to his work in the Bureau of Mines of Saxony.

Outside of Freiberg there is a substantial collection of copies of Werner manuscripts in the History of Science Collections in the University of Oklahoma, in the Oklahoma State University Library, and in my own possession.

REFERENCES

American Commission on Stratigraphic Nomenclature, 1948, Discussion of Nature and Classes of Stratigraphic Units: excerpts from a letter written by W.W. Ruby: American Association of Petroleum Geologists Bulletin, v. 32, pt. 1, p. 376-381.

American Geological Institute Glossary of Geology and Related Sciences, 1960, 2nd Edition: Washington 325 p.

Hutton, James, 1795, Theory of the Earth with Proofs and Illustrations, v. II: Edinburgh, Cadell and Davies, 567 p.

Meuder, H.W., 1802, Herrn Robert Jameson's mineralogische Reisen durch Schottland und die Schottischen Inseln. Aus dem Englischen übersetzt und von einem Auszuge aus Herrn Bergrath Werner's Geognosie die Lehre von den Gebirgsarten betreffend, als Einleitung begleitet: Leipzig, Siegfried Lebrecht Crusius, 256 p.

Ospovat, A.M., 1960, Abraham Gottlob Werner and his influence on Mineralogy and Geology: University of Oklahoma, Doctoral Dissertation: Ann Arbor, University Microfilms, Inc., 259 p.

Ospovat, Dov, 1971, Hutton and his critics: the failure of a theory: Harvard University, Unpublished Paper.

Playfair, John, 1802, Illustrations of the Huttonian Theory of the Earth: Edinburgh, Printed for Cadell and Davies, London and William Creech, Edinburgh, 528 p.

Taylor, K.L., 1969, Nicholas Desmarest and geology in the eighteenth century: *in* Schneer, C.J., ed., Toward a History of Geology, Cambridge, Mass., Massachusetts Institute of Technology Press.

Werner, A.G., 1786, Kurze Klassifikation und Beschreibung der verschiedenen Gebirgsarten: Abhandlungen der Böhmischen Gesellschaft der Wissenschaften, II.

Werner, A.G., 1971, Short Classification and Description of the Various Rocks: Translated with an Introduction and Notes by A.M. Ospovat: New York, Hafner Publishing Company, 194 p.

History of Concepts in Precambrian Geology, edited by W.O. Kupsch and W.A.S. Sarjeant, Geological Association of Canada Special Paper 19, 1979

THE PRIMAEVAL EARTH AS CONCEIVED BY SWEDISH 18th CENTURY NATURALISTS

Gerhard Regnéll
Department of Geology, University of Lund, Sweden

ABSTRACT

Speculations on the early state of the Earth by the Swedish 18th century naturalists U. Hiärne, C. Polhem, E. Swedenborg, C. Linnaeus, D. Tilas, J.G. Wallerius, and T. Bergman are briefly commented upon. If viewed in their contemporary European context, these authors may be characterized largely as exponents of the prevailing ideas of their times in this particular field. Bergman alone stands out as an independent thinker who exercised a certain influence upon the development of geology.

RESUME

Un bref commentaire est presenté sur les spéculations faites au sujet de l'état primordial de la terre par les naturalistes suédois du dix-huitième siècle: U. Hiärne, C. Polhem, E. Swedenborg, C. Linneaus, D. Tilas, J.G. Wallerius, et T. Bergman. S'ils sont considérés dans leur contexte contemporain-européen, ces auteurs peuvent être caracterisés comme étant des interprètes des idées dominantes de leur temps dans ce domaine particulier. Seul Bergman se distingue comme penseur indépendant qui exerca une certaine influence sur le développement de la géologie.

EIGHTEENTH CENTURY CONCEPTS

Understanding the nature of the early phases in the development of the Earth is a major task which has never ceased to challenge earth scientists (e.g., Windley, 1976). As is well known, ideas on the origin and first stages of the Earth were for a long time dominated by the tales in the Holy Bible. Only gradually did concepts based on the actual observation of geological phenomena and processes win recognition. This change in attitude occurred mainly during the 18th century.

The early history of geology in Sweden has not received much attention internationally. For example, Linnaeus is the only one of the figures to be discussed briefly below whose name is even mentioned in such a general work – admittedly one focussed on Great Britain – as Gillespie's *Genesis and Geology* (1951, 1959).

The situation in learned circles during the 18th century was analysed very skilfully and placed in its European context by T. Frängsmyr (1969) in his thesis *Geologi och skapelsetro (Geology and the Doctrine of the Creation).* It is to be regretted that this important and comprehensive work on the early history of geology in Sweden has been published only in Swedish, true with an English summary but one of little more than seven pages. This work is a rich source of information on the subject to be dealt with below; had it been more readily available to the international public, a simple reference to Frängsmyr would have been adequate in many instances.

The general philosophical background against which the later developments took place was summarized as follows (Frängsmyr, 1969, p. 12-13):

"In the middle of the 17th century a mechanistic universe began to take shape. The Aristotelian eternal world was put under debate and was replaced – though certainly not without a bitter fight – by Cartesianism. According to the new philosophy, the Universe was a big machine operated by rigid natural laws. Although this doctrine gave no place for a personal God who would interfere in the machinery, the Cartesianists did not wish to offend the Christian authorities. On the contrary, they were anxious to emphasize that the complicated machinery of the Universe was set in operation by divine intervention. God was the initiator and originator but, after having been started by him, all processes were governed by natural laws. According to Boyle, ever an apologist, the Creation meant that God set the particles moving. If he had continued acting beyond this, God would have denied the laws set by himself. However, theological ideas would prove in the length of time not compatible with a strictly mechanical universe." [Author's translation]

Direct observations made in the field led to the conclusion that the surface of the Earth had undergone progressive changes that required far longer time than the span of 6000 years allotted to the Earth, according to the Scriptures. However, certain authors still adhered as a matter of course to traditional lines of thought, notably J.G. Wallerius as late as 1776.

It is brought out by Frängsmyr's investigations that Urban Hiärne (1641-1724), a richly faceted personality who was primarily chemist and physician (e.g., Lindroth, 1952), should be regarded as a pioneer in Swedish geology, a fact which had previously been largely overlooked. Hiärne was the first one in Sweden to give a tolerably comprehensive description and explanation of the origin of the Earth and of the processes which are incessantly changing its face; he can therefore be counted among the early champions of actualism. Hiärne's picture of the Earth is very much in agreement with the model given by Athanasius Kircher in his *Mundus subterraneus* (1664). He speculated at length about the origin of mountains. He was not of the opinion that they were formed during the Deluge; rather he imagined that the primary mountains came into existence by the act of Creation.

There is another figure whose place in the history of geology has been recognized only lately, due in part to the fact that his geological speculations were not published but were instead set forth in letters to learned correspondents. I am referring to Christopher Polhem (1661-1751), who was a renowned mechanical engineer but whose name may be completely unknown to most historians of geology. Frängsmyr (1969) has given a penetrating analysis of Polhem as a geological thinker. Polhem ventured to regard the Book of Genesis as a symbolic representation of the origin of the Earth. He advanced the opinion that all the changes which are so evident would require a span of several hundred thousands of years, a notion which is highly remarkable at the time it was expressed. In a letter of 1710 (referred to by Frängsmyr, 1969, p. 119), Polhem commented upon the structure and size of the Earth. He assumed that at the centre there is a cavity, the diameter of which he computed as 10,000 km and which he believed to be filled with fire. This cavity he thought to be surrounded by a shell of pumice with a thickness of 500 km. Equally thick, he believed, was the outermost shell which consists of different rocks, water and soil. Minor cavities lodged in this shell contain fire, air or water. These ideas are hardly original but rather in line with hypotheses current at his time.

If the name of Polhem does not mean very much to people outside Sweden, the reverse is true of that of Emanuel Swedenborg (1688-1772), though for reasons other than for his geological writings. These are, however, important enough to have been discussed at length in several papers (Nathorst, 1906, 1908; Regnéll, 1949; Frängsmyr, 1969). In the field of natural science Swedenborg was in many respects an original thinker. As a cosmologist he was a Cartesianist yet, at the same time, in his great cosmogonic work the *Principia rerum naturalium* (1734), he adhered to the Christian tale of creation. As a geologist Swedenborg may be ranged into the tradition from Hiärne and Polhem. His belief in the Scriptures was counter-balanced by his rational mind and the witness of his eyes. Observations in the field led him to the conviction that changes of the relief of the Earth's surface were due not only to the Mosaic Flood but also to forces acting more recently. To Swedenborg there was clear evidence that Sweden had in the past been covered by the sea, a notion that added fuel to the lengthy discussion of the 'receding of the waters' vs. 'the elevation of the Swedish wall' carried on during the 18th century.

Much can be said about Linnaeus (1707-78) as a geologist; but much has already been said with profound insight by Frängsmyr (1969) and, prior to him, by Nathorst (1907, 1909). In several different contexts, Linnaeus advocated the opinion that the nature of the primaeval surface of the Earth remained unknown because it was obscured by the changes which had taken place over the length of time – and to him geological time did not fit into the compass of the biblical chronology. To illustrate his view of this matter I select a passage from the famous preface to *Museum Regis Adolphi Friderici* (1754), in which Linnaeus commented upon the importance of studying natural history and on the value of natural history collections. The original edition has parallel texts in Latin and Swedish; the preface, or part of it, was published separately somewhat later in German, French and English. The following quotation is from an anthology edited by A.H. Uggla (1939, p. 48):

> "*Varia saxorum* Strata *sub altissimis saepe alpium jugis latitantia, antiquissima sunt monumenta, quae prisci orbis vicissitudines ante oculos ponunt, easque, dum reliqua silent omnia, nobis in aurem loquntur.*
>
> *Innumera* Petrificata *peregrinorum atque a nemine unquam mortalium nostri aevi visorum Animalium, quae his lapidum acervis, sub celsissimus saepe montibus obruta jacent, unica sunt prisci orbis fragmenta, quae supra omnem Historicorum memoriam assurgunt.*
>
> *Horum aliorumque lapidum immensa congeries tanto cumulo tellurem operit, ut nemo hominum hactenus eum perrumpere, atque ad Terram primogeniam penetrare potuerit.*"

Or, in English:

"The several rock strata, in many cases beneath the highest mountains, are the oldest antiquities that display the changes of the former world, about which they alone are whispering while the rest is silent.

The countless fossils of strange and unknown animals – not seen living in present times by anybody – which are buried in these beds, frequently beneath the highest mountains, are the only ones to bear witness of the inhabitants of our former Earth in ages further back than any historian can follow.

The immeasureable quantity of these and all other rocks cover the Earth to such a height that nobody has yet been able to penetrate into the primaeval Earth."

A similar opinion was expressed in one of Linnaeus' autobiographies (cf. Frängsmyr, 1969, p. 222).

Most certainly Linnaeus had in mind the Lower Palaeozoic sequence so well exposed in Mt. Kinnekulle, in the Swedish province of Västergötland, which – erroneously of course – he took as a standard for layered sequences in general. The dolerite capping the bedded rocks in Kinnekulle was designated by Linnaeus as 'grey rock' (occasionally as 'sandstone'); strangely enough it was thought to be indurated ground moraine. Traditionally, the rocks termed 'grey rock' or 'ground rock' were interpreted as *creati in initio mundi* (A. Drossander, at the end of the 17th century, according to von Hofsten (1945, p. 36): see also below, under Tilas, 1742). Even toward the end of the 18th century a definition like the following one is met with (Retzius, 1895, p. 252):

"*Grey rock*, is among the primaeval rocks, the origin of which is far beyond our chronology and which in fact forms the structure referred to as the skeleton of the Earth in immense mountain ranges both beneath and upon the surface." [Author's translation]

Linnaeus (1747, p. 35) found it proper to state that "not all grey stones were created in the beginning but, like other rocks they are also *filii temporis*, children of time." [Transl.] The usage in the older literature of terms such as 'grey stone' and 'trap' is far from consistent (cf. Frängsmyr, 1969, p. 36 *et seq.*).

In order to give a general idea of Linnaeus' conception of the state of the primaeval Earth I cannot do better than to quote a passage from Torbern Bergman (1774). Bergman observed (p. 278, foot-note S) that "Mr Archiater [i.e., Linnaeus, after ennoblement von Linné] has been good enough personally to look over this outline of his thoughts. They were presented originally in a speech 1743, *De Telluris habitabilis incremento*, and were expounded further in his itineraries; in the system [i.e., *Systema Naturae*]; in *Amoenitates* [Linnaeus' collected dissertations]; and in other works" [Author's translation].

Bergman (1774, p. 274-278) writes as follows:

"The world-famous Archiater and Knight v. Linné figures the original dry land as an island under the line on which there was a high mountain and, accordingly, all kinds of climate, from the hottest to the coldest. . . . Where the Earth is concerned, sand is supposed to crystallize from sea-water, for when it rains the water becomes white and turbid, precipitation takes place and the particles grow together by crystallization. Clay is called *sedimentum maris viscidum*. Sea-water is always less clear than spring-water and, when not agitated, precipitates such slime as is found on all poles, plants, and stones in the sea. On drying, this is nothing but clay which because of *viscidum* does not exhibit a constant crystal pattern under a magnifier. Mould is formed by plants and is nothing but their skeletons, but lime is formed by animals.

Sand crystallizes to sandstone by influence of percolating water; but after many hundred [years] it weathers when exposed and becomes silt. If sand is mixed up with other soil particles, preferably some ferruginous matter which is the most common on Earth, it grows together into grey stone when all pores which contain water are filled by quartz and spar crystals dependent on the nature of the soil matter dissolved by the water. Precious stones are crystals that are different in shape and composition. The shape is due to the salts, and the composition to the nature of the soil particles dissolved by the water. Clay becomes soapstone, hornfels, and mica... Firmly pressed mould turns in the end into shale which as a corollary of its origin burns when heated, and often exhibits imprints of animals and plants that throve in the mud which is nothing but coarse mould. Calcareous matter which has been attacked by the waves and silted up is called chalk, from which flint is generated in fissures under influence of water. Lime mixed with clay or shale causes these to change into a calcareous rock so that petrified shells are not the progeny of limestone mountains but rather give birth to such mountains.

Rocks are supposed to have originated in the following way. In places where sargasso exists in the open sea the surface becomes calm and smooth. Here an argillaceous sediment is deposited on the sand-covered bottom upon which mud of decomposed sea-plants is often found. Beneath the sargasso there lives a great quantity of peculiarly strange animals and sea-creatures. These die, become buried in the clay because of which the bottom rises gradually so that it approaches the surface of the water. Then the waves become forceful, expel the sargasso and, therefore, throw up sea-wrack, sea-weed, etc. and, finally, a top-layer of sand. The sand grows into sandstone and the mud into shale. The clay, by the intermingled shells, is converted into a bed of limestone replete with petrified animals originating from an abyssal depth. The decayed sea-weed and sea-wrack form a shaly layer, and on top of it there is a grey rock which has been formed by shingle growing together, in the same way as shingle by gravel and gravel by driftsand when all interstices had become filled by quartz and sparcrystals."

On resigning from the presidency in the Royal Swedish Academy of Science in 1742, Daniel Tilas (1712-72, Member of the Council of Mines, King-at-arms) presented a speech, the title of which might be given in English as *The history of the realm of rocks*. This 32-page paper remains little known, for it is not available in any international language. In fact Tilas' exposition is a rather interesting piece of work because it tries to combine faithfulness to the Scriptures with the critical attitude of an experienced mining engineer.

I shall permit myself to quote a certain number of paragraphs from Tilas (1742), for I am of the opinion that a direct quotation gives the flavour of the old writers which is lost if one just reports upon the contents. In this particular case

I can only hope that the quotation will achieve the effect desired, for what follows is, of course, a translation from the original Swedish:

"Although it has been, and still is, much disputed whether the origin of the Ground Rocks should be referred to the period of creation or to the destruction that occurred through the Flood, I have arrived at the conviction that the origin of a majority of the Ground Rocks must be traced back to the first hour of creation; however, quite a few of less importance and distinction I hold to be relics and memorials of the destruction that took place . . .

The first benefit afforded by the Ground Rocks was that, by their being solid and coherent, they were to give strength and firmness to the entire Globe, in no way differently from the bones of a living body which, without such a frame, could not possibly support its own weight . . . Finally, the Ground Rocks were to serve as a workshop for all Metals and Minerals where they could gradually, and by and by, grow and ripen; in the same way as the Soil is the womb of Grass, Plants, and Trees, so the Ground Rocks are the womb of Minerals.

In consequence, I imagine that in its original shape the Earth was tidy and fair enough, everything being magnificently and neatly arranged by the hands of the Supreme Master and not distorted and covered by horrible Screes and Fanglomerates as at the present day.

I cannot possibly agree with those who believe that the Globe was enveloped in a Rock-crust, so to say, which for some reason broke up by the sinking in of some parts and the rising of others to form the Ranges and Hills that we now see. Those who hold this opinion I would ask to take the trouble to visit some Ridges, not only in the Lowlands but also in the High Mountains, and to watch closely their shape, nature and conditions. They will find, apart from what I have said above, how the matter hard as stone in the very Ground-Rocks must once have been fluid; for everything that is called Saxum or Grey Rock, along with a couple of Limestone-quarries and even Sandstone-Mountains, clearly exhibits Ripples, Veins, and Stripes exactly as in a Mud-puddle in which a number of lighter and darker objects have been mixed . . .

Let us now investigate, Gentlemen, what kinds of Mountains and Rocks existed in the beautiful and splendid state of the Globe just indicated. I admit that, in examining this particular matter, the darkness of ancient times is most profound. However, we may rely on guess-work so far. To those times I refer the following kinds of Mountains and Rocks:

1) All Mountains which are composed of so-called *Saxum* or *Grey Rock* and contain the three particles *Felspar, Quartz* and *Mica* in manifold mixtures and varieties.

2) All those Limestone mountains, Marble and Alabaster quarries, the matter of which is made of Calcareous Spar . . .

3) A great many Sandstone mountains . . .

4) Porphyry-quarries . . .

5) All vertical and horizontal Hornfels . . .

6) All Mountains which contain a quantity of Horn-rock, Talcum, and Soapstone . . .

7) All Swedish Iron-ore-mountains, and Solid Rock-ores, which are in many

places so big and extensive that they could almost be considered the common Ground-Rock of the particular place.

All those kinds of rocks here enumerated I consider to be such as were formed in the first beginning of the world and which I now wish to designate as *primaeval*, and further such rocks as have subsequently not shown any obvious growth or increase, in particular the kind of *Saxum* first mentioned. I believe that many additional kinds of rocks were formed at the first Creation, but, then, it is doubtful whether only their matter was created and they, so to speak, matured in the course of time, or whether they were generated in a perfect state right from the beginning; . . .

My principal reasons not only for assigning the rocks just enumerated to the oldest time and holding them to be primaeval, but also for considering them to belong with those rocks which do not show any appreciable growth, I elicit from the fact that, although I have been on the look-out for it, I have not yet been able to establish that any rock of this kind has been formed anew since the time of the Flood but, on the contrary, I have seen it evidenced a thousand-fold times that they are broken down unceasingly every day."

Tilas' paper contains many other interesting statements but I cannot possibly go into details. I must pass rapidly over the work of J.G. Wallerius (1709-85), whose name I have mentioned above. Wallerius was a renowned mineralogist and agricultural chemist and, besides, an odd personality as it seems. His book called *Tankar om verldenes i synnerhet jordenes, danande och ändring (Thoughts on the creation and changes of the world, particularly of the Earth)*, 1776, had a wide circulation since it was translated into Latin, German, French and Italian – perhaps more than it merited!

"He maintained that the earth and all that originally existed on it had developed out of water; he tried to verify this by his own experiments as well as by citing old authorities. God created first heaven and earth, and this meant to Wallerius that two elements had been created, light and water; out of light did He make the sun and the stars on the fourth day, out of water the planet of earth. Mountains did He create on the third day, but the face of the earth had assumed its present form as a result of the Deluge." (Frängsmyr, 1969, p. 350).

Quite honestly, the book did little to promote science, for Wallerius was an apologist of the Biblical tales of Creation (e.g., Zenzén, 1952, p. 97). Wallerius' treatise is rather an echo of ideas that belonged in the past.

Torbern Bergman (1735-84) finally merits full comment, considering his position as a truly outstanding member of the scientific society not only of Sweden but of all Europe. However, since Hedberg (1969 a,b) has published a study of Bergman as a geologist I can restrict myself to making only a few remarks, the more so as Frängsmyr (1969) has done full justice to the achievements of that brilliant scientist.

When Bergman died at the early age of 49, a life of restless scientific activity in many fields, but first and foremost in chemistry, was ended prematurely. His contributions to geology are found in *Physisk beskrifning öfver jordklotet (A physical description of the Globe)*, the first edition of which appeared 1766 and the second, revised and much enlarged, in 1773-74. The work was made available to the international public by translations into most European languages; however, it is remarkable that no English edition was ever published, as far as known (cf. Hedberg, 1969 a, p. 20). Large portions of the book are purely descriptive. Bergman's acumen is apparent in the first place in the section treating with the formation of the Earth, how its surface came into being and how it is still acted upon by various agencies. To put it briefly, Bergman set the Scriptures aside and demonstrated how everything has been brought about gradually by the operation of natural laws. He introduced a genetic classification of the mountains and gave an account of the formation of bedded sequences.

Bergman recognized three classes of rocks, namely in order from oldest to youngest: (1) *Uraldrige* or "primaeval" (2) *Flolägrige* or "bedded strata" (3) *Hopvräkte* or "thrown together". The primaeval rocks were characterized (Bergman, 1774, p. 292-293) as follows:

> "The first kind makes up the interior and the very body of major mountain ranges, which extend in connexion over the Earth and give it strength like the bones of the body of animals. They widen out downwards and probably meet at depths and constitute a firm lump, or maybe a coherent shell . . . The mountain roots probably unite, because they extend gradually at depth and because they have never been penetrated into loose ground; in addition, the laws of equilibrium postulate a reasonably homogeneous core which to judge from the behaviour of the compasss needle must be magnetic. But let us return to rocks of the first kind. They differ from all others by the fact that they rise more gently and with no terraces; further, in that they are composed generally throughout of one material. Although by changes of colour or grain some layering is shown of different parallel, vertical or strongly dipping beds they do seldom fracture along the seeming bedding planes but fracture equally readily in all other directions. The grain is frequently characterized by angular or scaly particles. They are cut by veins . . . Finally, traces of petrified animals or plants are never found in these mountains. They are called primaeval because they are probably a little younger than the Earth and have at least been in existence prior to the appearance of animals and plants. They are also classed as *Gängberg* because they are cut by dykes".

More or less the same section was cited by Hedberg (1969, p. 34-35).

Bergman, like any other scientist, despite his independence of thought, nevertheless built his scientific edifice on foundations laid by numerous predecessors. The relation between Bergman and the famous German mineralogist and chemist J.G. Lehmann has been commented upon by Frängsmyr (1969, p. 303 *et seq.*). There is no doubt Bergman's treatment of a number of questions recalls ideas expounded by Lehmann in his classic *Versuch einer Geschichte von Flötz-Gebürgen* (1756). The formation of mountains and bedded rocks was explained in much the same way by both authors; however, there are significant differences. While Lehmann commented very vaguely on the chemical precipitation of matters dissolved in the water of the primaeval sea, Bergman's account of crystallization is, in contrast, thorough, lucid and to the point. Furthermore, Lehmann was of the opinion that bedded rocks had been deposited in the Flood; this was explicitly denied by Bergman. However, in the first edition of Bergman's book there is no reference to Lehmann, although otherwise Bergman presented a very extensive list of references. This must certainly mean that Lehmann's treatise was not known to Bergman when he prepared the original edition of *A Physical Description of the Globe.*

Both Frängsmyr (1969, p. 341) and Hedberg (1969 a, p. 46; 1969 b, p. 190) have observed that Bergman merits to be ranked as a pioneer in the establishment of the principles of Earth History, whose fundamental importance was to be recognized subsequently.

The state of geology in 18th century Sweden was given in a nut-shell by Frängsmyr (1969, p. 351-352):

"In the period between Hiärne and Bergman hardly any significant progress was made in terms of concrete geological questions, yet some noteworthy forward steps were taken. The actualistic perspective had been noticed by Hiärne, but it was impossible to break the time barrier established by orthodox commentators. The decreasing importance of the Bible as a natural history text is exemplified by the fact that none of the later geologists, except Wallerius, seriously considered the Deluge as a geological agent. With Bergman a new era can be discovered. His methodical distinction bore witness to a new attitude towards the delicate question of the history of the earth: empirical and experimental investigation, free from biblical influences. This was a new ground upon which modern geology would build."

REFERENCES

Bergman, T., 1766, Physisk beskrifning öfver jord-klotet, pa Cosmographiska Sällskapets vägnar författad: (-Werlds-Beskrifning, II), Uppsala, 432p.

————, 1773-74, Physisk beskrifning ofver jord-klotet, pa Cosmographiska Sällskapets vägnar författad: (-Werlds-Beskrifning, II), Uppsala, J. Edman, 2nd Edition: Vol. 1 (1773), 470p.; Vol. 2 (1774), 536p.

Frängsmyr, T., 1969, Geologi och skapelsetro. Föreställningar om jordens historia frän Hnärne till Bergman: Lychnos-Bibliotek, v. 26: Almqvist and Wiksell, Uppsala, 379p.

Gillispie, C.C., 1951, Genesis and geology, a study in the relations of scientific thought, natural theology and social opinion in Great Britain, 1790-1850: Harvard Historical Studies, v. 58, Cambridge, Mass., Harvard University Press, 315p. (2nd ed. 1959, New York, Harper Torchbooks, Harper and Brothers, 306p.)

Hedberg, H.D., 1969 a, Influence of Torbern Bergman (1735-1784) on stratigraphy: Stockholm Contributions in Geology 20, Stockholm, p. 19-47.

————, 1969 b, The influence of Torbern Bergman (1735-1784) on stratigraphy: a resumé: *in* Schneer, C.J., ed., Toward a History of Geology: Cambridge, Mass. and London, Massachusetts Institute of Technology Press, p. 186-191.

Hofsten, N. von, 1945, Inbjudningsskrift till ahörande av de offentliga föreläsningar med vilka professorn i latinska spräket och litteraturen Josef Svennung och professorn i geologi, särskilt petrografi och mineralogi Erik Norin tillträda sina ämbeten: Uppsala, Almqvist and Wiksell, 59p.

Lindroth, S., 1952, Urban Hiärne 1641-1724: *in* Lindroth, S. ed., Swedish Men of Science: Stockholm, The Swedish Institute (Almqvist and Wiksell), p. 42-49.

Linnaeus, C., 1747, Wästgöta-resa, pa riksens högloflige ständers befallning förrättad ar 1746: Med anmärkningar uti oeconomien, natural-historien, antiquiteter etc., Stockholm, L. Salvius, 284p.

————, 1754, Museum S:ae R:ae M:tis Adolphi Friderici etc. Holmiae, Stockholm.

Nathorst, A.G., 1906, Emanuel Swedenborg sasom geolog: Geologiska Föreningens i Stockholm Förhandlingar 28, Stockholm, p. 357-400.

——————, 1907, Carl von Linné sasom geolog: *in* Carl von Linnés betydelse sasom naturforskare och läkare. Skildringar utgifna af Kungl. Vetenskapsakademien i anledning af tvahundraarsdagen af Linnés fodelse: Uppsala, p. 1-80.

——————, 1908, Emanuel Swedenborg as a geologist: *in* Stroh, A.H., ed., Emanuel Swedenborg as a scientist, 1, Stockholm. sect. 1., p. 1-47 (special pagination in separate print).

——————, 1909, Carl von Linné als Geolog: *in* Carl von Linnés Bedeutung als Naturforscher und Arzt: Jena, Gustav Fischer, p. 1-80.

Regnéll, G., 1949, On the position of palaeontology and historical geology in Sweden before 1800: Arkiv för Mineralogi och Geologi (K.Sv.Vet.-akad) 1, Stockholm, p. 1-64.

Retzius, A.J., 1975, Försök til mineral-rikets upställning. I en handbok at nyttja vid föreläsningar: Lund, Berlingska Boktr., 374p.

Tilas, D., 1742, Stenrikets historia: Utford i det vid praesidii aflaggande haldne talet inför Kongl. Swenska Wetenskaps-Academien den 14.april 1742: Stockholm, L.L. Grefing, 32p.

Uggla, A.H., ed., 1939, Carl von Linné. Fyra skrifter, illustrerade av Harald Sallberg: Stockholm, Lindfors, 160p. (An offset edition was published 1964.)

Wallerius, J.G., 1776, Tankar om verldenes, i synnerhet jordenes, danande och ändring. Stockholm, Henr. Fougt, 183p.

Windley, B.F., ed., 1976, The Early History of the Earth: London, New York, Wiley-Interscience Publications, 630p.

Zenzén, N., 1952, Johan Gottschalk Wallerius 1709-1785. Axel Frederik Cronstedt 1722-1765: *in* Lindroth, S., ed., Swedish Men of Science: Stockholm, The Swedish Institute (Almqvist and Wiksell), p. 92-104.

History of Concepts in Precambrian Geology, edited by W.O. Kupsch and W.A.S. Sarjeant, Geological Association of Canada Special Paper 19, 1979

URICONIAN AND LONGMYNDIAN: A HISTORY OF THE STUDY OF THE PRECAMBRIAN ROCKS OF THE WELSH BORDERLAND

W.A.S. Sarjeant[1] and A.P. Harvey[2]
[1]Department of Geological Sciences, University of Saskatchewan, Saskatoon, Saskatchewan S7N 0W0
[2]Librarian, Department of Palaeontology, British Museum of Natural History, London, England

ABSTRACT

After a brief review of the overall geology of Shropshire, the history of the earliest geological observations in the County is detailed. Particular stress is given to the pioneer work of Robert Townson, not mentioned in earlier accounts of the development of Shropshire geology. The work of Charles Callaway in establishing the Precambrian age of this sequence and in naming its major divisions ('Longmyndian', 'Uriconian', etc.) is detailed and his disputes with J.F. Blake about the sequence in the Long Mynd are discussed. An account is thereafter given of the widening of knowledge resulting from work by Lapworth, Watts, Cobbold and Whittard and the geologists of H.M. Geological Survey, between 1900 and 1938, and of the recognition of Precambrian inliers in Radnorshire and Herefordshire. The controversial conclusions of James (1956), concerning the sequence and structure of the southern Long Mynd, and of Taylor (1958), concerning the sedimentary environment of the Longmyndian, are outlined and contrasted with the alternative ideas put forward by Survey geologists in 1968. Radiometric ages assigned to these rocks appear to contradict palaeomagnetic evidence and to refer to tectonic events rather than to the formation of the rocks themselves. After a brief account of the Uriconian outside Shropshire, the many remaining problems presented by the Shropshire Precambrian are stressed.

RESUME

Après une brève revue de l'ensemble de la géologie du Shropshire, nous présentons un historique détaillé des premières observations géologiques sur le Comté. Une attention particulière est accordée aux travaux pionniers de Robert Townson, qui n'est pas mentionné dans les premiers rapports sur le développement de la géologie du Shropshire. Le travail de Charles Callaway établissant l'âge précambrien de cette séquence et nommant ses divisions majeures ('Longmyndien', 'Uriconien', etc.) est examiné en détail, et ses querelles avec J.F. Blake au sujet de la séquence dans le Long Mynd sont discutées. Nous présentons ensuite un compte rendu de l'étendue des connaissances résultant des travaux de Lapworth, Watts, Cobbold et Whittard et des géologues du Service Géologique de Sa Majesté entre 1900 et 1938, et de la reconnaissance de venêtres tectoniques précambriennes dans le Radnorshire et le Herefordshire. Nous donnons les grandes lignes des conclusions sujettes à controverse de James (1956) concernant la séquence et la structure du Long Mynd méridional, et de Taylor (1958) concernant l'environnement sédimentaire du Longmyndien, et nous montrons les différences avec les idées proposées par les géologues du Service en 1968. Les âges radiométriques assignés à ces roches semblent contradictoires aux évidences paléomagnétiques et référeraient plus aux événements tectoniques qu'à la formation des roches. Après un bref compte rendu de l'Uriconien à l'extérieur du Shropshire, nous faisons ressortir les nombreux problèmes qui subsistent sur le Précambrien du Shropshire.

INTRODUCTION

Of all the inland counties of England, Shropshire offers the greatest variety of natural scenery. Through the successive effects of at least four main phases of earth movements – one or more phases in the Precambrian, the Taconian (Ordovician), the Caledonian (Silurian-Devonian) and the Armorican (Carboniferous-Permian) – rocks ranging in age from Precambrian to Lower Jurassic have been brought to the surface in a series of structures extensively modified by movements on major fault-lines. Quaternary deposits – glacial, glaciofluvial, or fluvial – blanket many valleys to varying depths and cover much of the lowland area north of Shrewsbury. The combination of geological structure and variable lithology, with its concomitant effects on erosion and the growth of plants, not only gives variety to the landscape of the county but has also been the dominant influence throughout human history, determining the pattern of settlement, agriculture and pastoral farming and the movement of raiders and armies. Even today, the geography of Shropshire is still controlled by its geology to a degree unusual in industrialised, over-populated England. This is true even of the region around Coalbrookdale, the cradle of the Industrial Revolution; and only the recent expansion of Wellington and the development of Telford New Town are at last beginning to obliterate the old patterns.

The Caledonian orogenic phase has been the most striking in its effects on the Shropshire landscape, for it has imposed a strong NE-SW trend, as a result of compression from west and north against the Precambrian basement. Shropshire is, as a consequence, divisible into three broad structural units. In the north is a broad syncline (the Prees Syncline), modified by two NE-SW faults (the Hodnet and Ercall Mill Faults) and bounded to the southeast by the great Church Stretton Fault and its northward continuation, the Brockton Fault. This is essentially a lowland region, underlain by younger rocks (Carboniferous to Jurassic).

The southern part of the county is dominantly characterised by more ancient rocks (Precambrian and Lower Palaeozoic), the outcrops of younger rocks being few. It comprises two broad divisions; the Long Mynd-Wrekin anticlinal uplift on the west, separated by the Church Stretton fault axis from the Clee Hills Basin (a part of the East Shropshire-West Staffordshire syncline) to the east. However, the structure of both these regions is complicated by further faults and folds. The Long Mynd-Wrekin area consists, in its northern part, of a succession of synclines and anticlines, each with a NE-SW axis: the Long Mountain Syncline, the Shelve Anticline, the Ritton Castle Syncline and the fault-bounded, anticlinal Precambrian rocks of the Longmyndian Massif, forming the Long Mynd and subsidiary hills to the north. To the south of these structures, the associated syncline of the Clun Forest

THE PRE-CAMBRIAN ROCKS OF THE SHROPSHIRE AREA

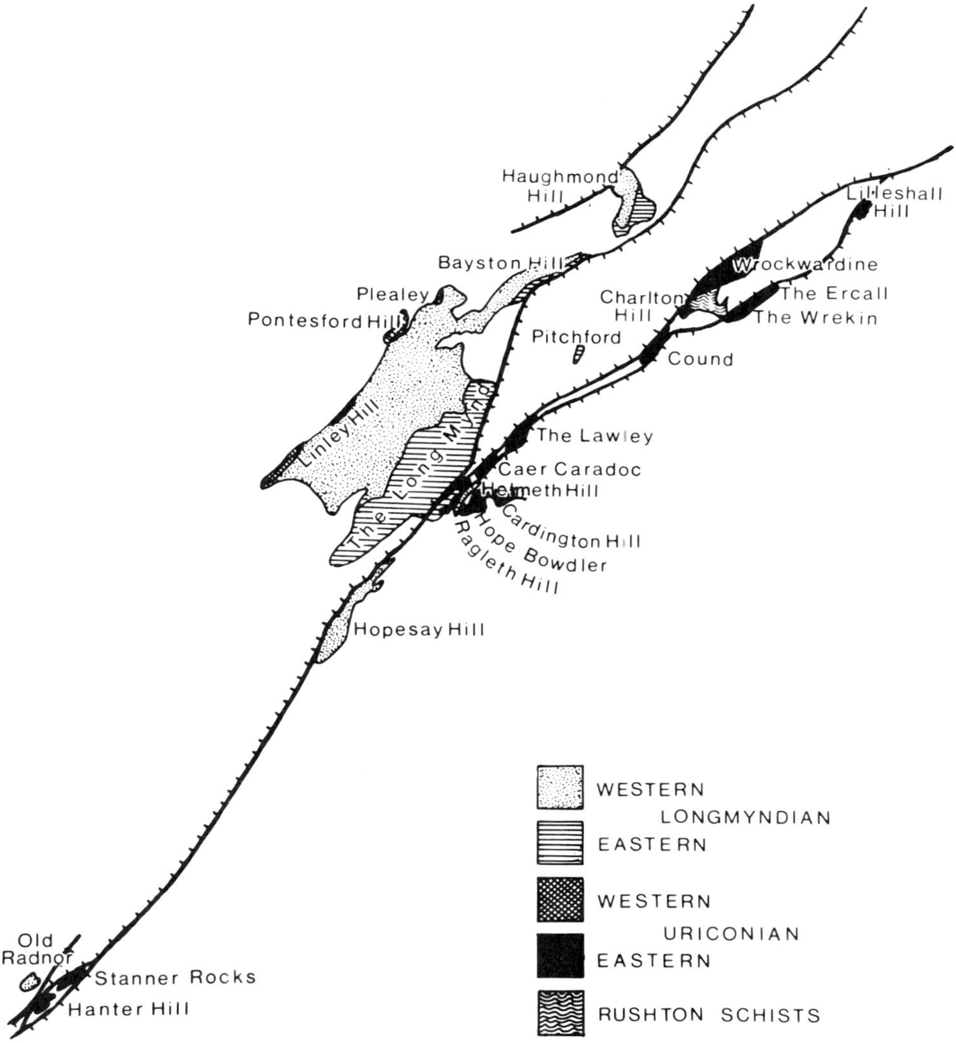

Figure 1. Sketch map of the Precambrian rocks of Shropshire and Radnorshire, with the major fault pattern (greatly simplified) indicated, ticked on the downthrow side. (The rocks of Stanner Rocks and Hanter Hill do not necessarily correlate with the Eastern, rather than the Western, Uriconian).

Basin forms the southwesternmost part of the county.

The Clee Hills Basin, which occupies the east and southeast of the county, is bounded on its western margin by the irregular line of Precambrian hills brought to the surface by the Church Stretton and Brockton Faults – the Stretton Hills, The Wrekin and its neighbours and, ultimately, the farthest-flung outlier, Lilleshall Hill, near the Staffordshire border. Its structure is complicated by the Ludlow Anticline, whose axis swings in an arc from south to west (separating the Brown Clee Syncline from the Titterstone Clee Basin) and by faulting on a general east-west line.

Precambrian rocks dominate the landscape on either side of the main south-north road through Shropshire, from Ludlow through Church Stretton to Shrewsbury and ultimately to Market Drayton. To the west is, first of all, Hopesay Hill, then the rather formless bulk of the Long Mynd with its abrupt valleys (locally called "batches" or "beaches"), and Pontesford Hill; a northward, less elevated extension intersects the road at Bayston Hill and the farthest outlier of this chain, Haughmond Hill, lies on the north bank of the Severn to the west of the main road, north of Shrewsbury. These rocks are dominantly of a sedimentary character, though they are fringed to the west by volcanics (which form Pontesford Hill). On the eastern side of the road, the landscape is more broken but more spectacular, with a discontinuous line of steepsided hills, the Stretton or Caradoc Hills, formed of Precambrian volcanics – Hope Bowdler and Ragleth Hills and their neighbours, next Caer Caradoc with its profile like a crouching lion, and finally the long whaleback of The Lawley. Farther to the east, The Wrekin forms a prominent landmark; it and its lower neighbours to the north (The Ercall and Lawrence Hill) and south (Primrose Hill) are formed of volcanics and metamorphics. In the dimmer distance is to be seen the last outlier of this series, Lilleshall Hill, again formed of volcanics.

Despite the blanket of Quaternary drift, the many upstanding ridges and the sections cut by rivers permit the cropping out of a more varied sequence of pre-Mesozoic rocks than is seen anywhere else in England – a varied array of Precambrian rocks and an equally varied, often highly fossiliferous, sequence of Palaeozoic rocks. With the single exception of Dorset, this is probably the English county most visited by geologists. Nonetheless, the growth of knowledge of the Shropshire Precambrian has been surprisingly slow; even today, many problems and difficulties subsist and much remains to be learned.

EARLY STUDIES

The earliest writer to comment on the geology of what were later to be recognised as the Precambrian hills of Shropshire was Dr. Thomas Beddoes (1760-1808), who is principally remembered for his researches in pneumatic medicine but who also undertook wide-ranging chemical investigations. In his *Observations on the Affinity between Basalts and Granite* (1791), Beddoes makes it clear that he must have visited Shropshire, for he noted that:

> "Beyond Shrewsbury . . . we have . . . a number of insulated, and generally rugged, points rising over the face of Shropshire and the adjacent counties. Were the plains covered with water a few yards in depth, these eminences would appear from distance to distance like so many stepping stones. They all, except the Malvern Hills which, though composed of granite, I consider as part of the same

system, consist of whinstone. Among these stepping stones I reckon the basaltic hills near Welshpool, the Wrekin, Lilleshall Hill, . . . In a south or southwest direction from the Wrekin, a number of craggy eminences arise. They are basaltes, and form a striking contrast with the smooth, rounded, and lumpish swells of schistus in their neighbourhood" (*ibid.*, p. 64-65).

(This last phrase clearly refers to the Long Mynd).

On a later page, Beddoes noted:

"As limestone is sometimes said to rest immediately on granite, so at the foot of the Wrekin, and at Lilleshall Hill, no slate is interposed between the limestones and basaltes; so that the analogy extends even to the exceptions" (*ibid.*, p. 66).

Beddoes' paper deserves a wider readership, for it is surely one of the earliest to suggest a metamorphic origin for granites. It is not even mentioned, however, in any of the standard histories of geology.

The next geological visitor to Shropshire about whom we have definite information was Arthur Aikin (1773-1854), a Warrington-born chemist who was a founder-member of the Geological Society of London. He published in 1797 a book describing a visit to North Wales and the Borderlands. In Chapter 18 "On the Plain of Salop" (p. 187-212), he noted that "Hagmond Hill, about two miles from Shrewsbury" [Haughmond Hill] was composed of "primitive argillaceous schistus interspersed with mica, and based upon porphyry" and that "The Wreakin [*sic*] itself, with two other smaller hills on the north and south of it, consists of a coarse dark grey whin, red on the surface, owing to the oxidation of its iron". Aitken recognised that The Wrekin was the northernmost extremity of a ridge which appeared again on the south side of the Severn and there comprised "the hills of Acton-Burnel, Frodsley [Frodesley], the Lawley, Caer-Caradoc, and Hope-Boudler Hills". He commented:

"Each of these, like the Wreakin, has the long diameter from north to south, and the direction of the perpendicular strata is the same with the short diameter; they are craggy at the top, and ascend from the plain of Salop very abruptly at an angle of about 60°. They abound in whin, porphyry, green earth, fragments of whin etc. in a clay cement, and are based upon granite."

He further noted that the "Longmont" (Long Mynd) was composed of "a very shivery kind of schistus", but gave no further geological particulars of this hill.

Robert Townson's Observations

However, the most important early geological visitor to the country was Robert Townson (1763-1827), an enthusiastic and versatile enquirer who was later to emigrate to New South Wales and to become deeply embroiled in the politics of that colony. In his *Tracts and Observations in Natural History and Physiology* (1799) are to be found the first extended observations on Shropshire geology. He commented that these were "the result of several little excursions made at different times; sometimes, in true style, on foot, with hammer in hand, other times on horseback" (Townson, 1799, p. 158). Since this work is almost forgotten and difficult of access, it is appropriate to quote him at length:

"Near the Severn, and two or three miles N. E. of Shrewsbury, is Haughman-

Figure 2. Robert Townson; from an oil painting by Augustus Earle (Mitchell Library, New South Wales).

Hill [Haughmond Hill]. This is of compound sandstone, and the strata where I observed them were vertical. About 10 miles east of Shrewsbury, the Wrekin, the most celebrated hill in the county, rises from the plain near the banks of the Severn; it is a ridge hill, about two miles long, running nearly northeast and southwest. The land is considerably higher on its eastern side, as the country here begins to rise into hills. It is chiefly composed of reddish coloured chert, which in some parts has the spotted and variegated appearance of Breccia. I recollect to have seen in one part a bed or an immense mass of hard close grained siliceous sandstone. In another part wacke or basalt is found, and in another a granitoid rock, composed almost wholly of small grained red fieldspar [*sic*], with a little quartz and some minute specks of a greenish matter, probably of the nature of hornblende. This hill, or mountain, as is may be called, belongs to the primative [*sic*]. It has no regular stratification, is without any mineral veins, and produces nothing valuable. To the N. E. of the Wrekin lie the Arcall [Ercall] Hills. That nearest the Wrekin is chiefly composed of the hard close grained siliceous sandstone, with some wacke or basalt. The other appears, from a quarry that is opened, to be chiefly of chert, yet the siliceous sandstone is found likewise here, and some of it has been used in a pottery at Coalport, in this district. Wacke or basalt is again seen here. The quarry of chert is covered by a bed of a very singular kind of red gravel about four or five feet thick, composed of small fragments and the small debris or sand of the granitoid rock" (*ibid*., p. 162-164).

Townson had thus recognised how very diverse in character were the rocks of The Wrekin and The Ercall, noting also that they were very ancient, "Primitive" in the Wernerian classification. By calling the rocks of Haughmond Hill "compound sandstone", he was recognising their comparability to the rocks of the Long Mynd. After discussing the Shropshire coal measures at length, he noted:

> "Towards the northern end of the coal district is Lilleshall Hill. Judging from some specimens I have seen it is of jasper, flanked with limestone on its north end and eastern side" (*ibid*., p. 179).

He then proceeded to consider the rocks of the hills south of Shrewsbury:

> "Near Pontesbury, the hilly district begins. Here I only examined the Ponsort Hill [Pontesford Hill] about nine miles from Shrewsbury. This is a hill of very considerable height and steep ascent. Ascending by its most gentle slope, I found at its base different kinds of chert, as Isabella coloured, greenish white, reddish white; and one of a glandular structure, having brownish red glands, rather cellular at the centre, filled up with quartz, which are united together by greenish grey chert; likewise a liver coloured chert or jasper, with flesh coloured streaks. But the hill itself, as far as my observations extend, is composed of dark greenish grey wacke [This fossil is rather lighter, softer and of a more earthy texture than basalt], greyish black basalt, and finely granulated wacke, or rather grau-wacke, for though carelessly examined this appears to be a homogeneous stone and has the usual colour of wacke, yet, when examined with a lens, it appears to be an aggregate. The aggregate or sandy nature is very visible on the surface when decomposition has commenced: there is little or no quartzous sand amongst it. I have two specimens which contain fragments of reddish brown petrosilex as big as peas, and one with a rounded pebble, of reddish very fine grained granitello, an inch in diameter . . .

"I found likewise an amygdaloid rock composed of a liver coloured ground, (of wacke?) with glands of white compact zeolite, of a flesh coloured lamellar substance, resembling barytes, but probably decomposed zeolite; and streaks of a pistachio green substance, like the lapis nephriticus, but which, from its swelling under the blowpipe, and melting into a greenish black glass, . . . I suppose to be compact actynolite" (*ibid.*, p. 181-183).

Having thus bravely attempted to describe the complex character of these rocks in the terms of his day (and indeed his description is quite adequate to enable recognition today of the rocks he was dealing with), Townson dealt at length with the rocks of the west Shropshire lead-mining district before turning back to describe the Long Mynd:

". . . The Longment . . . is composed of what I call compound sandstone, that is a stone which, instead of being formed by grains of quartz, is formed of grains of very minute fragments of other kinds of stone. These here, seem to be of an argillaceous and jaspideous nature, mixed with a few grains of feldspar . . . here and there I have found seams of reddish indurated clay or shistus [*sic*] amongst it; and on the eastern side, about Church Stretton and All-Stretton, the hill is skirted for about two miles by common grey argillaceous shistus. Near the latter place there is a quarry of coarse whin, which is used for repairing the road, and in another quarry, opened for a similar purpose nearer the village, the shistus is pervaded by short thick irregular veins of grey chert . . .

"The Longment is only separated by a narrow valley from a beautiful chain of high hills . . . The Lawley, which is the most northern, is about two miles long, but very narrow, with very smooth sides of rapid ascent, clothed with verdure, and uniting at the top in a sharp ridge. It is composed of a dark blueish grey, hard, heavy and compact wacke. In some parts a stone similar to No. 5 of the Caradoc . . . is found. And in many places, rocks of light reddish brown jasper and reddish white petrosilex protrude from the soil. Quite at the north end, where the rock has been broken for material to mend the roads, an imperfect or ill characterised granit or granitoid rock is found. It is composed of red feldspar, white quartz and blackish green hornblend; but only in the most perfect specimens. In general it is rather of the nature of granitic sandstone, and has some appearance of being formed by deposition; a little mica lies between the small beds which were just indicated. I have some specimens where the components are not distinct or granular; where the quartz and feldspar, mixed with a blackish green earthy substance, flow into one another. This granitoid stone seems to form only a wedge or patch among the wacke. At the south end of this hill I found the same as No. 8 of the following article.

"The Caradoc or Quardock is only separated from the last hill by a narrow valley. It is likewise a ridge hill and very similar in form to the preceding, of the same length, and runs in the same direction, but exceeds it in height, and rises considerably in the middle. The north end is composed of greyish coloured sandstone, in which grains of red feldspar and other matter are mixed with the quartzous sand. In some specimens, when examined with a lens, something like pitch is seen to fill the interstices between the grains. – Further south, I found

(No. 5) a greenish grey kind of whin or wacke. It is not a homogeneous stone, but rather an intimate mixture of a greenish black substance, probably of the nature of hornblende, with a greenish white substance which is easily scratched with a knife, and makes no effervescence with acids: then amygdaloid rocks prevail. – First I found (No. 8) a blackish grey compact wacke, very nearly approaching the true compact basalt, with glands, about the size of a large pea, of fasciculated and stellated-fibrous pistachio green actynolite (actinotus). These glands easily melt under the blowpipe into a blackish slag with ebullition; this fossil I don't recollect ever to have read of or seen before. It may be called the *Amygdalites actinotus*. – In the same part of the hill there is a blackish grey-wacke very cellular, with small hemispheres of haematites, the size of a pin's head, scattered about the sides of the cells. Here likewise small veins of blood red jasper, and some agate are found. – Further south, on each side of the little dell which divides this hill into the Little Caradoc and the Caradoc, great blocks of hard white siliceous sandstone abound. Then the amygdaloid rocks are found again. One kind is a dark brownish red wacke, thickly strewed with very small glands of white zeolite and white calcareous spar; and another is a blackish grey wacke with cells, only in part filled, with flesh-coloured and white zeolite and calcareous spar, all mixed together. Some of this red substance melted without intumescence or phosphorescence more like feldspar than zeolite; these three different kinds of fossils are mixed together that it is difficult to investigate them. – The rocks on the top and those still more southward, are of jasper. I have from thence a reddish brown compact jasper. The same but fine-cellular with the sides of the cells invested with green earth. A light brick coloured jasper, dull, and rather softer than a true jasper, with small oblong greenish black spots; and a liver-colour jasper with small oblong glands of white calcareous spar. I have from some part of this hill a black-brown jasper, with two little cavities in which are minute rock cristals, and minute rhomboidal cristals of flesh-coloured feldspar. Wherever there is any appearance of stratification, the strata cross the hill and rise towards the north. It is at the foot of this hill on the western side, that I lately said both coal and limestone were formerly got.

"To the south of the Caradoc lies the Elmlyth [Helmeth Hill]. This hill is much inferior in height and length to either of the preceding. Ascending it on the west side I found a singular kind of compound rock, a kind of granitic sandstone, and at the top a dark grey wacke, which in some places is intimately mixed and blended with reddish chert and jasper.

"To the east of this, separated by a valley, is the Hope Bewdlar [Hope Bowdler] Hill, of which I only examined the south end; this is of porphyry, of a liver-coloured ground of hard wacke with reddish feldspar.

"To the south of the Elmlyth lies the Hazelar [Hazler] Hill. At its base, near the high road, there is a quarry of reddish sandstone; but of a very curious kind. If examined with a lens it is seen to be wholly composed of small grains of flesh-coloured feldspar and grains of quartz, with scarce a particle of mica; however, on a nice examination, a few spangles of black mica are discovered. It is quarried for mending the roads. The rocks above, which protrude through the soil are of liver-coloured wacke.

"To the south of this is the Raglyth [Ragleth Hill]. The rocks on the very summit are of reddish grey, in another part of light flesh-coloured chert. But on the western side I found the rock to be a compound or aggregate stone, very difficult to be described. At first sight it appears as if small particles of reddish feldspar were embedded in a coarse grey earth basis; but, examined with a lens, it appears to belong to the compound sandstones; it is composed of grains of feldspar, grains of quartz, and some earthy matter. This hill terminates the chain of primitive hills, which began at the Lawley; and, south of the Raglyth, to a considerable extent, the stratified rocks prevail, which may be seen in the turnpike road to Ludlow. The strata I have observed are of the nature of argillaceous sandstone and flagstone."

In these lengthy quotations, I have endeavoured to show the meticulous care with which Townson examined and described the rocks of the Shropshire hills. His observations were unfortunately destined to be forgotten; it was not until almost 80 years later that these rocks were to be again studied in comparable detail. The complexity of the rocks he was examining is still all too apparent to the modern student; certainly Townson's geological studies were outstanding at this date and deserve to be remembered as constituting not only the first important study of what were later to be recognised as Shropshire's Precambrian rocks, but as one of the earliest thorough examinations of Precambrian rocks anywhere in the world.

Geological Visitors of the Early Nineteenth Century

In 1811, Aikin presented to the Geological Society of London the results of his own further studies of the Shropshire rocks. Although he referred to Townson's studies in passing, he employed a much less searching classification, grouping all the igneous rocks either as feldspar rocks of non-magnetic character or as greenstone rocks. He concluded:

"From these facts, it appears that the line of elevated ridgehills from Lilleshall to Ragleath [*sic*] is an unstratified mass of rocks of trap-formation incumbent on highly elevated strata of transition slate . . . The trap-formation itself does not seem to correspond with any of those described by mineralogical writers, and its essential characters are, its unconformableness with the transition slate on which it rests, the great abundance of claystone, both massive and vesicular, which it contains and the presence of actynolitic amygdaloid" (Aikin, 1811, p. 212).

In his pioneer geological map of England and Wales published in 1815, William Smith (1769-1839) shaded in grey the area between Shrewsbury and Church Stretton; this colour symbol, "Coal districts", was in part appropriate, since a small coalfield around Leebotwood was briefly productive during the nineteenth century. Much of the rest of Shropshire was shaded as "Red rhab and Dunstone".

Thomas Walford, the author of *The Scientific Tourist* (1818), noted "Greenstone, amygdaloid, feltspar, siliceous schistus" from The Wrekin, "Shivery schistus" from the "Longmont Hills" and "Basalt ironstone" from "Eastward of the Wrekin and Kinlet, 8 m. s. from Bridgenorth".

In his revised edition (a reduction) of the map of England and Wales (1820), Smith included a continuous strip of green shading for "Trap rocks" from the area of The Wrekin to a point south of Church Stretton, thus recognising the approximate line of the igneous outcrop but not its discontinuity. Although Smith visited Shropshire to examine estates along the Severn, he was probably most concerned with coal and does not seem to have examined the rocks of the Shropshire hills.

At about the same time, George Bellas Greenough (1778-1855), the first President of the Geological Society of London, published a map founded in part on Smith's and in part on original observations by himself and others. Shropshire is here dealt with in greater detail. A shading of green with red dots, "Rocks continually varying in their composition, often approaching to the character of Granit Greenstone & Conglomerate" occupies the area of Caer Caradoc, Hope Bowdler Hill etc., a short east-west strip south of Pontesford, and the region of The Wrekin. Not only the discontinuity of the outcrop, but also the variable character of the rocks present, was thus recognised and it is clear that Greenough had Aikin's work very much in mind. To the west, including the area of the Long Mynd, a uniform shading of pale yellow indicates: "Argillaceous slate in laminae varying in thickness, often calcareous. Grey Wacke, Culm."

Murchison and the Geological Survey

The most lastingly important examination of the geology of the county was made by Roderick Impey Murchison (1792-1871), during the field studies which were to lead to the establishment of his "Silurian System". His field work in Wales and the Welsh Borderland was principally undertaken in the summers of 1832 and 1833, the results being presented in a series of papers to the Geological Society beginning in March, 1833. On April 17th of that year, he came to the sixth and most ancient division of the rocks he recognised below the Old Red Sandstone:

> "Red Conglomerate, Sandstone, and Slaty Schist. This is a vast deposit of several thousand feet in thickness, consisting of very coarse, quartzose conglomerates, which alternate with some schistose beds and much purple-coloured sandstone (Compound Sandstone of Townson), the strata of which in Haughmond, Pulverbatch and Linley Hills, Shropshire, are highly inclined or vertical, in conformity with those of the preceding formations. No organic remains have been observed, by which, and by its very peculiar mineral structure, this formation is shown to be entirely distinct from the preceding groups." (Murchison, 1833, p. 476).

He noted also that:

> "In those districts where parallel ridges of all these formations are brought to day within a zone of small breadth, rocks of trappean or igneous origin are usual accompaniments, as in the neighbourhoods of Wrekin and Caer Caradoc, Shropshire . . ." (*ibid.*, p. 477).

A further commentary followed in a paper presented on January 22nd, 1834 when Murchison stated that, at the bottom of what he termed the "Upper Grauwacke Series" of the "Transitional Beds", there were many thousand feet of "Hard, close-grained, grey, greenish and purple sandstone. Red and grey quartzose conglomerate. Slate-coloured and purple schists. Coarse slates; little or no calcareous matter." This thick sequence of sediments, which occupied "the hilly region of the Longmynd, Linley, the Stiperstones, etc.; extending to the north-east in Lyth and Haughmond Hills" was now named the "Longmynd Rocks" and correlated (wrongly) with beds at Gwstaden in Breconshire (Murchison, 1834a, p. 14 and folding table). The occurrence of trap rocks in The Wrekin and Caer Caradoc was again noted.

A fuller account of the igneous rocks was contained in his address on May 22nd, 1834:

> "Lilleshall hill consists chiefly of compact felspar rock, having in parts a sienitic structure. The Wrekin, which has been described by Mr. Aikin in the Geological Transactions, is of nearly similar composition. Lea and other rocks near Wrockwardine have the same base, but pass into porphyry and clinkstone. In Charlton Hill, porphyries and greenstones occur where they have not previously been noticed.
>
> These rocks mark parallel axes of different lengths – the north bank of the Severn, ranging from north-east to south-west, and piercing through beds of grauwacke, chiefly those of the Horderley and May Hill formation, the sandstones of which, on the sides of the Wrekin and Arcal [Ercall] and at Charlton Hill, are converted into quartz rocks at their point of contact with the trap.
>
> The line of disturbance occasioned by the protrusion of the zone of Charlton Hill, is traced in certain trap rocks which appear in the bed of the Severn, near Cound, extending thence to the south-west.
>
> *Caer Caradoc.* – This remarkable ridge, formerly described by Mr. A. Aikin, is on a line of eruption parallel to that of the Wrekin. It includes the hills of Lawley, Great and Little Caradoc, Helmeth, Ragleath, etc., together with a large *contrefort* on its south-eastern face. At Cardington and Hope Bowdler it consists of many varieties of felspar rock, sienite and greenstone. The beautiful amygdaloid with actynolite first pointed out by Mr. Townson, and supposed to be peculiar to one spot, is shown to be of frequent occurrence in several of these hills . . ." (Murchison, 1834b, p. 85-86).

Thus Murchison recognised, in this study, the whole line of the volcanic Precambrian hills and perceived their structural relationships, though his conclusion that the overlying quartzites were of volcanic origin (repeated in his account of the Stretton hills) was incorrect. He also made new observations concernings the more westerly Precambrian rocks:

> "The ancient grauwacke system of the Longmynd and Linley Hills, or mineral axis of Shropshire, is penetrated by a vast number of points of eruptive trap, chiefly greenstones. Pontesford Hill, amid varieties of greenstones, contains also a porphyry and a remarkable amygdaloid. This zone is reproduced at intervals in Sharpstone Hill, and has been observed . . . in Haughmond Hill, four miles northeast of Shrewsbury, where it passes into a coarse sienite. Vertical, veined, and indurated strata appear at different points along these lines of eruption." (Murchison, 1834b, p. 86).

Although Murchison again described the Longmyndian sediments and their associated igneous rocks in his *Silurian System* (1838), he recognised from the outset that these rocks were more ancient than the Silurian, even in his enlarged vision of that system; indeed, he considered them to be the most ancient rocks in England and placed them into the Cambrian. He considered that the lowest strata in the "Longmynd Rocks" occurred along the western side of the Stretton valley, dipping to the W. N. W., noting that they could be seen in order of superposition

Figure 3. Roderick Impey Murchison, Bart. (1792-1871). Engraving by Joseph Brown after a painting by Pickersgill.

The Lawley. Caer Caradoc. Hope Bowdler. Broccard's Castle. Ragleath.

THE CARADOC RANGE (sketched by Mrs. Stackhouse Acton).

Figure 4. The Stretton Hills; from an illustration in Murchison's *Siluria* (1867).

in the Carding-mill valley and parallel gullies. The lowest rocks were "glossy schists or clay-slate", with two or three minute layers of siliceous limestone scarcely an inch thick: these were overlain by a vast and regular series of hard, purple or plum coloured, greenish and grey schistose flagstones and siliceous grits with occasional quartz veins. Eruptive bosses penetrated this sequence. The rocks of the Stiperstones were now recognised to be of younger (Silurian) age.

The mapping of Shropshire by H. M. Geological Survey began around 1850: the Survey officers concerned in it were William Talbot Aveline (1822-1903), Edward Hull (1829-1917), David Hiram Williams (? - 1849) and Warington Wilkinson Smyth (1817-1890), under the general supervision of Andrew Crombie Ramsay (1814-1891) who was later to succeed Murchison as Director-General of the Survey. The first relevant maps to appear, both in 1850, were Sheets 61SW (Church Stretton, Longmynd, Brown Clee), for which Aveline was primarily responsible, and 55NW (Ludlow, part of Titterstone Clee) mapped by Aveline and Smyth. Both sheets were revised in 1855, after further fieldwork by Aveline and a Survey palaeontologist, John William Salter (1820-1869). In that year, Sheet 61NW (Shrewsbury, Pontesbury, Shineton), mapped by Williams, Aveline and Hull, appeared; and three years later, the publication of Sheet 61NE (Wellington, Shifnal, Newport, The Wrekin) mapped by Williams, Smyth, Aveline and Hull, completed coverage of the relevant areas. (Though Ramsay's name appears on all these sheets, we have been unable to find evidence that he undertook any mapping in Shropshire). The treatment of Shropshire's ancient rocks in these maps is not detailed; the igneous rocks are shown in maroon and are variously captioned as "Greenstones", "Basalt Greenstone" and "Basalt. Igneous Rocks. Greenstone" in different quarter sheets; the adjacent quartzites are shown as "Caradoc sandstone". (On quarter-sheet 61 NE, "Altered Caradoc sandstone" is shown surrounding The Wrekin, following Murchison's concept). The rocks of the Long Mynd are shown as "Cambrian rocks", but further details are written onto the maps themselves, the rocks of the northern Long Mynd and its northeastward extension being labelled "Purple and green sandstones and conglomerate" and those of the southern Long Mynd "Purple and green sandstones and slaty beds, disturbed and cleaved." These maps provide the first published indication of the complexities of the Stretton fault system; this is especially apparent on quarter-sheet 61 NE, where the faults west of The Wrekin are well shown. Many new discoveries concerning the distribution of the ancient rocks are apparent; in particular, the presence of "Cambrian" in Ragleth and Helmeth Hills, east of Church Stretton and the occurrence of igneous rocks around Wrockwardine, to the northwest of The Wrekin.

In the course of his work with Aveline, Salter discovered the first indications of life in the "Longmynd Rocks", in ripple-marked greenish sandstones of the Carding Mill valley and Callow Hill. These evidences were principally trace-fossils (paired holes presumed to be formed by the Lobworm, *Arenicola didyma,* and annelid tracks), together with what Salter considered to be a trilobite cephalon; he proposed the new name *Palaeopyge ramsayi* for this (1856). In a later paper (1857), he described anew the ripple marks and recorded the occurrence of rain-drop impressions and suncracks; the paired holes were given a new name, *Arenicolites sparsus*, "young" and "adult" forms being distinguished on the basis of their size (they are now recognised as being the twin mouths of U-shaped tubes).

Salter's discoveries appeared to afford strong support for attributing a Cambrian age to the "Longmynd Rocks"; the life-forms present were primitive indeed but seemed quite comparable to those known from the type Cambrian of Wales. Murchison, who had succeeded Sir Henry de la Beche as Director-General of the Survey in 1855, mentions having requested Salter and another palaeontologist, John Rhind (?-1857) to undertake further searches for fossils in these sediments (Murchison, 1867, p. 29), but no new discoveries resulted.

In 1862, "A Lady" published a small book entitled *The Rocks of The Wrekin* – which, however, gave no description whatsoever of the geological constitution of that hill!! Seven years later the same authoress, this time under her own name of Charlotte Eyton, published a second work, *Notes on the Geology of North Shropshire* (1869), mentioning "Lower Cambrian grits and interstratified lavas" in Haughmond Hill and noting that "The Wrekin and its little northern neighbour, Lawrence's Hill, consist of dark red lava, with veins of several different minerals" (Eyton, 1869, p. 17). She further commented that the Ercall Quarry showed the nucleus of that hill "to be formed of pink syenite – a Plutonic rock, or one supposed to have been cooled before it arrived at the surface, instead of flowing out in a liquid state. A similar rock forms the nucleus of Primrose Hill, at the southwestern extremity of the [Wrekin] chain" (p. 18).

In a paper presented to the Geological Society of London on May 23rd, 1877, the petrologist Samuel Allport (1816-1897) made a more extended examination of the igneous rocks of The Wrekin, noting that in a quarry at the south end of Lawrence Hill:

> "thick beds of volcanic ash are seen dipping in a northerly direction, at an angle of 55° . . . They here consist of several alternations of coarse and fine material, the whole of which have been highly indurated and otherwise altered. These beds are covered by masses of altered pitchstone and felsites, which are well exposed on the south slope of the hill, on its steep northern face, and also on the opposite crags of Ercall Hill" (Allport, 1877, p. 450).

He mentioned that the ashbeds and overlying rocks were also exposed in The Wrekin itself, the altered pitchstone and felsites being the "compact felspar" of Murchison's account. The presence of a dolerite intrusion on the summit of the ridge of The Wrekin, where it formed a rounded elevation, was recorded, as was the presence in the Lawrence Hill Quarry of two highly altered basalt dykes. He concluded that the rocks clearly evidenced formation by subaerial volcanic activity.

RECOGNITION OF PRECAMBRIAN AGE

By 1878, the distribution of the ancient rocks of Shropshire was known with fair clarity and the existence of a sedimentary sequence and a volcanic sequence recognised, but their relative ages remained to be determined, the order of succession within the beds was still to be worked out and the structure was not yet determined. Nor was the full range of beds yet discovered.

The major breakthrough was made by a Bristol geologist, Charles Callaway (1838-1915). On March 11th, 1874, he communicated to the Geological Society the discovery of an outcrop of fossiliferous shale near Shineton, to the east of The Wrekin; a number of new species of trilobites were listed, which Callaway considered to be of Tremadocian (then considered Upper Cambrian) character. The stratigraphic position he attributed to the beds was regarded with doubt by some Fellows and the paper was published only in abstract. However, three years later Callaway addressed the Society at length on his discovery, establishing the Tremadocian age of the fossils beyond doubt and noting that between the Shineton Shales and The Wrekin there were micaceous, green sandstones; these yielded the brachiopod *Kutorgina cingulata* and were likewise considered of Cambrian age, Callaway suggesting their correlation with the Hollybush Sandstones of the Malverns. Similar sandstones cropped out on the southeast flank of Caer Caradoc and formed an inlier at Lilleshall; at The Wrekin and at Caer Caradoc, the fossiliferous sandstones were separated from the hills by Murchison's "metamorphosed" quartzites.

The proposed age for the rocks gained rapid acceptance. In 1877, the Reverend W.S. Symonds (1818-1877), Rector of Pendock, Worcestershire, and an enthusiastic amateur geologist, noted in a contribution to the *Transactions* of the Shropshire Archaeological and Natural History Society that:

> "We find . . . on the westward flanks of The Wrekin a Syenitic gneiss, which I have long considered to be the oldest rock of The Wrekin, and believe that a tunnel through the hill would shew an ancient syenitic nucleus similar to that of the Malvern Hills and Malvern Tunnel. I believe also that a similar axis underlies the upheaval of the Longmynds. The Malvern syenitic axis is traversed in all directions by infiltrated volcanic rocks. This is probably the case that The Wrekin, and the old synitic rock owes its elevation to volcanic movements, which also elevated the other rocks which surround The Wrekin."

Symonds discussed the rocks of the Long Mynd, accepting them as being of Cambrian age.

The age of Murchison's "metamorphosed" quartzites was discussed by Callaway in a second paper presented to the Geological Society, on 5th June, 1878; he demonstrated that the quartzites were older than his "Hollybush Sandstone" and indeed considered a "great discordance" to separate them. The age of the Longmynd series was taken to be Lower Cambrian (presumably on the basis of Salter's "trilobite") and, since no comparable strata occurred in that series, the quartzites were thought to be yet older. Though Callaway considered that the quartzites were bounded above and below by faults, he was nonetheless confident that the igneous rocks were earlier than the quartzites, since the latter contained at their base "small rounded or unrounded pieces of felstone greatly decomposed, but in some cases showing distinctly the banded structure characteristic of some of the Wrekin felstones" (1878b, p. 760). Their Precambrian age was thus established.

Summaries of both papers were published by Callaway in the *Midland Naturalist* (1878a, 1879c); no new information was given. However, this journal also gave a summary of a lecture presented to the Birmingham Natural History and Microscopical Society in 1879, in which Callaway suggested that the Shropshire igneous rocks corresponded with the Pebidian series of Pembrokeshire and with the Huronian of Canada (1879a). Callaway's receipt of an award of £50 from the Government Fund for the Endowment of Scientific Research, to enable further researches into the relations of these Precambrian rocks, was announced in a subsequent issue of the *Midland Naturalist* (1879b, p. 103). The first fruits of this research were soon forthcoming for, in the following issue, Callaway published a letter announcing the discovery of a *second* Precambrian group in Shropshire:

"In Primrose Hill, the south-westerly spur of The Wrekin, I have come upon an exposure of hornblendic gneiss, with a high dip to the NE. Associated with this bedded rock are a red granitoidite and a well-crystallised diorite. The identity of these rocks with some of the common Malvern types in my cabinet is undoubted, and the strike precisely corresponds with that of the Malvernian series. This metamorphic group is unconformably overlaid by the tuffs and rhyolites of The Wrekin. I have just returned from a visit to St. Davids, where I was struck by the close resemblance between the lower part of the Pebidian and the ashy slates and halleflintas of Lilleshall Hill. These Lilleshall rocks I have also detected on the flanks of the Malvern Hills. Thus light begans to dawn upon the obscurity which has hitherto enveloped these ancient and difficult rocks." (Calloway, 1879b, p. 158-159).

This new discovery seems to have resulted from a suggestion made to Callaway by Thomas George Bonney (1833-1923), the newly-appointed Professor of Geology at University College, London. Bonney had himself first visited The Wrekin in the company of Samuel Allport around 1877 and, deciding afterwards to attempt a microscopical examination of the rocks he had collected, had visited Callaway to obtain supplementary specimens. On a joint excursion to the hill, he had suggested to Callaway that the area south of The Wrekin might repay detailed study; and Callaway's resultant studies revealed the presence of the metamorphic rocks (see Bonney, 1879). Bonney's petrological analyses of the Shropshire Precambrian rocks appear as supplements to each of Callaway's major addresses to the Geological Society.

On June 11th of 1879, Callaway presented to the Geological Society the first instalment of a detailed account of Shropshire's Precambrian rocks. He discussed in turn the whole series of outcropping masses – Lilleshall Hill, The Wrekin group, Wrockwardine and Charlton Hill, the small outcrop at Cound, The Lawley, Caer Caradoc, the group of hills between Cardington and Church Stretton (Helmeth, Hope Bowdler, Cardington and Ragleth) and the southernmost exposures at Stanner Rocks, Wartle Knoll and Hanter Hill in Radnorshire – and demonstrated that they could no longer be grouped as "intrusive greenstones". The Primrose Hill metamorphic rocks were briefly noted, but stress was laid on the igneous rocks; these were shown to consist primarily of extrusive rocks (tuffs, volcanic breccias, pitchstones, felstones and hornstones) penetrated by minor intrusions of basalt and dolerite. Callaway's conclusions concerning their Precambrian age were reinforced by his discovery of a conglomerate in the Longmynd series of Haughmond Hill, in which many recognisable pebbles of the igneous rocks were present (1879d, p. 661-662).

Callaway also briefly noted (*ibid.*, p. 662) that "on the N. W. flank of the Ragleth and further to the S. W., the Longmynd series is represented by a felspathic sandstone." This was the first report of the occurrence of Longmyndian rocks to the east of Church Stretton.

Also during 1879, Callaway wrote a popular scientific article summarising the geology of The Wrekin and advancing the claim that this was "the oldest mountain in England". He visualized its origin as resulting from the accumulation of layers of tuff and lava, which were afterwards in part thrust upwards as a horst between two parallel faults to form an island whose erosion produced surrounding sand beaches (afterwards altered into quartzites). He thought that this island was buried by sediments during the Caradocian (Ordovician), but rose again above the sea in Old Red Sandstone times, to be buried again beneath the terrestrial New Red Sandstone deposits. He further considered that the island was eventually re-exhumed after the retreat of the seas of the Glacial (Pleistocene) period which he believed had left a deposit of marine shells on its lower slopes (1879e). Clear sketches of the successive stages in the geological history of The Wrekin were presented. Though the account contains a number of errors, it shows the clarity of Callaway's thinking and his ability to formulate, from his observations, a lucid and logical reconstruction of past geological events.

Callaway's discovery of extrusive igneous rocks in the Precambrian of the Malvern Hills (reported to the Geological Society on May 26th, 1880) strengthened the correlations with Shropshire. The second instalment of Callaway's work on the Shropshire Precambrian was not, however, presented to that Society until December 21st, 1881; again, this was a reinterpretation of rocks considered by Murchison to be "greenstones" – this time, of those exposed in Pontesford Hill and in a series of outcrops on the western flank of the Longmynd:

> "The prevailing rock-types are purple rhyolites at the north end of the chain, purple hornstone (or halleflinta) in the middle, and purple and green hornstone (or halleflinta), with some indurated grits, partly derived from a gneissic source, at the south" (1882, p. 123).

Their close lithological affinity with the rocks of The Wrekin-Caer Caradoc chain was clear. The presence of gneissic fragments was the only indication of metamorphics in this area; such rocks did not crop out in this area. That the volcanics were overlain by "Longmynd" sediments was rendered beyond doubt by the presence of fragments of purple rhyolite, "almost universal" in the latter.

A further complication was introduced by Callaway's discovery (1884) of some poor exposures of quartzose mica schists at Rushton, immediately to the west of The Wrekin. The rocks, which Callaway named the 'Rushton Schists', cropped out in the roadways and in farmyards at the western end of the village; they were overlain by Cambrian beds to the south and obscured by drift to the north, so that, whilst it was clear that these also were of Precambrian date, their relationships to the Primrose Hill gneisses and to the extrusive rocks was not evident.

However, careful study soon enabled Callaway to begin reconstructing the sequence. In a paper presented to the Geological Society on June 23rd, 1886, he began by proposing the name "Uriconian" for the Precambrian igneous sequence. He noted that the conglomerates and grits of the Longmynd series were not only "crowded with pebbles of purple rhyolites" but also contained pieces of a pale-green felsite, and that the grits contained similar fragments in variable proportions with quartz; fragments of mica-schists and of gneissic rocks of Primrose Hill type indicated that a metamorphic terrain was also undergoing denudation at this time. Study of a conglomerate at Charlton Hill, which Callaway believed to have formed contemporaneously with the Uriconian rocks, showed it to contain fragments of quartz and mica schists and of gneissose rocks, confirming the order of succession already deduced. The overlying quartzites were seen to rest unconformably on the Uriconian rocks; although Callaway nowhere commented directly on Murchison's concept that the quartzites were metamorphic in origin, his observations made this no longer tenable.

The origin of the conglomerates was again considered in a paper presented to the Shropshire Archaeological and Natural History Society in 1888. Callaway clarified the fact (left uncertain in his earlier paper) that the Charlton Hill conglomerate was interstratified with the ashy and felsitic rocks of the Uriconian, so that its age was undoubted. He considered it "perhaps, amongst the oldest sedimentary rocks in the earth's crust." In this paper, the term "Longmyndian" was first proposed for the sedimentary sequence.

Callaway had already recognised that the green sandstones flanking the Uriconian rocks, and overlying the quartzites, were of Cambrian age, but he had been unable to be more precise. Charles Lapworth (1842-1920), recently appointed first Professor of Geology at the University of Birmingham, now published the results of a study of the fragmentary fossils in these beds (henceforth called the Comley Sandstones), showing that they included the Lower Cambrian trilobite *Olenellus* (not hitherto reported from Britain). This discovery, announced in two journals (1888a, b), gave a strong reinforcement to Callaway's conclusions on the age of the Uriconian. Lapworth suggested also that "a Precambrian age for the Longmyndian was a matter of fair probability" (1888a).

The Blake-Callaway Controversy

In the same year, the Lower Cambrian age of the Longmyndian rocks had also been questioned by John Frederick Blake (1839-1906), who was then Professor of Natural Science at University College, Nottingham. In presenting an account of the "Monian" sequence, a series of rocks of presumed Precambrian date exposed in Anglesey and eastern Ireland, he pointed out the similarity between the sediments of the Upper Monian at Bray Head, County Wexford, and those of the Longmyndian, suggesting also that the Uriconian volcanics might correlate with the Middle Monian (1888, p. 542-543).

Stimulated by this idea and having time available following retirement from his Nottingham appointment, Blake forthwith undertook a study of the problematic Shropshire sequence. The results were presented to the Geological Society on March 12th, 1890. He noted the problem presented by the enormous apparent thickness of the Longmyndian sequence – computed at some 26,000 feet – and mentioned how synclinal structures had been envisaged to account for this. On the basis of his studies he concluded that the Longmyndian was divisible into two distinct units, separated by an unconformity. The younger unit, forming the western part of the Long Mynd, included the conglomerates containing Uriconian pebbles and was

Figure 5. Charles Callaway (1838-1915).

Figure 6. John Frederick Blake (1839-1906). From a photograph in P.F. Kendall and H.E. Wrott: *Geology of Yorkshire* (1924, p. 302).

thus definitely post-Uriconian. The older unit, forming a narrower strip on the eastern flanks of the hill, might well predate the Uriconian volcanic episode. He now considered the western sequence to be probably of Cambrian date, the eastern sequence as probably Precambrian (Upper Monian).

The supposed trilobite, *Palaeopyge ramsayi,* described by Salter in 1857, was considered by Blake to be definitely inorganic (Blake, 1888, p. 354). Most later authors (e.g., Watts, 1925, p. 334; Greig *et al.*, 1968, p. 41) have accepted Blake's conclusion and this genus is excluded from the list of valid trilobite genera in the *Treatise of Invertebrate Palaeontology* (Harrington *et al.*, 1959). However, Pocock *et al.* (1938, p. 43) did not entirely discount the possibility of its authenticity and Dr. A.W.A. Rushton (pers. commun. to W.A.S. S., 1973) comments: "In my opinion, it would be difficult to prove or disprove whether the remains are of organic origin. Perhaps their stratigraphic position implies that they are *Lusus naturae.*"

The Uriconian rocks, which were also considered at length by Blake, were now presumed to be attributable to the Upper Monian rather than the Middle; he stated that they had been "extruded from the midst of the Longmyndian rocks" (Blake, 1888, p. 470), a comment whose ambiguous phraseology was to puzzle subsequent workers.

Blake's conclusions were immediately challenged in the discussion following his lecture, by Henry Hicks (1837-1899), a Welsh-born London doctor who had made extensive studies of the Pembrokeshire and Anglesey Precambrian rocks. Hicks (1890) endorsed the earlier view that the Longmyndian sequence was of Cambrian date; he considered that the rocks of Blake's Lower Long Mynd Series "had undoubtedly formed an arch over the Caer Caradoc rocks, that the limb on the west side has been greatly broken by reversed faults and that the limb on the east had been broken by great thrusts which had caused newer rocks to be pushed over and hide the lower." Blake brusquely retorted that he "had seen no evidence for any of the thrusts imagined by Dr. Hicks" and, although admitting that the summit and base of his system were not defined, stated flatly that "the system itself was there" (*ibid.*, p. 420).

Blake's opinions did not appeal at all to Callaway, who swiftly responded in a paper read on 7th January, 1891. He attacked Blake on all major counts. Tackling Blake's statement on the extrusion of the Uriconian rocks from the midst of those of the Longmyndian, he reported that he could find no evidence for their intrusion into the sedimentary sequence: in the few instances where these rocks cropped out in close proximity, there was no evidence of thermal metamorphism of the sediments. Nor could he discover any evidence to support Blake's supposed intra-Longmyndian unconformity: indeed, he had found some fragments of volcanic rock in the supposed "Monian" sequence, though these were uncommon. Callaway therefore reiterated that there *was* an unconformity beneath the Longmyndian, separating it from the schists and gneisses; the evidence for this was again adduced. All in all, whilst recognising that the age of the Longmyndian was still *"sub judice"* (Callaway, 1891, p. 118), he saw no reason to modify his earlier views on the order of superposition.

Predictably, a spirited discussion followed this presentation. Blake stated that he had not regarded the volcanic rocks as intrusive, "but as having burst out from among the slates" (see discussion of Callaway, 1891, p. 124) – a comment which did little to clarify his earlier statement. He reaffirmed that he had traced the unconformable junction between the two Longmyndian divisions "across the district" and, whilst admitting that the presence of volcanic fragments in the lower group was puzzling, made no comment thereon. If the whole Longmyndian was to be called Longmyndian, where then was the Cambrian of the western part of the district? After Bonney and Hicks had spoken in support of Callaway's ideas, the latter, in reply, commented that he "had difficulty in understanding Prof. Blake's theory of *extrusion* as distinguished from *intrusion*" (see discussion of Callaway, 1891, p. 125) and, after discussing several other matters, pointed out that undoubted Cambrian sediments did indeed occur in the western districts, the Shineton Shales cropping out on the west side of the Long Mynd, under the Stiperstones, the Lower Cambrian being cut out by a fault.

Lapworth was meanwhile continuing his examination of the fossils of the Comley series; in 1891 he published a description and theoretical restoration of the trilobite *Olenellus* (*Holmia*) *callavei* based on fragments from the Comley quarry and, in a footnote, recorded the finding of a trilobite of Middle Cambrian type, *Paradoxides broomi*, in the conglomerate beds lying above his *Olenellus* limestone. He noted that there was intergradation between the lowest Comley beds and the underlying quartzites, now known as the Wrekin Quartzite; a Lower Cambrian date must therefore be assigned also to this unfossiliferous sedimentary unit, which was henceforth viewed as the earliest sediment formed in a Lower Cambrian marine transgression.

Three years later, when Lapworth prepared an account of south Shropshire geology in association with William Whitehead Watts (1860-1947), for a Geologists' Association field trip they jointly led, he felt unwilling to decide on the relative ages of the Pre-Cambrian rocks, merely listing them and indicating alongside the orders of superposition envisaged respectively by Callaway (A) and Blake (B):

	A	B
Longmyndian Red Sandstone Series	5	4
Longmyndian Flagstone and Shale Series	4	2
Uriconian Lavas and Ashes	3	3
Rushton Schists	1 & 2	1?
Primrose Hill Gneiss	1 & 2	3?

Later in the paper (1894, p. 307), Lapworth and Watts refer to Blake's "Upper Monian" as the "Western Longmyndian" and his presumed "Lower Cambrian" division as the "Eastern Longmyndian". These terms, originally coined by Callaway but hitherto little used, were to be widely employed in later papers.

Though Callaway, in the preamble to a paper establishing the existence of Longmyndian outliers at Old Radnor, Radnorshire and Huntley, Gloucestershire, stated firmly that "The rocks of the Longmynd Hills . . . originally supposed to be Lower Cambrian, are now admitted to be of pre-Cambrian age" (1900, p. 511), the controversy was far from being resolved. Callaway was awarded the Murchison Medal of the Geological Society in 1903 as a tribute to his Precambrian researches

Figure 7. Charles Lapworth (1842-1920). Photo: F. Raw.

Figure 8. Edgar Sterling Cobbold (1851-1936). Photo: C.E. Resser.

in England, Wales and Scotland; but neither he nor Blake published anything further on Shropshire geology.

An amateur geologist living in Church Stretton, Edgar Stirling Cobbold (1851-1936), had by now begun to follow up the work of Callaway and Lapworth, becoming a friend and frequent field companion of the latter and of Watts (see Watts, 1936). In addition to greatly adding to knowledge of the Cambrian faunas on the county in a series of papers published between 1910 and 1936, Cobbold also acquired a thorough knowledge of the Precambrian outcrops. When he contributed a long chapter on geology to a local history, Cobbold (1900) suggested that the Eastern Longmyndian passed downwards into the Eastern Uriconian, the connecting link being the buff and blue grey grits, with their bright red grains, cropping out on Ragleth and Helmeth Hills. These had been originally reported by Callaway (1879d): Cobbold now applied to them an unpublished name formulated by Lapworth – the "Helmeth Grits".

DEVELOPING IDEAS SINCE 1900

On June 22nd, 1904, William S. Boulton, Professor of Geology at University College, Cardiff, presented to the Geological Society the results of a detailed study of the Uriconian rocks of Pontesford Hill; these were selected for study because they were thought to exemplify most of the Uriconian lithological types. He described the hill as "a 'plagioclinal' ridge, bounded on all sides by faults" (1904, p. 483) and distinguished two groups of constituent rocks:

> "1. A Bedded Group, consisting of the Northern and South-Eastern Rhyolite respectively, differing in composition and structural characters. Of these, the Northern Rhyolite, of Uriconian type, is probably bedded, but may be intrusive in the Andesite Group, whilst the South-Eastern Rhyolite, which is associated with breccias and grits of the same com osition, is certainly bedded. Between these acid rocks intervenes a thick series of basic andesite-tuffs (palagonite, halleflinta, crystal and gritty), interbedded with basic augite-andesite lava . . .
>
> 2. Olivine-Augite Dolerites (granulitic, ophitic, and coarsely amygdaloidal), having a chemical composition and mineralogical characters similar to those of the dolerites of the Carboniferous districts of Staffordshire and Shropshire, are intruded among the bedded rocks in laccolite-like masses." (Boulton, 1904, p. 483).

He noted also that "some of the fine tuffs and volcanic grits show unmistakeable signs of deposition in water."

After presentation of the paper, a written contribution from Lapworth was read which offered a new approach to the correlation problem:

> "The Longmyndian formation itself (which was made up of a Lower division of grey, green and red shales and grits, reminiscent of the Charnian of Leicestershire; and an Upper division of red sandstones, grits and conglomerates, reminiscent of the Torridonian of Scotland, with an intermediate group – the Bayston Group – combining characteristics of both) was certainly sedimentary throughout. But the materials of which the beds were made up appeared to have been largely pyroclastic in origin, and were often suggestive of simultaneous volcanic action at no great distance outside the Longmyndian area. Broadly speaking, the amount of this ashy or felspathic material increased as the succession was

Figure 9. William Whitehead Watts (1860-1947). Photo reproduced by permission of the Royal Society.

ascended", and the typical volcanic rocks of the so-called Uriconian marked apparently the local incoming or culminating phase of this volcanic action, connected perhaps with the movements which brought the Longmyndian to a close and prepared the way for the Cambrian" (*ibid.*, p. 485).

Thus, whilst accepting the view of Blake that the extrusion of the Uriconian sequence occurred during the time that the Longmyndian sediments were accumulating, Lapworth concurred with Callaway in recognising particulate igneous material as present in varying proportions throughout those sediments and in viewing the age of the whole Longmyndian and Uriconian as Precambrian.

Subdivision of the Longmyndian

In 1910, when he and Watts again collaborated in preparing the account of "Shropshire" for the Geologists' Association jubilee volume, *Geology in the Field,* Lapworth presented the results of his work more fully. He now recognised seven divisions within the Longmyndian:

Wentnor Group (Western Longmyndian System)

7. *Ratlinghope Series*: red and purple grits and shales, with conglomerates.
6. *Bayston Series*: red and grey grits, with the three zonal conglomerates of Stanbatch, Darnford and Haughton [in descending order], the last rich in pebbles of Uriconian volcanic rocks.

Stretton Group (Eastern Longmyndian Series)

5. *Portway Series*: purple, grey and green slates and flags, with few shales.
4. *Lightspout Series*: massive grey and green grits and flags, with few shales.
3. *Synalds Series*: purple shales and occasional flaggy grits and green shales, with the *Carding Mill Grit* at the base.
2. *Burway Series*: grey-green flags and shales, having as a basement band the siliceous *Buckstone Grit*.
1. *Stretton Shale Series*: consisting of
   b) *Brockhurst Shales*: hard grey-blue and dark green laminated shales, with rare calcareous nodules.
   (a) *Watling Shales*: green shales with occasional purple mudstones, flaggy beds, and rare calcareous bands.

The Narnells Rock Grit corresponded with the horizon at which Blake had placed his unconformity: but Lapworth, although recognising that there were indeed marked lithological differences between the two major groups he now distinguished, neither placed the boundary between them at this level nor found evidence for the existence of such a discordance.

Though not hitherto published, the names proposed by Lapworth in this paper were already in widespread use and had indeed been employed by two Liverpool amateur geologists, Thomas Mellard Reade (1832-1909) and Philip Holland, when presenting the results of a series of analyses of Longmyndian rocks to the Liverpool Geological Society in 1909. They were impressed by the concentration of calcareous material in the Stretton Series and its virtual absence elsewhere, noting that it was accompanied by the development of "cone-in-cone" structure and the presence of phosphates.

In 1912, the discovery of an inlier of Longmyndian rocks around the hamlet of

Pedwardine and the village of Brampton Bryan, in Herefordshire, was reported by Arthur Hubert Cox. The rocks, given the name "Brampton Grits and Conglomerates", appeared to be overlain by Tremadocian shales with *Dictyonema,* though the nature of the junction was not clear.

The outcrop of Longmyndian sediments at Old Radnor, noted earlier by Callaway (1900), was described in greater detail by Edmund Johnston Garwood (1864-1949), Professor of Geology at University College, London, and Miss E. Goodyear, in a paper presented to the Geological Society on June 6th, 1917 but not published till 1919. They showed that the sediments were attributable to the Bayston Series, with its two higher conglomerates, and expressed their view that the Brampton Bryan outcrop was of similar character.

Subdivision of the Uriconian

In 1925, a new account of the geology of south Shropshire was prepared for the Geologists' Association by Watts who, since Lapworth's death in 1920, had assumed his mantle as doyen of Shropshire geologists. Two series of Uriconian rocks were now recognised (p. 7-8):

> "Linley (or Pontesford) Volcanic Series (Western Uriconian of Callaway). Green shales, purple and green grits, andesitic and rhyolitic lavas and tuffs.
> Cardington Volcanic Series (Eastern Uriconian of Callaway) of the Cardington and Caradoc Hills, etc. Andesitic and rhyolitic lavas and intrusions, tuffs, volcanic grits and shales, pierced and locally overlain by dolerites and basalts."

Watts considered that the formation of the Cardington Series preceded, and that of the Linley Series succeeded, the accumulation of the Longmyndian; but he brought forward no evidence to support this conclusion.

In 1935 (one year before his death), Cobbold, in collaboration with Walter Frederick Whittard (1902-1966) of Imperial College, London, published the results of many years of intermittent study of the Helmeth Grits. A petrological study of the massive beds which made up the bulk of the "Grits" showed the groundmass to consist of a chlorite-epidote aggregate, carrying abundant angular and subangular fragments of rhyolite and trachyandesite, together with quartz grains, fragments of granitic and granophyric rocks and of shales. Rapid accumulation from a local source was indicated and the authors recognised that these "Grits" were in reality lithic tuffs formed from already consolidated Uriconian volcanics. The shale fragments were thought to be derived from sediments contemporary with the vulcanism. Associated with the massive horizons were thinner beds of sandy shale, formed during intervals in the tuff accumulation. The authors pointed out the great similarity between these interbedded shales and the Watling Shales of the Longmyndian. They considered "that the tuffaceous beds, which are outstanding rocks of the Helmeth Grit Group, may be taken to represent the decadent phases of the vulcanicity of the Uriconian" (1935, p. 354) which, they concluded, was older than the Stretton Series of the Longmyndian, the shale beds in the Helmeth Grits representing the first beginning of Longmyndian sedimentation.

Remapping of Shropshire by the Geological Survey had commenced in 1914-15 and further work had been undertaken in 1919-20 and 1926-28. The District Geologists were Thomas Crosbee Cantrill (1867-1931) and Charles Hawker Din-

Figure 10. Caer Caradoc from near Leebotwood. Photo: W.A.S. Sarjeant, 1961.

Figure 11. The Lawley from the south, near Shootrough. Photo: W.A.S. Sarjeant, 1961.

ham (b. 1883): other Survey officers who had worked in this region during this period were Roy Woodhouse Pocock (1887-1954), Thomas Robertson (b. 1892), Charles Bertie Wedd (1868-1945), Talbot Haes Whitehead (1892-1966) and Disney Alexander Wray (b. 1888). In 1932, solid and drift editions of the new Shrewsbury Sheet (152 New Series) were simultaneously published; this was thereafter to be a favourite map for testing the interpretative abilities of University geological classes (the sheet to the south, 166, was not destined to be published until almost 40 years later).

In 1938, the accompanying *Memoir* appeared. In it, Whitehead suggested (p. 13) that the Western Uriconian might be the oldest division and that the Longmyndian might be intermediate in date between it and the Eastern Uriconian; this idea, advanced very tentatively, was an inversion of earlier views and was apparently formulated in ignorance of the work of Cobbold and Whittard (1935). A great deal of new detailed information was incorporated in this *Memoir*: in particular, the first thorough studies of the Eastern Uriconian of The Lawley (1938, p. 27-30) and of the intrusive dykes in the Longmyndian (*ibid.*, p. 57-58) were presented.

The subdivisions proposed by Lapworth for the Eastern Longmyndian were retained, but a three-fold subdivision of the Western Longmyndian was now envisaged, paralleling that proposed by Blake in 1890:

Wentnor Series

Bayston Group: purple grits and conglomerates, with some hard shales: thickness about 8,000 ft.

Bridges Group: dark purple shales and sandy mudstones, with beds of grit and sandstone: maximum thickness, in the northern area, about 4,000 ft.

Oakswood Group: red-purple grits and conglomerates; thickness about 5,000 ft.

On the basis of disharmonious folding between competent and incompetent beds, it was thought that the Wentnor Series might be locally inverted (*ibid.*, p. 12, 53). The idea that the Longmyndian might be of Cambrian date, even in part, had by now been abandoned and has had no subsequent advocates.

In 1941, the exposures of igneous rocks in Stanner Rocks, Worsell Wood and Hanter Hill, near Kington in Radnorshire, were discussed at length by Norman Holgate and K.A. Knight Hallowes. These rocks had been attributed to the Uriconian by Callaway (1879d) and, though this attribution had been challenged by later writers on the area, Holgate and Knight Hallowes entirely accepted it. Moreover, they demonstrated that pebbles of rocks of Stanner-Hanter type were present in the Longmyndian conglomerates of Old Radnor Hill, a further reinforcement for the concept that the Uriconian antedated the Longmyndian.

The Developing Longmyndian Controversy

The wartime opening of a new quarry into Haughmond Hill provided a major new section through the Longmyndian. John Challinor (b. 1894), distinguished geological historian, concluded that representatives of both the Eastern and Western Longmyndian sequences were present; on the evidence of graded bedding in the former, he determined an eastward dip and concluded that the latter sequence was overlain by the former. On this basis, he suggested that the succession in the Long Mynd itself was an inverted one (1948, p. 108). His views were promptly challenged by Whitehead who, whilst accepting Challinor's conclusions on "way up", believed that the two sequences in Haughmond Hill were separated by a fault so that their order of succession was not evident. Whitehead further considered that, while the eastward dip of the Eastern Longmyndian rocks of Haughmond Hill certainly contrasted with their general westward dip on the Long Mynd, both might well nonetheless be right way up, since the structure and stratigraphy of these rocks were still problematical (1948, p. 182).

The first postwar postgraduate thesis devoted to the Precambrian of Shropshire was begun around 1950 by John H. James (b. 1924) at the University of Bristol, where Whittard had been Head of Department since 1937. His initial publication (1952) was a report of field-mapping in the Linley district. Here, evidence from graded bedding in tuffs within the Wentnor "Series" (Western Longmyndian) showed them to be inverted; the lavas of the Western Uriconian, lying above the Wentnor "Series" with marked angular discordance, must also be inverted and must in consequence be *older* than the Wentnor "Series", thus affording some support for Whitehead's theory.

On the basis of James' work, Whittard (1952) suggested the possibility of a deep, synclinal overfold in the Wentnor Series, whose presence would account for the equivalence in character of the Oakswood and Bayston Groups; the Bridges Group would form the core of this fold. He stressed, however, that this was highly conjectural. Support for this idea was shortly forthcoming, for in 1955 Whitehead reported the observation, at three places within a radius of five miles from Shrewsbury – at Haughmond Hill, at Sharpstone Quarry 2½ miles south of Shrewsbury, and in a disused quarry at Stapleton Common, five miles SSW of the city – of sedimentary structures in the Western Longmyndian rocks, each of which gave evidence that the base of the sequence lay towards the east or southeast. In contrast, Whitehead believed that the base of the Oakwood Group lay towards the west, as James had demonstrated. Final proof was unlikely to be forthcoming, however, since the position of the "nose" of the presumed fold lay beneath the Triassic sediment cover somewhere north of Shrewsbury.

The full results of James' work were given in an address to the Geological Society on November 2nd, 1955 and published in the following year (James, 1956). His work was confined exclusively to the southern part of the Long Mynd. He recognised the following sequence:

Wentnorian

Bridges Group: Purple banded shales and siltstones with occasional green shales. Thickness, 2,000-3,900 feet.

Oakswood Group: Purple massive sandstones containing three conglomerates: Lawn Hill 500-1,300 feet thick, Oakswood 100-200 feet, and Radligh 100-140 feet, with green laminated shaly tuffs, siltstones and sandstones near the base. Thickness, 3,500-5,500 feet.

Bayston Group: Purple, grey, greenish grey and dark grey sandstones with thin shales, and containing two conglomerates: Darnford 340 feet and Stanbatch 400 feet thick. Thickness, 5,600-6,300 feet.

Unconformity

Mintonian

Portway Group: Purple and green shales and conglomeratic sandstones with the Huckster Conglomerate at the base. Thickness, 1,600-3,800 feet.

Unconformity

Strettonian

Lightspout Group: Grey and greenish grey flaggy siltstones with purple siltstones and shales above, having the marker horizon of the Haddon Hill Grit near the base. Thickness, 2,100-2,800 feet.

Synalds Group: Purple shales with subordinate sandstones and containing the marker horizons of the Batch Volcanic beds, with the Upper Cardingmill Grit near the base. Thickness, 2,000-2,800 feet.

Burway Group: Greenish grey, finely laminated flags with massive siltstones and sandstones, and having the Lower Cardingmill Grit as the topmost bed and the Buckstone Rock as the basal bed. Thickness, 2,200-2,400 feet.

Stretton Shales: Grey-blue Brockhurst Shales and green and purple Watling Shales, with the Helmeth Grit as the basal bed. Thickness, 3,400 feet (about).

As was already apparent from Whittard's earlier paper (1952), James visualised the Wentnorian as a major isoclinal structure, overturned on the western limb, its core formed by the Bridges Group. The Uriconian was believed to underlie the whole of this structure, its western and eastern divisions being merely the surface expressions of the two limbs of an overall syncline (*ibid.*, Fig. 4) and not, as had been repeatedly suggested hitherto, the products of two unrelated igneous episodes.

James' stratigraphic views have by no means secured universal acceptance. In 1958, James Haward Taylor (1909-1968), in a review of Precambrian sedimentation in England and Wales, accepted that the Uriconian was indeed uniformly earlier than the Longmyndian but preferred to use the subdivisions of the Longmyndian recognised by the Survey geologists in 1938. Taylor noted that almost all the Precambrian sediments of the counties considered were the products of incomplete weathering and incomplete sorting and that indeed many of the sandstones were microbreccias, in which volcanic fragments were prominent. The Eastern Longmyndian sequence was considered typically geosynclinal, the Stretton Shales representing quiet argillaceous sedimentation prior to the orogenic phase and the Portway Group marking the change from synorogenic to post-orogenic conditions.

The Western Longmyndian, however, with its thick and commonly structureless sandstones, exhibited none of the characters of synorogenic sediments. Sedimentary structures (false-bedding and ripple marking) occur and, whilst the sediments are still poorly sorted and contain a variety of rock fragments and unstable constituents, there is less interstitial clay-grade material than in flysch and the larger grains tend to be more rounded. The facies was considered typical of the post-orogenic molasse and its conglomerates comparable to the *Nagel Fluh* of the Alps: it was considered to extend over the foreland, beyond the limits of the geosyncline. Taylor stressed that this was the most widespread facies, present also at Old Radnor, at Huntley, and at Brampton Bryan. All in all, the English and Welsh Precambrian sediments showed little trace of epicontinental sedimentation and showed every evidence of deposition in an orogenic zone where there was considerable volcanic activity.

James' views, however, continued to be accepted by Whittard (1958) in a field guide to South Shropshire. Furthermore, in a review of the tectonic history of Shropshire, David Lawrence Dineley (b. 1927; also of the University of Bristol) noted that geological parties from the Preston Montford Field Centre had been able to establish that the major synclinal overfold in the Longmyndian, detected by James in the southern Long Mynd, was continued northward (Dineley, 1960, p. 8).

The first geophysical studies of Shropshire were published in 1957. The results of an aeromagnetic survey of the Church Stretton area by Frederick Arthur Henson (1919-1957), of the University of Nottingham, demonstrated the very major character of the Church Stretton fracture line. Henson (1957) also showed that a strong magnetic anomaly was associated with the Uriconian of the Caer Caradoc area and used it to suggest that Uriconian volcanics continued eastwards from this region at a shallow depth under the Palaeozoic cover. Further work on this topic was brought to a close by his premature death (Evans, 1957). In the same year, palaeomagnetic readings taken in the Wentnor Series were mentioned in a general account of palaeomagnetism in the British Isles (Creer *et al.*, 1957).

In an incidental discussion of the Uriconian prefacing a description of the Ordovician sediments of Pontesford Hill, William Thornton Dean and Dineley (1961) stated that they had been unable to find evidence for an unconformity between the Wentnorian and Western Uriconian at Lydd Hole, such as had been reported by James; instead, they considered this to be a faulted junction. The fault appeared to be normal, hading at only a degree or two from the vertical; but its northward extension could not be mapped (1961, p. 368). In his revision of the field guide prepared years earlier by Whittard, Dean departed from the views of his predecessor by reverting to the subdivisions of the Longmyndian proposed in 1938.

In a review of the stratigraphy of the Precambrian of the British Isles, John Graham Comrie Anderson (b. 1910) stressed that correlation remained a problem of particular difficulty because the outcrops, outside northern Scotland, were relatively small and fault-bounded (1965, p. 91). He distinguished the English Precambrian region as "The south-east Kratogen" (*ibid.*, p. 28). He considered that the Rushton Schists, together with similar rocks in the Malvernian, were the oldest rocks of the Kratogen *(ibid.,* p. 98-99) and accepted Taylor's views that the Uriconian and Eastern Longmyndian were connected with a geosyncline and orogeny (or orogenies) of fairly late Precambrian date and that the Western Longmyndian was a postorogenic sediment of molasse type (*ibid.,* p. 98).

Renewal of Mapping by the Geological Survey

Mapping of Shropshire by officers of H. M. Geological Survey had recommenced between 1952 and 1959, successively under the direction of Victor Ambrose Eyles (d. 1978) and Vernon Wilson. The first fruits of this work was recognition of the presence of ignimbrites. Raymond Dearnley reported that at Earl's Hill, one mile S. E. of Pontesbury:

> "Thoroughly welded vitric tuffs occur . . . which have suffered extreme compression and distortion, and the resulting (eutaxitic) structure may resemble flow banding although the individual primary shards are well preserved" (1966, p. 1).

Fragments of ignimbritic material were noted also in tuffs near Lydd Hole, on the flanks of the Long Mynd, and in the Eastern Uriconian southwest of Uppington and at Lawrence Hill. Sillars (non-welded, pumice and shard tuffs) were present on The Ercall and in the Stretton Hills (Hope Bowdler to Willstone Hill) east of Church Stretton. At Lilleshall Hill, the tuffs contained fragments of both welded and non-welded tuffs and sillars. In the Longmyndian conglomerates (Haughmond, Oakswood, Lawn Hill), these rock types were also represented. From these observations, it is apparent that ash-flow tuffs formed quite widely in the Uriconian (as a result, presumably, of deposition of hot fragmental material from *nuées ardentes*), which in turn implies that eruptions were occurring, at least locally, under terrestrial conditions.

In 1967, the long-awaited New Series Sheet 166 of the Geological Survey was at last published in solid and drift editions. More units of Uriconian and Longmyndian were discriminated than in the earlier Shrewsbury sheet. The officers primarily responsible for mapping in this area were D.C. Greig, J.E. Wright, B.A. Hains, and G.H. Mitchell, and these four geologists were the principal authors of the accompanying *Memoir,* published in 1968. The chapters on the Uriconian and Longmyndian are undoubtedly the most important contribution to knowledge of the Shropshire Precambrian forthcoming this century, providing as they do a great volume of additional information about the petrology, order of superposition, and faulting of the southern Salopian outcrops of the Uriconian (the southern part of The Lawley, Caer Caradoc and the adjacent hills, and the Wart Hill inlier) and of the southern Longmyndian sediments. Beyond a note of the similarity of the rock types (Greig *et al.*, 1968, p. 30), the possible identity between the Eastern and Western Uriconian is not commented on.

The authors took issue with James over the existence of a major unconformity at the base of the Portway Group, for which they could find no adequate evidence (*ibid.,* p. 43-44): for this reason, they rejected his 'Mintonian' division. Nor could they find definite confirmatory evidence for his theory of a major synclinal structure in the Longmyndian. However, since available evidence did not directly contradict the idea, his theory was felt to be at least tenable (*ibid.*, p. 50).

The Survey officers also took issue with Taylor's ideas on the sedimentary environment of the Longmyndian. They reported a variety of sedimentary structures, in particular in the Stretton Series, which were indicative of a shallow-water environment with intermittent exposure, such as might occur in a tidal estuary or delta (*ibid.*, p. 66-73, 75). They considered the sandstones throughout the Longmyndian to be of subgreywacke type and, although recognising that the precise conditions of formation of such sediments were not understood, they felt the grain-size distribution to be consistent with formation as river-transported sediments or as shallow-marine, tidal-flat deposits (*ibid.*, p. 74-75). Thus they could not accept the model of a "geosynclinal" Eastern series and a "molasse" Western series envisaged by Taylor. Instead, they commented:

> "The whole sequence appears to be a product of Molasse-type (paralic) sedimentation, showing similarities of tectonic setting, lithology and thickness with the Old Red Sandstone of Scotland.
>
> It is considered that the Longmyndian was deposited in a narrow crustal depression which originated by subsidence of the underlying basement between major fault lines. It is possible that the Pontesford-Linley and Church Stretton lines of disturbance were the two controlling fault systems and the apparent absence of the Stretton Series to the south-east of the Church Stretton line may be due to original non-deposition. This picture invites comparison with the taphrogeosyncline of Kay (1951, p. 61) which arises from block faulting in the basement rather than from broad crustal downwarp. The form of the Longmyndian trough is not identical with that illustrated by Kay, the latter being wedge-shaped in cross section and fault-bounded on one side only. The shallow-water character of the Longmyndian sediments may be compared with the non-marine, fluviatile and estuarine deposits of the Triassic taphrogeosyncline of North America . . . The interpretation of an unconformity below the Wentnor Series required slight, pre-Wentnorian folding of the Stretton Series, followed by limited erosion and subsequent deposition of the Wentnor Series over an area more extensive than that of the original trough. It is further suggested that the deep overfolding of the whole Longmyndian sequence was restricted approximately to the limits of the original depositional trough and was due largely to renewed compressive block movements in the underlying basement. Similar intense folds, apparently resulting from movements of crustal blocks, are known from other areas, e.g. the overfolding and thrusting of the Palaeozoic rocks to the west of the Malvern line of disturbance." (R. Dearnley and J.E. Wright *in* Greig *et al.*, 1968, p. 175-76).

In 1970, the first author published his only contribution to the elucidation of the Shropshire Precambrian – the discovery, 11 years earlier, of a typically spindle shaped volcanic bomb from among the very poorly exposed, down-faulted rhyolitic tuffs on The Lawley's western flanks (Sutton and Sarjeant, 1972, p. 299). Although this was not *in situ,* it strongly indicated the local presence of coarse, agglomeratic material and the proximity of a vent.

In a review of the Proterozoic history of southern Britain, J.W. Baker (1971) reconsidered briefly the rocks of the Stanner-Hanter region of Radnorshire. He noted that these rocks were certainly pre-Longmyndian and suggested that they were the product of four unrelated events:

1) Intrusion of early basic magmas.
2) Amphibolitisation, aplitic veining and pegmatites, consistent with an amphibolite facies regional metamorphism.
3) Acid porphyries and granophyres related to Uriconian vulcanicity.
4) Unmetamorphosed dykes related to a regional swarm.

This sequence of events in part paralleled that presumed to have occurred in the formation of the ancient rocks of the Malvern Hills, farther to the east (*ibid.*, p. 256-257).

Baker also noted that the Rushton Schists and Primrose Hill Gneisses should be considered, not as separate units, but as part of "one continuous tract extending beneath a Cambrian and Pleistocene cover and bounded on the north by the same post-Uriconian pre-Cambrian fault" (*ibid.*, p. 256). Greig *et al.* (1968) considered the Rushton Schists to be regionally metamorphosed rocks of the greenschist facies, containing quartz, mica, chlorite and soda felspars, with local garnet clusters; Baker noted that on Primrose Hill, similar schists pass into garnetiferous acid gneisses, in which alkali-felspar was important. He believed that such rocks were formerly more extensive, serving as the main source for the "Monian-like" lithic fragments in the Uriconian and Longmyndian.

In 1972, R.S. Thorpe published the results of geochemical analyses of igneous rocks from Shropshire, the Malvern Hills and Charnwood Forest. He observed that the Uriconian rocks appeared to be composed dominantly of andesites and dacites, noting that, from the recently published Geological Survey map, the relative thicknesses were basalt 27 per cent, andesite/dacite 61 per cent and rhyolite only 12 per cent. He found that the andesites and dacites had low chromium and nickel contents and that the rhyolites, although having suffered the effects of alkali migration during metamorphism, nevertheless had contents of zirconium, yttrium, lanthanum and cerium that were less variable and relatively high in comparison with the intermediate and basic rocks. He concluded that the Uriconian is a calc alkaline, basalt-andesite-rhyolite association and that a primary andesite lava was important in its evolution.

The high yttrium content, he noted, was a feature shared with the Warren House volcanic suite of the Malvern Hills. The high content of this element relative to lanthanum and cerium was an unusual characteristic shared also with the Charnian volcanics; though the formation of chemical complexes during fractionation might have increased this ratio, the high yttrium content might well be a primary feature.

Thorpe concluded that the data he presented indicated that the three magmatic suites were not to be regarded as lateral equivalents and that, although they might be contemporaneous, the evidence did not favour that possibility. He considered that the Uriconian complex probably predated the Malvernian igneous rocks and was a late- to post-orogenic volcanic group, either post-orogenic relative to a middle Proterozoic tectonic cycle or syn-orogenic relative to a late Proterozoic orogeny.

Radiometric Age Determinations

Only during the last decade have age determinations for Shropshire Precambrian rocks started to become available. In their review of age determinations on Precambrian and early Phanerozoic rocks from Wales and the Welsh Border, Fitch *et al.* included in a list of apparent ages "not clearly too low" (1969, p. 33-34):

"Ercall rhyolite, Buckatree Glen .... 632 ± 32 m.y.
Basalt, Leaton Quarry, Wrockwardine .... 638 ± 81 m.y.
Rhyolite, Pontesford Hill, Salop .... 677 ± 81 m.y."

These results supported earlier conclusions that the Uriconian rocks were extruded in the latest Precambrian; however, since the authors noted that all the rocks they examined "show clear evidence of subsequent alteration and/or deformation" (*ibid.*,) their value is questionable. Indeed, J.W. Baker considered them to be the "result of a post-formational radiogenic event" (1971, p. 256).

Four years later, Adrian H. Bath reported the first rubidium-strontium isotope data for the Longmyndian sediments. The samples examined were from the older part of the sequence. For the Burway Group, the mean age obtained was 529 ± 6 m.y.; for the Synalds Group, 452 ± 31 m.y.; and for the youngest beds studied, the Lightspout Group, 529 ± 23 m.y. In view of past stratigraphical interpretations, these results were surprising, for they would imply that the Longmyndian rocks were of Lower Palaeozoic date, an unexpected harking-back to the early ideas of Murchison! However, Bath stressed that the results should be viewed with caution; he concluded only that the Longmyndian rocks were deposited no earlier than around 600 million years ago, in the very latest Precambrian or Cambrian (1974, p. 573).

Even this conclusion was swiftly challenged by P. Toghill (1975, p. 235), who suggested that the age obtained was that of a late Cambrian metamorphic episode affecting the Eastern Longmyndian sediments. Bath responded by pointing out that ratios of strontium^{87} to strontium^{86} had also been reported and that these were a factor in deciding the maximum probable depositional date for the sequence. He did not, however, discount the possibility that the rubidium-strontium ages corresponded to a later metamorphic episode.

The most recent review of the Precambrian of the Borderland is by Frederick Weir Dunning (1975). After again noting that the Rushton and Primrose Hill metamorphics and the Hanter-Stanner rocks strongly recalled those of the Malvernian sequence, he suggested that all these rocks were younger than 1400 m.y. but presumably not younger than the Uriconian (c. 620 m.y.). His review of information concerning the Uriconian rocks, though useful, contains no new ideas, though he stresses that their age remains uncertain and might prove as old as 900-1,000 m.y.

In a discussion of the Longmyndian rocks, Dunning (*ibid.*) stressed that the 600 m.y. age obtained for this sequence is also the potassium-argon age computed for a widespread overprinting event in the British Precambrian basement and a highly likely age for the folding, rather than the deposition, of these rocks. Certainly it might have been possible for the Longmyndian sediments to have been deposited in a period as brief as ten million years and for folding to have immediately followed deposition; but Dunning feels that the date is incompatible with the palaeomagnetic work of Creer *et al.* (1957) and more recent, unpublished work of J.C. Briden:

> "The Longmyndian pole directions are seemingly too far different from Cambrian directions for the Longmyndian to be as young as 600 m.y., unless there were large and rapid movements at this time, which, of course, is perfectly possible." (Dunning, 1975, p. 93).

He notes that, if Bath's computation is indeed correct, the only possible British correlations for the Longmyndian would be the Argyll Group (Middle Dalradian) of Scotland and perhaps the highest formations of the Mona Complex and Arvonian. The earliest date (900-950 m.y.) would give it no correlatives in the Scottish Torridonian, but would mean that the movements which tilted the Stoer Group of the Torridonian might have coincided with the taphrogenic movements which formed the Longmyndian trough and initiated sedimentation therein. An equating of the Longmyndian sediments with a part of the Moinian succession of Scotland, which may well have formed during the interval between sedimentation of the Stoer and Torridon Groups, would then be possible.

THE URICONIAN OUTSIDE SHROPSHIRE

Though this history is concerned essentially with Shropshire, it should be noted that volcanic rocks, of a type sufficiently comparable in general features to be loosely designated "Uriconian" and of presumed or proven Precambrian age, occur not only in the immediately adjacent counties of Herefordshire and Radnorshire (already discussed) but also quite widely in other parts of England. To the eastward, they crop out as components in the Malvernian of the Malvern Hills (Groom, 1900; Raw, 1952); at Barnt Green, in the Lickey Hills of Worcestershire just south of Birmingham (Lapworth, 1898); at Caldecote and Hartshill in Warwickshire (Allen, 1957); and in the Charnian sequence of Charnwood Forest, Leicestershire (see paper by T.D. Ford, 1979). Subsurface occurrences have been reported from boreholes at Woo Dale, near Buxton, Derbyshire (Cope, 1950); at Glinton, near Peterborough (Kent, 1962); farther south in Northamptonshire, at Orton and Great Oxenden (Smith, 1936; Kent, 1962); at Upwood and Warboys, Huntingdonshire (Dunning, 1975); and even at North Creake, Norfolk (Kent, 1947). To the westward, the presence of comparable rocks in the Pebidian of Pembrokeshire and in the Monian of Anglesey has already been mentioned; but it should be noted in addition that Precambrian volcanic rocks of Uriconian type occur also in the Arvonian of Caernarvonshire, from which ignimbrites have been reported (Morris and Fearnsides, 1926; Dearnley, 1966).

THE REMAINING PROBLEMS

After some 180 years of intermittent study, the distribution and character of the Shropshire Precambrian rocks are now known with fair clarity, allowing for the problems presented by inadequate exposure in many areas. However, the sequence of geological events, although now agreed in general outline, is still not fully understood and even the overall structure remains a matter for dispute. More extensive absolute age-dates, obtained by radioactive means, should soon be forthcoming for some of these rocks. However, the structural complexities are not likely to be resolved in detail until a great deal of further work has been done, by instrumental as well as by conventional means. Unquestionably, this Precambrian sequence of the Welsh Borderlands is destined to continue to pose problems to geologists for many years to come.

ACKNOWLEDGEMENTS

The authors are indebted to Mrs. Joan M. Eyles (Great Rissington, nr. Cheltenham, Glos.), Dr. C.P. Hughes (Sedgwick Museum, Cambridge), Dr. John F. Lewry and Dr. Geoffrey Parslow (University of Regina) and Dr. A.W.A. Rushton (Institute of Geological Sciences, London) for their help and advice. They would also like to thank Dr. Colin McGregor and Mrs. D.M. Sutherland (Head of Library Services) of the Geological Survey of Canada for arranging the loan to W.A.S. Sarjeant of all relevant one-inch maps of Shropshire.

REFERENCES

In this section, only those works directly cited in the text are listed. It should be stressed, however, that many others – in particular, the standard stratigraphic works on England and Wales, works on the history of geology, and biographies and obituary notices of geologists – have been consulted in addition.

Aikin, Arthur, 1797, Journal of a tour through north Wales and part of Shropshire, with observations in mineralogy and other branches of natural history: London.

——————, 1811, Observations on The Wrekin, and the great coal-field of Shropshire: Geological Society of London Transactions, v. 1, p. 191-212.

"A Lady", 1862, The rocks of The Wrekin and what is written upon them: Wellington, Salop, for the Author by Hobson: London, Houlston and Wright, 48 p.

Allen, John R.L., 1957, The Pre-Cambrian geology of Caldecote and Hartshill, Warwickshire: Leicester Literary and Philosophical Society Transactions, v. 51, p. 16-31.

Allport, Samuel, 1877, On certain ancient devitrified pitchstones and perlites from the Lower Silurian district of Shropshire: Geological Society of London Quarterly Journal, v. 33, p. 449-460.

Anderson, J.G.C., 1965, The Precambrian of the British Isles, *in* Rankama, K., ed., The geologic systems: The Precambrian, v. 2, p. 25-111.

Baker, J.W., 1971, The Proterozoic history of Southern England: Geologists' Association of London Proceedings, v. 82, p. 244-266.

Bath, Adrian H., 1974, New isotopic age data on rocks from the Long Mynd, Shropshire: Geological Society of London Journal, v. 130, p. 567-574.

Beddoes, T., 1791, Observations on the affinity between basalts and granite: Philosophical Transactions of the Royal Society of London, v. 81, p. 48.

Blake, John Frederick, 1888, On the Monian system of rocks: Geological Society of London Quarterly Journal, v. 44, p. 463-547.

——————, 1890, On the Monian and basal Cambrian rocks of Shropshire: Geological Society of London Quarterly Journal, v. 46, p. 386-420.

Bonney, T.G., 1879, The Pre-Cambrian rocks of Great Britain: Birmingham Philosophical Society Proceedings, v. 1, p. 140-159.

Boulton, William S., 1904, The igneous rocks of Pontesford Hill, (Shropshire): Geological Society of London Quarterly Journal, v. 60, p. 450-486.

Callaway, Charles, 1874, On the occurrence of a Tremadoc area near The Wrekin in south Shropshire, with description of a new fauna (Abstract): Geological Society of London Quarterly Journal, v. 30, p. 196.

———, 1877, On a new area of Upper Cambrian rocks in south Shropshire: Geological Society of London Quarterly Journal, v. 32, p. 652-672.

———, 1878a, Recent discoveries in the geology of Shropshire. I: Midland Naturalist, v. 1, p. 205-209.

———, 1878b, On the quartzites of Shropshire: Geological Society of London Quarterly Journal, v. 34, p. 754-763.

———, 1879a, Report of lecture by C. Callaway on "The pre-Cambrian rocks of Shropshire", delivered to the Birmingham Natural History and Microscopical Society: Midland Naturalist, v. 2, p. 81.

———, 1879b, Two Precambrian groups in Shropshire: Midland Naturalist, v. 2, p. 158-159.

———, 1879c, Recent discoveries in the geology of Shropshire. II. The quartzites of Shropshire: Midland Naturalist, v. 2, p. 39-44.

———, 1879d, The Precambrian rocks of Shropshire. Pt. I. With notes on the microscopic structure of some of the rocks, by T.G. Bonney: Geological Society of London Quarterly Journal, v. 35, p. 643-669.

———, 1879e, The oldest mountain in England: Popular Science Review, new series, v. 3, p. 15-24.

———, 1880, On a second Precambrian group in the Malvern Hills: Geological Society of London Quarterly Journal, v. 36, p. 536-539.

———, 1882, The Precambrian (Archaean) rocks of Shropshire. Pt. II. With notes on the microscopic structure of some of the rocks, by T.G. Bonney, Geological Society of London Quarterly Journal, v. 38, p. 119-126.

———, 1884, A new metamorphic area in Shropshire: Geological Magazine, dec. 2, v. 1, p. 362.

———, 1886, On some derived fragments in the Longmynd and newer Archaean rocks of Shropshire: Geological Society of London Quarterly Journal, v. 42, p. 481-485.

———, 1888, On some ancient Salopian conglomerates: Shropshire Archaeological Society Transactions, v. 9, p. 239-243.

———, 1891, On the unconformities between the rock-systems underlying the Cambrian quartzite in Shropshire: Geological Society of London Quarterly Journal, v. 47, p. 109-125.

———, 1898, On the metamorphism of a series of grits and shales in northern Anglesey: Geological Society of London Quarterly Journal, v. 54, p. 374-381.

———, 1900, On the Longmyndian inliers at Old Radnor and Huntley (Gloucestershire): Geological Society of London Quarterly Journal, v. 56, p. 511-520.

Challinor, J., 1948, New evidence concerning the original order of deposition of the Longmyndian rocks: Geological Magazine, v. 85, p. 107-109.

Cobbold, E.S., 1900, The geology of the district, *in* Campbell-Hyslop, C.W., ed., Church Stretton: Shrewsbury: Wilding, v. 1, p. 1-115,

———, 1927, The stratigraphy and geological structure of the Cambrian area of Comley (Shropshire): Geological Society of London Quarterly Journal, v. 83, p. 551-573.

Cobbold, E.S. and Whittard, W.F., 1935, The Helmeth Grits of the Caradoc Range, Church Stretton; their bearing on part of the Pre-Cambrian succession of Shropshire: Geologists' Association, London, Proceedings, v. 46, p. 348-359.

Cope, F.W., 1949, The Woodale Borehole, near Buxton: Proceedings of the Geological Society of London, no. 1446, p. 24 (also in Geological Society of London Quarterly

Journal, v. 105, p. iv).
Cox, Arthur Henry, 1912, On an inlier of Longmyndian and Cambrian rocks at Pedwardine (Herefordshire): Geological Society of London Quarterly Journal, v. 68, p. 364-373.
Creer, K.M., Irving, E. and Runcorn, S.K., 1957, Geophysical interpretation of palaeomagnetic directions from Great Britain: Philosophical Transactions of the Royal Society of London, series A, v. 250, p. 144-156.
Dean, W.T., 1968, Geological itineraries in South Shropshire, with a section, "The Clee Hills", by D.L. Dineley: Geologists' Association of London, Guide no. 27, 48 p.
Dean, W.T. and Dineley, D.L., 1961, The Ordovician and associated Pre-Cambrian rocks of the Pontesford district, Shropshire: Geological Magazine, v. 98, no. 5, p. 367-376.
Dearnley, R., 1966, Ignimbrites from the Uriconian and Arvonian: Bulletin of the Geological Survey of Great Britain, London, v. 24, p. 1-6.
Dineley, D.L., 1960, Shropshire geology: an outline of the tectonic history: Field Studies, v. 1, no. 2, p. 1-23.
Dunning, Frederick W., 1975, Precambrian craton of central England and the Welsh Borders, *in* A.L. Harris *et al.*, eds., A correlation of the Precambrian rocks in the British Isles: Geological Society of London Special Reports, no. 6. London: Geological Society of London, p. 83-96.
Evans, W.D., 1957, Obituary notice: Frederick Arthur Henson: Geological Society of London Proceedings, no. 1554, p. 146.
Eyton, C., 1869, Notes on the geology of north Shropshire: London: Robt. Hardwicke, 88 p.
Fitch, F.J., Miller, J.A., Evans, A.L., Grasty, R.L. and Meneisy, M.Y., 1969, Isotopic age determinations on rocks from Wales and the Welsh border, *in* Wood, A., ed., The Pre-Cambrian and Lower Palaeozoic rocks of Wales: Cardiff: University of Wales Press, 461 p. (p. 22-45).
Ford, T.D., 1979, The history of the Precambrian rocks of Charnwood Forest, England, in Kupsch, W.O. and Sarjeant, W.A.S., eds., History of Concepts in Precambrian Geology: Geological Association of Canada, Special Paper 19, p. 65.
Garwood, Edmund J. and Goodyear, Miss E., 1919, On the geology of the Old Radnor district, with special reference to an algal development in the Woolhope Limestone: Geological Society of London Quarterly Journal, v. 74 (for 1918), p. 1-30.
Greig, D.C., Wright, J.E., Hains, B.A. and Mitchell, G.H., 1968, Geology of the country around Church Stretton, Craven Arms, Wenlock Edge and Brown Clee: Geological Survey of Great Britain, London, 379 p.
Groom, T.T., 1900, On the geological structure of portions of the Malvern and Abberley Hills: Geological Society of London Quarterly Journal, v. 56, p. 153-197.
Harrington, H.J., *et al.*, 1959, Treatise on Invertebrate Palaeontology, Part O. Arthropoda: New York: Geological Society of America, 560 p.
Henson, F.A., 1957, An aeromagnetic survey of the Church Stretton area, Shropshire: Geologists' Association Proceedings, v. 68, p. 107.
Hicks, Henry, 1890, Discussion of Blake, John F., 1890: "On the Monian and basal Cambrian rocks of Shropshire": *in* Geological Society of London Quarterly Journal, v. 46.
Holgate, N. and Knight Hallowes, K.A., 1941, The igneous rocks of the Stanner-Hanter district, Radnorshire: Geological Magazine, v. 78, p. 241-267.
James, J.H., 1952, Notes on the relationship of the Uriconian and Longmyndian rocks near Linley, Shropshire: Geologists' Association, London, Proceedings, v. 63, p. 198-200.
——————, 1956, The structure and stratigraphy of part of the Pre-Cambrian outcrop

between Church Stretton and Linley, Shropshire: Geological Society of London Quarterly Journal, v. 112, p. 315-337.

Kent, Peter E., 1947, A deep boring at North Creake, Norfolk: Geological Magazine, v. 84, p. 2-18.

———, 1962, A borehole to basement rocks at Glinton, near Peterborough, Northamptonshire: Geological Society of London Proceedings, no. 1595, p. 40-42.

Lapworth, Charles, 1888a, On the discovery of the *Olenellus* Fauna in the Lower Cambrian rocks of Britain: Nature, v. 39, p. 213.

———, 1888b, On the discovery of the *Olenellus* Fauna in the Lower Cambrian rocks of Britain: Geological Magazine, dec. 2, v. 5, p. 485.

———, 1891, On *Olenellus Callavei* and its Geological Relationships: Geological Magazine, dec. 3, v. 8, p. 529-536.

———, 1898, A sketch of the geology of the Birmingham district: Proceedings of the Geologists' Association, London, v. 15, p. 313-416. (Reprinted, with additions, 1907: Birmingham, Cornish Brothers, 104 p.)

———, 1903, Speech on presentation of the Murchison Medal to Charles Callaway: Proceedings of the Geological Society of London, 1903, p. xl-xli.

———, 1904, [Written communication in discussion of Boulton, W.S., 1904, q.v.]

———, 1910, Geology in the Field, Shropshire: *in* Geologists' Association Centenary Volume, Geology in the Field, pt. IV: Geologists' Association, London, p. 739-769.

Lapworth, Charles and Watts, W.W., 1894, The geology of South Shropshire, with special reference to the district to be visited during the long excursion: Geologists' Association, London, Proceedings, v. 13, p. 297-355.

Morris, T.O. and Fearnsides, W.G., 1926, The stratigraphy and structure of the Cambrian Slate-Belt of Nantlle (Caernarvonshire): Geological Society of London Quarterly Journal, v. 82, p. 250-299.

Murchison, Roderick I., 1833, On the sedimentary deposits which occupy the western parts of Shropshire and Herefordshire, and are prolonged from N.E. and S.W., through Radnorshire, Brecknockshire and Caermarthenshire, with descriptions of the accompanying rocks of intrusive or igneous characters. II: Geological Society of London Proceedings, v. 1, no. 3, p. 474-477.

———, 1834a, On the structure and classification of the Transition Rocks of Shropshire, Herefordshire and part of Wales, and on the lines of disturbance that have effected that series of deposits, including the Valley of Elevation of Woolhope: Geological Society of London Proceedings, v. 2, no. 34, p. 13-20.

———, 1834b, On certain trap rocks in the Counties of Salop, Montgomery, Radnor, Brecon, Caermarthen, Hereford and Worcester; and the effects producced by them upon the stratified deposits: Geological Society of London Proceedings, v. 2, no. 36, p. 85-93.

———, 1838, The Silurian system, founded on geological researches in the counties of Salop, Hereford, Radnor, Montgomery, Caermarthen, Brecon, Pembroke, Monmouth, Gloucester, Worcester and Stafford; with descriptions of the coalfields and overlying formations . . .: London, John Murray, 2 vols.

———, 1867, Siluria. A history of the oldest rocks in the British Isles and other countries; with sketches of the origin and distribution of native gold, the general succession of geological formations, and changes of the earth's surface. London, John Murray, 566 p.

Pocock, R.W., Whitehead, T.H., Wedd, C.B. and Robertson, T., 1938, Shrewsbury District including the Hanwood Coalfield (One-inch Geological Sheet 152 New Series):

Memoirs of the Geological Survey of Great Britain, London: H. M. Stationery Office, 297 p.
Raw, Frank, 1952, Structure and origin of the Malvern Hills: Geologists' Association, London, Proceedings, v. 63, p. 227-239.
Reade, T.M. and Holland, P., 1909, Analyses of Longmyndian Rocks: Liverpool Geological Society Proceedings, v. 10, p. 276-287.
Salter, J.W., 1856, On fossil remains in the Cambrian rocks of the Longmynd and North Wales: Geological Society of London Proceedings, v. 12, p. 246-251.
——————, 1857, On Annelide-burrows and surface-markings from the Cambrian Rocks of the Longmynd: Geological Society of London Proceedings, p. 199-206.
Smith, B., 1936, On a boring at Oxendon, Northamptonshire: Summary of Progress of the Geological Survey of Great Britain and the Museum of Practical Geology, London, for 1934, pt. 2, p. 28-31.
Sutton, Ian D. and Sarjeant, W.A.S., 1970, Weekend excursion to Shropshire: Mercian Geologist, v. 3, no. 3, p. 291-303.
Symonds, Rev. W.S., 1877, On the geology of Church Stretton and Ludlow: Shropshire Archaeological and Natural History Society Transactions, v. 1, p. 395-429.
Taylor, J.H., 1958, Pre-Cambrian sedimentation in England and Wales: Eclogae Geologicae Helvetiae, v. 51, p. 1078-1092.
Thorpe, R.S., 1972, The geochemistry and correlation of the Warren House, the Uriconian and the Charnian volcanic rocks from the English Pre-Cambrian: Geologists' Association, London, Proceedings, v. 83, p. 269-285.
Townson, Robert, 1799, Tracts and observations in Natural History and Physiology: London, The Author, 282 p.
Toghill, P. and Bath, A.H., 1975, Discussion of Longmyndian dates: Geological Society of London Journal, v. 131, p. 235-236.
Walford, Thomas, 1818, The scientific tourist through England, Wales and Scotland: London; Booth, 2 vols.
Watts, W.W., 1925, The geology of South Shropshire: Geologists' Association, London, 73 p.
——————, 1936, Edgar Sterling Cobbold D.Sc., F.G.S.: Caradoc and Severn Valley Field Club, Transactions, Shrewsbury, for 1936, p. 1-13.
Whitaker, W. and Watts, W.W., 1889, List of works on the geology, mineralogy and palaeontology of Shropshire: Shropshire Archaeological and Natural History Society Transactions, v. 12, p. 33-62.
Whitehead, T.H., 1948, Longmyndian stratigraphy: Geological Magazine, v. 85, p. 181-2.
——————, 1955, The Western Longmyndian rocks of the Shrewsbury district: Geological Magazine, v. 92, p. 465-470.
Whittard, W.F., 1952, A geology of south Shropshire: Geologists' Association, London, Proceedings, v. 63, p. 143-197.
——————, 1958, Geology of some classic British areas: Geological itineraries for South Shropshire: Geologists' Association Guides, no. 27, 23 p.

History of Concepts in Precambrian Geology, edited by W.O. Kupsch and W.A.S. Sarjeant, Geological Association of Canada Special Paper 19, 1979

GEOLOGICAL CONCEPTS DEVELOPED IN THE PRECAMBRIAN OF SUDBURY, ONTARIO

John S. Stevenson
Department of Geological Sciences, McGill University,
Montreal, Quebec H3C 3G1

ABSTRACT

Although Alexander Murray of the Geological Survey of Canada described nickel in "magnetic trap" in 1856 near the present site of the Creighton mine, Sudbury, Ontario, little further attention was given to the occurrence until the location of the Murray mine in 1883. Since then, the nickel-copper orebodies of Sudbury have been extensively developed, their relationship to the enclosing norite and micropegmatite of the Sudbury Nickel Irruptive has been carefully studied and varying concepts of their origin proposed.

The "classical" interpretation of the geological history of the area, espoused by Coleman (1905) and developed by Collins (1934), views the ore-bearing Nickel Irruptive, Proterozoic and pre-Keweenawan in age, as a single body intruded in much the form it has today. The ores were thought to be derived by magmatic segregation from the Nickel Irruptive magma. Knight (1917) believed hydrothermal processes to be important in the formation of the Sudbury ores; he first enunciated the ring-dyke concept which pictures the composite Irruptive as intruded around the edges of a "ring-graben". This idea was later developed by Thomson (1957) and Williams (1957). In 1962, Hawley reaffirmed the Coleman-Collins concept, with certain modifications based on detailed mineralogical work.

The relative ages of the norite and micropegmatite of the Nickel Irruptive have long been a subject of much discussion; geochronologic work, only possible in recent years, indicates a younger age for the micropegmatite when compared with the norite.

Interest in lunar geology has been reflected in the proposal that the Sudbury Basin is a meteoritic crater, first suggested by Dietz in 1962 and developed by Guy-Bray and Geological Staff (1966), French (1967) and others. Basic to this concept is that of the origin of the Onaping formation within the Basin. Some workers such as French (1972) and Peredery (1972) favour an origin related to meteoritic impact for the rock in this formation, while others, such as Stevenson (1972), interpret the Onaping as an ash-flow deposit of volcanic origin. Thus Sudbury continues to be a very active area for the testing and study of geological concepts relating to Precambrian rocks.

RESUME

Bien qu'Alexander Murray du Service Géologique du Canada ait décrit en 1856 du nickel en "trappe magnetique" près du site actuel de la mine Creighton, Sudbury, Ontario, on n'accorda que peu d'attention à l'occurrence jusqu'à la localisation de la mine Murray en 1883. Depuis lors, les gisements de nickel – cuivre de Sudbury ont été très développés, et leur relation avec la norite encaissante et la micropegmatite du "Sudbury Nickel Irruptive" a été étudiée avec attention et divers concepts sur leur origine furent proposés.

L'interprétation "classique" de l'histoire géologique de la région, adoptée par Coleman (1905) et développée par Collins (1934), considère le "Nickel Irruptive" minéralisé, d'âge Protérozoïque et pré-Keweenaw, comme un corps unique mis en place pratiquement dans la forme qu'il a aujourd'hui. On pensait que les minerais étaient dérivés du magma du Nickel Irruptive par ségrégation magmatique. Knight (1917) croyait à l'importance des processus hydrothermaux dans la formation des minerais de Sudbury; il énonça le premier le concept du ring-dyke qui décrit l'éruptif composite comme une intrusion sur les bords d'un "graben en anneau". Plus tard, cette idée fut développée par Thomson (1956) et Williams (1957). En 1962, Hawley repris le concept de Coleman-Collins, avec certaines modifications basées sur un travail, minéralogique détaillé.

Les âges relatifs de la norite et de la micropegmatite du Nickel Irruptive ont longtemps fait l'objet de beaucoup de dissensions; les travaux géochronologiques, qui furent possibles seulement ces dernières années, indiquent pour la micropegmatite un âge plus jeune que celui de la norite. L'intérêt pour la géologie lunaire se reflète dans l'hypothèse selon laquelle le Bassin de Sudbury est un cratère météoritique, hypothèse suggérée pour la première fois par Dietz en 1962 et développée par Guy-Bray (1966) French (1967) et d'autres. A la base de ce concept on retrouve le problème posé par l'origine de la formation Onaping occupant l'intérieur du Bassin. Certains auteurs tels French (1972) et Peredery (1972) favorisent pour les roches de cette formation une origine reliée à un impact météoritique tandis que d'autres, comme Stevenson (1972) interprètent la formation Onaping comme un dépôt de coulée de cendres d'origine volcanique. Ainsi Sudbury continue d'être une région très attirante pour tester et étudier les concepts géologiques reliés aux roches précambriennes.

INTRODUCTION

The Nickel Irruptive at Sudbury (Fig. 1) contains the largest known amount of nickel-sulphide ore in the world and one must conclude that it occupies a unique spot in Precambrian geology. Hence it is not surprising that the geology of the Sudbury Nickel Irruptive has been studied ever since the first major discoveries in 1883, by a great many mineralogists, and the geological concepts developed here have influenced concepts developed elsewhere in the Precambrian. The geological work has been enhanced by the tremendous amount of underground development, but much remains to be done. As Collins (1934, p. 123) has said: "The Sudbury Nickel Irruptive furnishes a most interesting case of the painfully slow, caterpillar-like, yet logical way in which we grope our way to an understanding of big and intricate geological bodies".

EARLY EXPLORATION

In the Sudbury area the first nickel was found, along with copper, in "magnetic trap" by Alexander Murray in 1856 on Salter's Meridian line, a line that followed along the west line of Snider township (Murray, 1857). The find was apparently along the norite contact just west of the Creighton pit (Barlow, 1904). No quantity of ore was found at that time and no prospecting done on the showing. It was not until the 1883 discovery of the Murray orebody (Fig. 2) in a cut on the Canadian Pacific Railway, then under construction, that prospecting, exploration and develop-

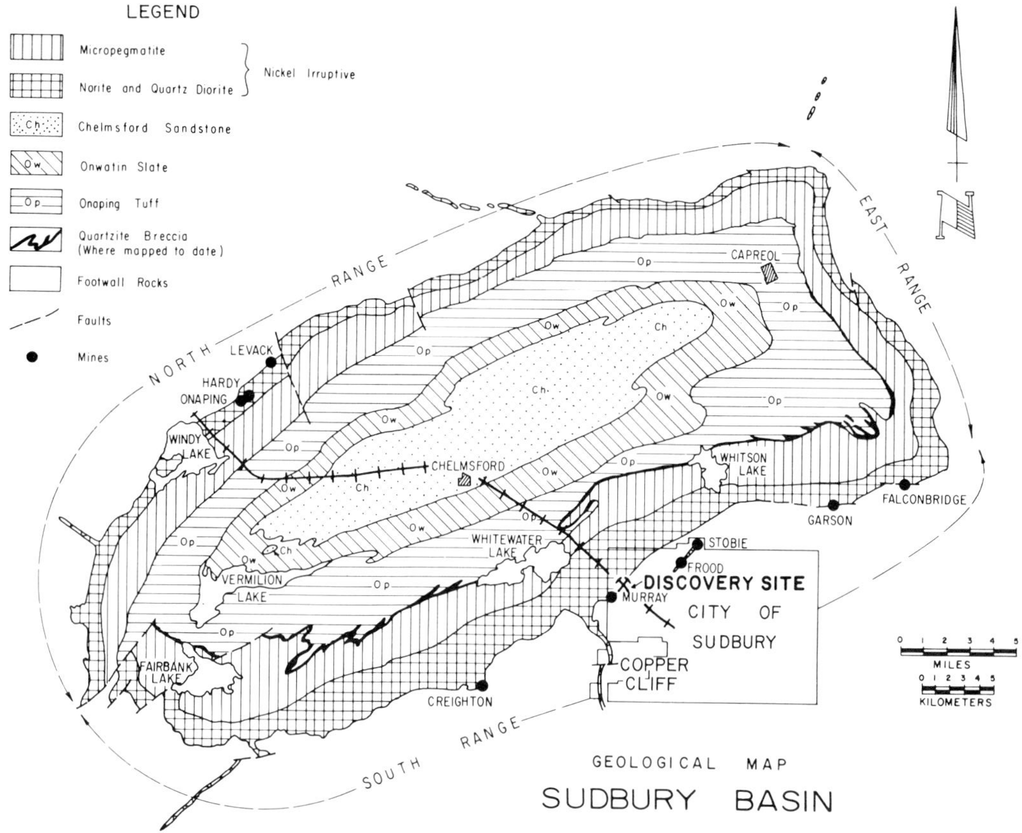

Figure 1. Geological Map of Sudbury Basin modified by J.S. Stevenson after several published maps.

Figure 2. The discovery site, Murray mine, commemorated by a plaque erected by the Ontario Archaeological and Historic Sites Board.

ment really got under way at Sudbury. It was as a result of this activity and the need by the mining industry for geological information about copper-nickel orebodies that the first systematic geological work in the area was begun.

In response to the need, the Geological Survey of Canada sent Robert Bell, accompanied by his assistant, A.E. Barlow, into Sudbury in 1888 to begin systematic mapping. After three seasons of field work Bell submitted his report on the Sudbury mining district in 1891 (Bell, 1891). This report and accompanying maps, with appended petrographic notes on Sudbury rocks by G.H. Williams of Johns Hopkins University, constituted the first comprehensive treatment of Sudbury geology. Early Bell recognized a distinct geological basin but only went so far as to define it as an area occupied by a group of less altered rocks of upper Huronian or possible lower Cambrian age, lying unconformably on the older Huronian rocks. He related the ores to isolated masses of diabase, diorite, and greenstone around his basin, but the detail and mapping coverage were not sufficient to bring out the continuity of the ore-associated basic rocks around the basin. He did not distinguish between areas of ore-bearing norite and varieties of greenstone (the term under which he included gabbros, diabases and diorites). The recognition of the ore-bearing rock as a hypersthene gabbro, norite, had not yet been made. Although Williams, in his petrographic description of rocks submitted to him by Bell, did describe a quartz-hypersthene gabbro from the Blezard mine, the field occurrence was reported to Williams as a dyke. As there are no such dykes at the Blezard mine, this specimen probably came from the main norite. A year later, in 1892, von Foullon reported orthorhombic pyroxene in gabbro from the Murray mine (Coleman, 1905). It is interesting to note that Williams was perhaps the first to identify micropegmatite from Sudbury, the identification being that of a specimen from Moose Lake in Levack township.

In respect of the origin of the ores Bell recognized that they were found principally at the contact of greenstones with older rock. He was impressed by their clear association with his greenstones, saying: "The general character of the deposits seems to indicate that they have originated primarily from a state of fusion." (Bell, 1891, p. 9F). One might therefore say that Bell was the first to espouse a magmatic origin for the Sudbury ore, an origin that is still much in favour.

More definitive papers and studies relating to the origin of the ores by magmatic differentiation were published shortly after Bell's work. Barlow (1904) notes papers by Adams, Kemp and Walker, the latter representing original research for a Ph.D. thesis at Leipzig. Walker had a feeling for the continuity of the nickel-bearing Irruptive. He divided the Irruptive around the basin into three separate general areas of rock types, the Whitson Lake, Traverse Mine and Windy Lake irruptives, and felt that although the mining of the time showed lack of continuity of irruptive rocks within these areas, future more-detailed mapping would show continuity. As a matter of fact his map, Fig. 3, approximately 1 inch to 7 miles, suggests the continuity of the nickel-bearing Irruptive completely around the basin (Walker, 1897). He identified the nickel-bearing rocks in his irruptive areas as norite, and the overlying granites as micropegmatite. Thereby he showed that the isolated specimens earlier identified as norite and micropegmatite by Williams and von Foullon did in fact represent rock types of considerable extent around the basin.

BARLOW'S MAPPING (1901-1904)

Because of a resurgence in prospecting and mining activity around 1900 after the first excitement of the 1880s, and the need for up-to-date geology, the Geological Survey of Canada sent A.E. Barlow into the district in 1901 to up-date and extend Bell's map of 1891. In view of his earlier experience with Bell in Sudbury, Barlow was an excellent choice for this job. After three years of intensive field work he produced a comprehensive report with accompanying maps of the region. He published two areal maps on a scale of 1 mile to 1 inch, both of the South Range, the Victoria Mines map that extended from the Victoria Mines on the west to the North Star Mine on the east, and, accompanying this map on the east, the Sudbury Map extending easterly to the then Cryderman mine, now the Garson mine. In addition, he published three detailed 400 scale maps; two sheets, a north sheet and a south sheet showed mines of the Copper Cliff offset, and a third sheet showed the Elsie and Murray mines. These areal maps indicated for the first time the continuity of the Nickel Irruptive along the South Range and, particularly the Sudbury map, the

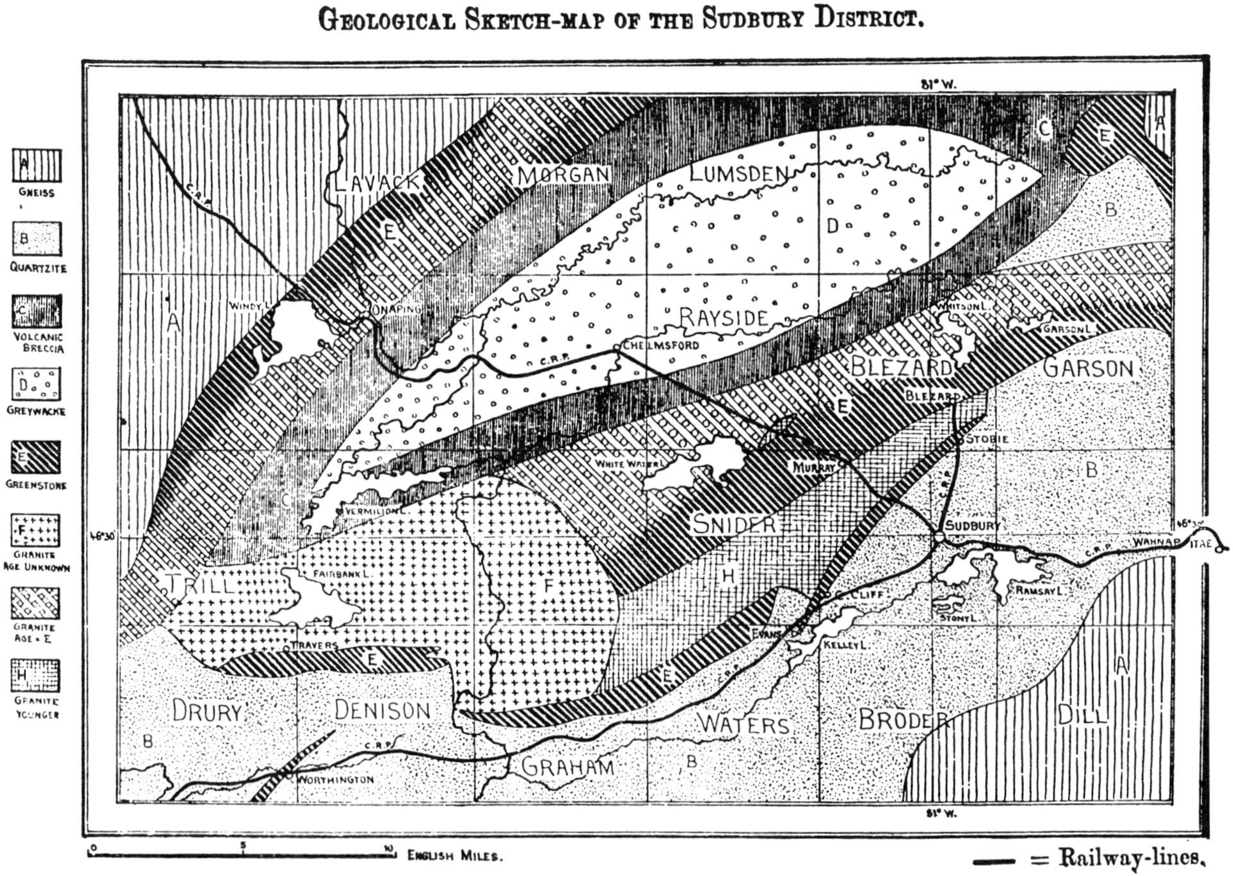

Figure 3. Walker's 1897 map suggesting for the first time the continuity of the nickel Irruptive, then referred to on his sketch as greenstone (E) and granite (Age=E).

localization of the orebodies along the lower edge of the nickel-bearing Irruptive, the norite. This map also indicated the two-fold nature of the Irruptive: a lower member, the norite, and an upper member, the micropegmatite. In his work on the Copper Cliff embayment or funnel and the related offset, Barlow was the first geologist to map and show the importance of embayments in the base of the norite and related offsets. His comment about this feature, "this main belt of norite sends off a narrow dyke-like extension or offset" (Barlow, 1904, p. 88), is rather cogent in view of the discussion that has ensued about the relation of the norite to the offsets, and the general acceptance today that the offset rock is part of the Irruptive.

Barlow did not get around to mapping the North Range because "at the time it was considered more important to ascertain definitely the continuity or otherwise of the southern or main nickel range" (Barlow, 1904, p. 7); and consequently, he did not show the continuity of the Irruptive completely around the basin. However his 1904 map of the South Range brought out the important features of Sudbury geology.

In a long discussion on the origin of the ore deposits, Barlow came out very strongly in favor of magmatic differentiation, but did point out that in some places, as at Victoria and some of the Copper Cliff mines, "various hydrochemical agencies accompanying dynamic action have been more active than others" (Barlow, 1904, p. 131). Of significance is his remark on pages 131-132, "The real practical side of the geological investigation, consisted in the outlining of the immense masses of intrusive norite or gabbro, with which the nickel and copper deposits of the region are alone associated."

However, even in these early days opinion differed on the origin of the ores and one finds Dickson (1903) concluding that the nickel-copper ores were deposited from hot circulating waters and that they replaced the norite. Dickson had spent the summers of 1901 and 1902 in the Sudbury district and in the mines, collecting material for a Ph.D. thesis at Columbia University under Kemp. Dickson was impressed by the amount of alteration and by the amount of ore-breccia in the mines he visited where much of the ore occurred as a cement for the rock fragments and along shear-planes. He concluded that the ores had not formed by magmatic differentiation but by late hydrothermal solutions resulting in "the secondary formation of the Sudbury ore-bodies as replacements along crushed and faulted zones" (Dickson, 1903, p. 515).

COLEMAN'S MAPPING (1902-1904)

In 1902, a year after Barlow had started his work, A.P. Coleman was sent to Sudbury by the then Ontario Bureau of Mines to examine the more important nickel discoveries, and a year later he was instructed to map the Northern Nickel Range, a job that had not been undertaken by Barlow in his mapping. During the summers of 1903 and 1904 Coleman did this and extended his mapping to the South Range as well, paying particular attention to the basic and acid edges of the Nickel Irruptive. On the map, 1½ inches to 1 mile, accompanying his report, Coleman (1905) showed for the first time the continuity of the Nickel Irruptive completely around the basin. It must be noted however that Barlow had said in his 1904 report

(p. 25): "the later geological work is tending to prove their connection (referring to belts of Irruptive) as one continuous ellipsoidal band." Coleman in his report stated that:

> "The ore deposits are at the margin of a huge laccolithic sheet of eruptive rock a mile and a quarter thick, 36 miles long and 17 miles wide. This sheet is now in the form of a boat shaped syncline, with its pointed end to the southwest and its square end to the northeast. The rock composing this sheet is norite at its outer (lower) edge, merging into granite or granodiorite at its inner (upper) edge. The ore bodies are around the margin of the norite or along dike-like offsets from it, and have evidently segregated from the rock while still molten, although they have undergone later rearrangement by circulating water" (Coleman, 1905, p. 2).

This map of Coleman's, published in 1905, showed the basin in its entirety for the first time. It has stood the test of time, the essential features of Sudbury geology having been but little changed in all subsequent mapping. It stands as a tribute to the careful work and geological perspicacity of one of our early Sudbury geologists. It also points up the value of doing geological work in a mining camp at the time of its active and widespread development.

During the summers of 1909, 1910, and 1911 Coleman did further work, particularly at the mines of the area; he published his findings along with an updated version of his earlier map (Coleman, 1913). With the publication of the 1913 report of Coleman, the early stage of systematic geological investigation at Sudbury came to an end.

EARLY WORK OF THE ONTARIO DEPARTMENT OF MINES

The next stage in the development of geological thought at Sudbury resulted from the establishment in 1915 of a Royal Commission by the government of Ontario to enquire into the feasibility of refining nickel in the province. As part of this enquiry C.W. Knight, Assistant Provincial Geologist, undertook in 1915 and 1916 the study of the Sudbury ore deposits for the Commission. With Dickson, Knight was one of the early hydrothermalists. He was led to his beliefs by first of all noting the late, by implication post-irruptive, age of much of the sulphide, particularly where it formed the matrix for breccias, and then by noting in these breccias what he considered to be extensive hydrothermal replacement of the fragments (Knight, 1917). In his concept of a hydrothermal origin for the Sudbury ores, Knight was later followed by Wandke and Hoffman (1924), Phemister (1925) and Yates (1948). As a result of his studies, Knight also formulated a new concept of the emplacement of the Irruptive. On the basis of the steep and reverse dips of the norite footwall contact on the South Range, he felt that the norite had been emplaced as a dyke, and because of the elliptical or ring-like shape of the outcrop of the Irruptive, he thought it had been emplaced around a down-faulted block of the earth's crust. He therefore was the first to apply the concept of the ring-dykes of the Scottish geologists to the emplacement of the Sudbury Irruptive.

In 1924, the Scottish geologist, T.C. Phemister, spent the summer in Sudbury studying the Irruptive and the ores for the Ontario Department of Mines. He made 11 traverses across the Irruptive at various places around the basin and made careful quantitative petrographic studies of the specimens collected. He also studied the rocks of the three principal offsets. His conclusions were at variance with those of many of the earlier workers. He concluded that the norite, which he preferred to call gabbro, and the micropegmatite were separate intrusions; that both the Creighton and Murray granites were younger than the Irruptive; that the quartz-diorite dykes which form the rock of the offsets were later than the granites and therefore later than the norite of the main mass of the Irruptive; and that the sulphides were formed by the action of ascending hydrothermal solution under high temperature and pressure, with replacement an important process (Phemister, 1925).

Phemister's views were so different from those of earlier field men that

> "the three of them remaining (Coleman, Moore and Walker), puzzled by this curious state of affairs, decided to revisit Sudbury last summer and make a fresh study of typical sections, so as to test once more the magmatic theory and see if their earlier conclusions were wrong" (Coleman *et al.*, 1929, p. 11).

Like Phemister they made systematic traverses across the Irruptive, collected specimens and made many chemical analyses. They continued to disagree with Phemister about his break between the norite and micropegmatite and reaffirmed their concept that the ores formed by magmatic differentiation from the magma of the Irruptive.

THE SECOND PHASE OF MAPPING

A second period of systematic mapping began when W.H. Collins of the Geological Survey of Canada entered the Sudbury field and during the period 1928-1934 mapped the whole Nickel Irruptive (Collins appears to have been the first Sudbury geologist to use the spelling, "irruptive", rather than the spelling, "eruptive", used by earlier writers.) and included in his mapping an area that extended from Espanola on the west to Wanapitei on the east. The results of his mapping were published in four 1-mile maps by the Geological Survey of Canada, and details of his work and concepts of Sudbury geology were published in four successive papers in the Royal Society of Canada (Collins, 1934, 1935, 1936, and 1937).

As a result of his detailed mapping of the norite and micropegmatite and of extensive petrographic and chemical studies along several traverses across the Irruptive, Collins believed that the norite and micropegmatite were transitional. To emphasize this on his maps, he showed a transition zone of variable width between the norite and micropegmatite. It must be noted here, however, that Phemister, who had worked in the field with Collins in 1930, later published a paper (Phemister, 1937, p. 1) in which he concluded that "the nature of the rocks of this intermediate zone does not support the theory of gravitative separation of the basic member but indicates rather the formation of hybrid types at the contact of two separate magmas." Collins believed "that the norite and micropegmatite were formed in place by gravitational separation from a single body of molten rock matter (magma) that was homogeneous when intruded" (Collins, 1934, p. 135). He concluded that by further differentiation of the norite layer, the sulphides collected at the bottom as crystallization proceeded, where some of the sulphide reacted with earlier crystallized Irruptive to form offset magma: final emplacement and cooling followed. Deformation and folding occurred principally after solidification. Collins was therefore a magmatist who believed in a folded, differentiated sill.

Collins also tackled two other important aspects of Sudbury geology, the granite problem and the history of the offsets. He concluded that the Creighton granite was older and the Murray granite principally younger than the norite (Collins, 1936). He studied in some details the offsets that Coleman had described and along which many important mines were located, and described them as quartz diorites. He believed, with Coleman, that the offsets were protrusions from the main Irruptive, grading, in the case of the Foy and Copper Cliff offsets at least, into the main norite.

Shortly after Collins began his work at Sudbury, A.G. Burrows and H.C. Rickaby commenced studies in the area for the Ontario Department of Mines. This work was continued from 1927 until 1932. Their first report (Burrows and Rickaby, 1929) dealt principally with the rocks inside the basin, and included for the first time a really complete description of the important Onaping tuff formation. Their second report (Burrows and Rickaby, 1934) dealt not only with the rocks to the south of the basin, but also more specifically with the relation between the ores and the norite and the origin of the ores. They believed, with Dickson, Knight and Phemister, that hydrothermal solutions and ordinary processes of replacement, structurally controlled, were dominant in the formation of the Sudbury ore deposits.

CONTRIBUTIONS BY COMPANY GEOLOGISTS

At about this time, in 1931, the International Nickel Company of Canada, Limited, established its geological department and began a detailed study of the Irruptive. Since then, both this company and the other major company in the basin, Falconbridge Nickel Mines, Limited, have actively pursued geological studies in the field and underground, all with follow-up laboratory work. Members of their geological staffs have published, and continue to publish, in scientific and technical journals and have made important contributions to an understanding of Sudbury geology.

In two consecutive papers by one of these geologists, Yates concludes that the Creighton and Murray granites and the quartz diorites are all later than the norite, although he does say that some quartz diorite may be a direct apophysis from the norite (Yates, 1938, 1948).

As one would expect, geologists of these companies were concerned early with the origin of the ores. Yates considered that the orebodies were in part hydrothermal and in part deep-seated metamorphic; he emphasized the close relationship of the quartz diorite and the ore. Davidson felt that there was no possibility of the ore at the Falconbridge mine being a magmatic segregation in situ, but that there was strong evidence that the orebodies were deposited by hydrothermal solutions, logically derived from the same magma source as the norite (Davidson, 1948). Lochhead (1955) was somewhat less definite, saying that the Falconbridge ores were emplaced after arrival and partial cooling of the norite, reaching their present position by travelling upward in an extremely mobile stage.

Mitchell and Mutch, in two papers on the Hardy mine in the North Range, concluded that the ore at this mine was definitely post-norite and that the ore deposition was by processes commonly accepted as hydrothermal (Mitchell and

Mutch, 1956, 1957). Zurbrigg *et al.* (1957), in describing the disseminated massive and breccia sulphide ores of the Frood-Stobie mine and noting the general lack of wall-rock alteration, considered these ores magmatic by implication. Clarke and Potapoff (1959), in describing the rather unusual ore occurrence at the McKim mine in the South Range (where the ore is in granite in the footwall adjacent to the norite), concluded that this ore was later than the granite which was in turn later than the norite. Cowan (1968), in a paper on the Strathcona mine in the North Range, concluded that the orebodies were the result of gravitational differentiation of sulphides in norite and injection of sulphide magma into favourable structural sites.

As has long been known and briefly referred to in many reports, the Sudbury district is characterized by a large amount of an unusual crush breccia particularly in the rocks immediately south of the Irruptive. Because of its extensive development east and west of the Frood mine, this has long been known as the Frood breccia. However, Speers (1957) found similar breccia at several points all around the basin and proposed the name, "common Sudbury breccia". The Sudbury breccia is generally considered to be younger than all the footwall rocks of the Irruptive but older than the Irruptive (Fairbairn and Robson, 1941). However, some geologists believe that it cuts the quartz diorite and is therefore younger than the Irruptive (Yates, 1948). Speers believed it to be pre-nickel in age, but believed that in places these breccias have been remobilized to intrude the Irruptive. Zurbrigg *et al.* (1957) have proposed that this breccia at Frood is younger than the Murray granite at Frood but older than the quartz diorite.

LATER WORK OF THE ONTARIO DEPARTMENT OF MINES

In 1953 J.E. Thomson of the Ontario Department of Mines began, and for several years continued, work in the Sudbury area. He visited key localities, made detailed studies in several Sudbury area townships, and prepared a compilation map on a scale of 1 inch to 1 mile. In referring to the difficulties besetting Sudbury geology, Thomson (1957, p. 3) notes Lapworth's advice by way of Knopf, "Map it, and it will all come out right". The Ontario Department of Mines has continued mapping in the Sudbury district, concerned principally with the Huronian belt. In 1969 the Department published the most up-to-date map of the area available, Map 2170, the Sudbury Mining Area, on a scale of 1 inch to 1 mile, a compilation and revision of previous maps, by K.D. Card.

Card in a recent publication (Card, 1978) includes the south-eastern corner of the Sudbury basin in his geological map of the Sudbury-Manitoulin area; some minor but no major changes have been made on the Sudbury portion of this map.

Because of the large amount of volcanic material represented by the Onaping tuff and its definite relevance to Sudbury geology, it was decided early in the course of Thomson's work to ask Howell Williams of the University of California, a geologist of wide experience in volcanology, to act as consultant for the Ontario Department of Mines on Sudbury volcanology. A report on his findings was published in 1957. At the same time, T.C. Phemister, long familiar through previous field work with Sudbury geology, was asked by the Department to study the controversial Copper Cliff rhyolite. This he did and in his report (Phemister, 1957) his principal conclusion was that the Copper Cliff rhyolite in the map area, previously regarded as of either sedimentary or volcanic origin, was interpreted as an acid intrusive.

The origin of the Onaping volcanics inside the basin and immediately overlying the irruptive has concerned Sudbury geologists for some time. Burrows and Rickaby (1929) made a detailed study of the formation and concluded along with many earlier writers that it consisted of volcanic ash or tuff, breccias and agglomerates principally of volcanic origin. Williams (1957), although agreeing to their volcanic origin, conceived of the Onaping tuffs as comprising a vast accumulation of glowing avalanches associated with rhyolite dykes and Pelean domes of rhyolitic composition. Stevenson (1961) later proposed that much of this ryholitic material consisted of varyingly altered quartzite; he modified and extended Williams' glowing avalanches concept to that of a single ash-flow of considerable extent for the Onaping tuff and breccia (Stevenson, 1972).

Further in respect of the Onaping tuff formation within the basin, Thomson and Williams (1959) concluded that glowing avalanches burst through fissures close to the margins of a pre-existing large depression on the site of the present Sudbury basin to form the Onaping formation, and that drainage of magma chambers caused the depression to subside still more, forming a volcano-tectonic sink. As a consequence of this concept, they also believed that the Nickel Irruptive apparently represents successive ring-like intrusions of norite and micropegmatite along fractures bordering the initial Sudbury basin. In contrast to the ring-dyke concept of Knight, Thomson and others for the form of the intrusion, Wilson (1956) modified the boat-shaped syncline concept of Coleman to that of a funnel-shaped intrusion, pointing out similarities to the Bushveld complex and suggesting that the Sudbury irruptive may have a zone of ultramafic rocks toward its base.

Another controversial feature of the Onaping formation has been a siliceous breccia that occurs at the base of the formation. Coleman (1905) referred to this as the Trout Lake conglomerate and showed it on his 1904 map as such. Burrows and Rickaby (1929) concluded that it was a volcanic breccia, included it with the Onaping, and did away with Coleman's Trout Lake conglomerate. Thomson and Williams (1959) conceived of the breccia as a brecciated rhyolite. Stevenson (1961) believes that this consists of a practically continuous formation of quartzite and quartzite breccia. This quartzite is of considerable interest because the only other known quartzites lie outside the basin, and its presence, as Hawley says, "introduces a complication which cannot be settled until its age can be determined as pre- or Lower Huronian" (1962, p. 21).

HAWLEY'S CONCEPTS (1962-1965)

Major contributions filling out our concepts of the origin of the Sudbury ores were made by Hawley (1962 and 1965). His familiarity with the problems of Sudbury geology was the result of extensive mineralogical and chemical studies, critical field work by himself and his graduate students, and discussions wtih many field geologists in the Sudbury area. He believed that the primary ores were magmatic, the result of segregation of immiscible sulphide liquid from the Nickel Irruptive, and that, superimposed on these ores, were later effects of minor importance involving remobilization due to intrusives and some hydrothermal mineralization and supergene deposition.

Hawley demonstrated that much of the ore had formed in place either as disseminated ore in the norite and in the quartz diorite of the offsets or as globules of silicate magma in a dominantly sulphide belt, resulting in his "immiscible-silicate-sulphide" ore of the Frood mine. He believed that as a result of differentiation by liquation from the magma, the abundant sulphide liquid so derived would collect in troughs at the base of the norite. Here it would form massive sulphide ores, or, in places, having engulfed and disrupted adjacent breccias, would form breccia ores.

In his 1965 paper Hawley stressed the very close relations between norite, quartz diorite and ore, pointing out that there was no need to separate the time of formation of the quartz diorite very greatly from that of the norite. In respect of the origin of the quartz diorite he amplified and showed his preference for a modification of an earlier concept of Collins' (1937) by noting the very common association of quartzz diorite and sulphides. He concluded not only that the ores were liquid at the time of the formation of the quartz diorite, but also that acquisitions from the ore fluids themselves possibly contributed to the formation of the quartz diorite from basic magma at the base of the Irruptive. He noted that the segregated quartz diorite and ore fluids might remain liquid at lower temperatures than that at which norite would crystallize.

Hawley contributed to the study of the form of the Irruptive by noting that Wilson's concept of a funnel-shaped intrusion is supported by an observation of H.F. Zurbrigg (Hawley, 1962, p. 24) that crystal layering has been seen at two places and, where measurable, is flatter than the footwall of the intrusive.

Between the time of Hawley's two papers, Kullerud and Yoder, working at the Geophysical Laboratory in Washington, D.C. on the Fe-Ni-S system and on sulphide-silicate relations, demonstrated experimentally that sulphur reacts readily with many rock-forming silicates to produce sulphides and oxides (Kullerud and Yoder, 1963). Applying these results to Sudbury, Kullerud (1963) thought that sulphur, introduced after intrusion and emplacement of the Irruptive at a time when possibly only the core of the Irruptive was still liquid, would extract iron, nickel and copper from the silicates and lead to the formation of immiscible silicate-sulphide liquids responsible for the orebodies. Hawley (1965), although admitting that this method of formation of the ores had several features that may be applicable to Frood, noted that there were still many requirements to be met and suggested that further field studies to test this idea were needed.

DEVELOPMENTS SINCE 1966

Recent detailed studies of the rocks at the base of the Irruptive have markedly developed our concepts of Sudbury geology by emphasizing the close relation between the ores and the marginal intrusions that are believed to be distinct from the main Irruptive but are still magmatically related to it.

In a paper given in 1966 at the Society of Economic Geologists' Symposium on Magmatic Ore Deposits, Souch *et al.* (1969) concluded that the ores are a part of a sulphide-bearing facies of the Irruptive from which they were products of crystallization of an inclusion-bearing and sulphur-rich silicate magma. They showed that the ores were part of a discontinuous sublayer of sulphide- and inclusion-bearing noritic rock. This sub-layer was divided into three main types, North Range, South Range, and the quartz-diorite offsets. Significantly, the inclusions were described as including abundant peridotite, pyroxenite and gabbro, as well as fragments of other rock types.

In this connection it should be noted that Wilson (1956) had combined his concept of the Irruptive as a funnel-shaped intrusive, with an association at depth of a zone of ultramafic rocks. Stevenson (1964) also noted evidence for an association of the Irruptive with ultramafic rocks.

In their studies of the Irruptive, new isotopic dates were obtained by Souch *et al.* (1969) in order to determine the relative ages of the sub-layer and other members of the Irruptive. From these studies they concluded that the norite and sub-layer may be contemporaneous but both may be older than the micropegmatite.

Further detailed work on the rocks at the base of the Irruptive includes a comprehensive study of the Strathcona mine in the North Range by Naldrett and Kullerud (1967). They described the four units, all norite, of the Nickel Irruptive that occur at the mine and brought forth evidence bearing on the emplacement of the ore. They speculated on the origin of the ore and extrapolated their results to other ore occurrences around the basin. In respect of their four norite units, Naldrett and Kullerud (1967) established an age sequence from oldest to youngest: felsic norite, mafic norite, xenolithic norite, and hanging wall breccia. In considering the origin of the Irruptive, they emphasized the significance of the many ultramafic xenoliths that occur in the xenolithic norite, not only at the Strathcona but associated with other orebodies around the basin. They concluded that these ultramafic xenoliths originated from an irruptive-related, well-fractionated and layered series in the lower portion of the funnel-shaped intrusion as proposed by Wilson. However, in a later paper, Naldrett *et al.* (1970) have suggested that, because the Irruptive magma was not capable of crystallizing olivine-bearing rocks, it is very possible that the ultramafics were derived from a deep-source chamber and not from the Irruptive itself.

In regard to timing, Naldrett and Kullerud (1967) concluded that the bulk of the ore sulphides were introduced with the xenolithic norite after crystallization of the main body of the Irruptive. From a calculation of settling rates, they have further concluded that the sulphides were transported upward as sulphide droplets held in suspension as an immiscible liquid in the uprising xenolithic norite magma, and that these droplets settled toward the base of the norite magma and farther into fractures in the underlying gneiss.

In discussing sulphide liquids and the genesis of magmatic sulphides of the Sudbury type, MacLean (1969) proposed that a rise in the oxidation state of a mantle-derived, sulphur saturated magma would release sulphur and produce a sulphide liquid. He noted (MacLean, 1969, p. 882) that this process could occur during emplacement of such a magma and that the "sulphide liquid would tend to lag and form part of a later phase of the intrusive sequence".

In reference to mineral assemblages in the ores, Naldrett and Kullerud (1967), as a result of their experimental work on sulphide-oxide systems, concluded that the sulphide ore crystallized as a magnetite-pyrrhotite-solid solution assemblage and that the mineral assemblage pyrite-monoclinic and hexagonal pyrrhotite-chalcopyrite-pentlandite formed during subsequent sub-solidus cooling.

In the Field Guide and in papers presented at the Special Session on Sudbury Geology at the Geological and Mineralogical Associations of Canada meetings in 1971 in Sudbury (Guy-Bray, 1972), several participants continued to stress the close relationship between the orebodies and the younger marginal or "sub-layer" intrusions. However, the relative ages of these rocks still appear to be uncertain. Hewins (1971) confirmed the younger age of the sub-layer compared to the main Irruptive, but felt that the mafic norite, originally regarded by Naldrett and Kullerud as a separate intrusion and therefore younger than the felsic norite, was part of a continuous sequence within the main Irruptive. Pattison and Phipps (1971), on the other hand, believed that, in at least two deposits in the North Range, sub-layer rocks (principally xenolithic norite) are overlain discordantly by and are therefore older than the felsic and mafic norites of the main Irruptive.

Considerable controversy exists and continues to exist about the relative ages of members of the Irruptive and a grey granitic breccia that is extensively developed in the North Range along the contact between the Irruptive and footwall gneiss. Naldrett and Kullerud (1967) concluded that this breccia is both post-xenolithic norite and post-ore. From a study of plagioclase compositions they noted that the breccia has been the locus of sodium metasomatism. They observed that much North Range ore is found in this breccia but that there is also much grey granite breccia (their footwall breccia) which does not contain ore. Souch *et al.* (1969) referred to this breccia as a leucocratic phase of their sub-layer rocks, in places gradational and in other places cutting these rocks. Greenman (1971), in discussing this breccia at Strathcona, confirmed its post-xenolithic norite age and concluded that it was intruded in a fluidized state shortly after the injection of the xenolithic norite, itself later than the main Nickel Irruptive. Naldrett *et al.* (1972) postulated essentially the same late timing for the granite breccia. However, Pattison and Phipps (1971) believed that these breccias are pre-Irruptive and might be related to the meteoritic impact which they presumed to be responsible for the Sudbury basin structure.

Perhaps one of the most detailed recent studies of the petrology of the Irruptive is one by Naldrett *et al.* (1970) in which they describe the textural and modal variations of the Irruptive over several traverses and in a bore-hole drill-section. Based on 188 electron microprobe analyses, they showed variations in the anorthite content of the plagioclases and variations in the Fe: (Fe + Mg) atomic ratios of hypersthene and augite. They described the norites as types of igneous cumulates, using the terminology first introduced by Wager *et al.* (1960) for the Skaergaard intrusion in Greenland, and they presented textural evidence that the South Range norite is a mesocumulate and that the felsic norite, the upper gabbro of the South Range and the oxide-rich gabbro of the North Range are orthocumulates. Their gabbros appear to comprise much of what was heretofore called transition zone. They emphasized cryptic variation and phase layering of the three principal cumulus minerals, plagioclase, hypersthene and augite. In respect of the initial Irruptive magma, they felt that the quartz-rich norite, a variety of norite that is continuous around the margin of the South Range, possesses a composition that has not been modified by crystal accumulation or adcumulus diffusion, and that the composition of the fine-grained extreme outer portion of this quartz-rich marginal zone should approximate the composition of the liquid part of the original Nickel Irruptive intrusion. However, in recent work it has been suggested that much of the quartz in the quartz-rich norite represents xenocrystic material from the assimilated, siliceous footwall inclusions (Stevenson and Colgrove, 1968).

In their study of the petrology of the Irruptive, Naldrett *et al.* (1970) demonstrated gravitational crystal settling and illustrated their concepts in cross-sections showing development of layering in the intrusion. They felt that although the idea of gravitational crystal settling was consistent with the concept of either a folded sill or a funnel-shaped intrusion, they preferred the latter concept.

In recent years not only the lowermost part of the Irruptive, but also the uppermost part, have been the subject of detailed studies. Stevenson (1963) suggested that the micropegmatite should be extended to include a fine-grained upper phase, and concluded that textural and mineralogical features such as quartz pseudomorphs after tridymite and spherulitic micrographic intergrowths indicated a high intrusion temperature and rapid cooling of the micropegmatite. Peredery and Naldrett (1975), in a detailed study of the same upper part of the micropegmatite, recognized two phases of this rock; these comprised an older plagioclase-rich phase intruded by a younger granophyric micropegmatite phase. Because of chemical similarities to an upper oxide-rich gabbro phase of the norite, they felt that the older plagioclase-rich rock was a phase of the gabbro split off from it by a later injection of granophyric micropegmatite.

The Sudbury Irruptive has been the subject of much geochronological study over the years and the papers on the subject relating to Sudbury in themselves reflect the development of this sub-discipline of geology. This work dates from the early pioneer work of Fairbairn and his colleagues (1957, 1958, 1960) and has continued to the present day, and includes such recent work as the contribution by Krogh and Davis (1974) and the comprehensive studies of Gibbins and McNutt (1975a, 1975b). As well, many papers on Sudbury geochronology have been published in the intervening years.

Dates on the age of the Irruptive have progressively changed over the years and the most recent dates are those by Gibbins and McNutt (1975a, p. 1970), who have reported that "Whole-rock samples of norite from North and South Ranges of the Sudbury Irruptive yielded a rubidium strontium isochron of 1956 ± 98 Ma age" and that "Four new determinations of South Range and six published analyses of North Range micropegmatite define an isochron of age = 1680 ± 30 Ma". Krogh and Davis (1974) have reported an age of 1844 ± 3 Ma. on various zircons separated from North Range felsic and mafic norite and South Range black norite.

It is apparent that the recent chronologic work is beginning to indicate a younger age for the micropegmatite in respect of the norite, a concept that conforms to recent geologic thinking about the relative ages of these two members of the Irruptive.

The study of the evolution of concepts of Sudbury geology has recently received an added stimulus in the presentation of the astrobleme or meteoritic theory, a concept that reflects current interest in lunar geology. This was first introduced by Dietz (1964) and supported by the evidence of shatter cones by Guy-Bray and Geological Staff (1966) and by petrographic and petrogenetic evidence by French (1967, 1969 and 1971), Dence (1972) and Peredery (1971). Other work suggests that the petrographic features may be explained by volcanic phenomena (Stevenson, 1972).

Regarding meteoritic phenomena in Canada, Currie (1970, p. 137) has stated: "Clear-cut evidence of the action of ultra-high pressures has not been found in Canadian occurrences of shock metamorphism (with the dubious exception of one coesite deposit)."

The continuing interest in the fundamental concepts of Sudbury geology was emphasized by the symposium held at Sudbury in 1971, at a joint meeting of the Geological Association of Canada and the Mineralogical Association of Canada, and the publication of the papers presented (Guy-Bray, 1972).

Concerning the problems of Sudbury geology as they are now seen, Card and Hutchinson (1972, p. 67) have well said:

> "Any attempt to explain the origin of the Sudbury structure must take into account its regional geological setting. Particularly to be considered are its unique location relative to regional structural elements, its age in relation to regional geological history, and its relationship to regional gravity, sedimentary, stratigraphic, petrographic and metallogenic factors."

REFERENCES

Barlow, A.E., 1904, Report on the origin, geological relations and composition of the nickel and copper deposits of the Sudbury mining district, Ontario, Canada: Geological Survey of Canada, Annual Report, no. 873, 236 p.

Bell, R., 1891, On the Sudbury mining district, Geological Survey of Canada, Annual Report, New Series, V, part 1, F5- p. 5-95.

Burrows, A.G. and Rickaby, H.C., 1929, Sudbury basin area: Ontario Department of Mines XXXVIII, Annual Report, part III, 55 p.

————————, 1934, Sudbury nickel field restudied: Annual Report, Ontario Department of Mines XLIII, part II, 49 p.

Card, K.D., 1978, Geology of the Sudbury-Manitoulin Area, Districts of Sudbury and Manitoulin: Ontario Geological Survey Report 166, 238 p. Accompanied by Map 2360, Scale 1 inch to 2 miles (1:126,720) and 4 charts.

Card, K.D. and Hutchinson, R.W., 1972, The Sudbury structure: its regional geological setting: Geological Association of Canada, Special Paper 10, p. 335-380.

Clarke, A.M. and Potapoff, P., 1959, Geology of the McKim mine: Geological Association of Canada, Proceedings, v. 11, p. 67-80.

Coleman, A.P., 1905, The Sudbury nickel region: Ontario Bureau of Mines XIV, Annual Report, pt. III, p. 1-188, maps.

————————, 1913, The nickel industry, with special reference to the Sudbury region, Ontario: Canada Department of Mines, Mines Branch, no. 170, 206 p.

Coleman, A.P., Moore, E.S. and Walker, T.L., 1929, The Sudbury Nickel Irruptive: Contributions to Canadian Mineralogy, 1929: University of Toronto Press, 54 p.

Collins, W.H., 1934, Life history of the Sudbury Nickel Irruptive. Part I. Petrogenesis: Royal Society of Canada, Transactions, p. 123-177.

————————, 1935, Life history of the Sudbury Nickel Irruptive. Part II. Intrusion and deformation: Royal Society of Canada, Transactions, 3rd series, Sec. IV, v. 29, p. 27-47.

————————, 1936, The Sudbury Nickel Irruptive. Part III. Environment: Royal Society of Canada, Transactions, 3rd series, Sec. IV, v. 30, p. 29-53.

————————, 1937, The Sudbury Nickel Irruptive. Part IV. Mineralization: Royal Society of Canada, Transactions, 3rd series, Sec. IV, vol. 31, p. 15-43.

Cowan, J.C., 1968, Geology of the Strathcona ore deposit: Canadian Institute of Mining and Metallurgy, Bulletin, 61, no. 699, p. 38-54.

Currie, K.L., 1970, Shock metamorphism structures: *in* Geology and Economic Minerals of Canada: Economic Geology Report No. 1, Geological Survey of Canada, p. 135-137.

Davidson, Stanley, 1948, Falconbridge mine: *in* Structural Geology of Canadian Ore Deposits: Canadian Institute of Mining and Metallurgy, Geology Division, p. 618-626.

Dence, M.R., 1972, Meteorite impact craters and the structure of the Sudbury Basin: *in* Guy-Bray, J.V., ed., New Developments in Sudbury Geology: Geological Association of Canada Special Paper 10, p. 7-18.

Dickson, C.W., 1903, The ore-deposits of Sudbury: Reprinted in 1913, American Institute of Mining Engineers, Transactions, New York, v. 34, p. 1-67.

Dietz, R.S., 1964, Sudbury structure as an astrobleme: Journal of Geology, v. 72, p. 412-434.

Fairbairn, H.W., Hurley, P.M. and Pinson, W.H., 1957, Age program on biotites in Sudbury district, Ontario: Fourth Annual Programme Report of the U.S. Atomic Energy Committee, Contract AT (30-1) 1381, p. 54-56.

————, 1958, Sudbury, Ontario, Age Program, Fifth Annual Program Report of the U.S., Atomic Energy Committee, Contract AT (30-1) 1381, p. 51-57.

————, 1960, Mineral and rock ages at Sudbury-Blind River, Ontario: Proceedings of the Geological Association of Canada, v. 12, p. 41-66.

Fairbairn, H.W. and Robson, G.M., 1941, Breccia at Sudbury: Annual Report of the Ontario Department of Mines, v. 50, part 6, p. 18-33.

French, B.M., 1967, Sudbury structure, Ontario: some petrographic evidence for an origin by meteorite impact: Publication X-641-67-67, National Aeronautics and Space Administration, Goddard Space Flight Centre, Greenbelt, Md., 56 p.

————, 1969, Possible relations between meteorite impact and igneous petrogenesis as indicated in the Sudbury structure, Ontario, Canada: Publication X-644-69-371, National Aeronautics and Space Administration, Goddard Space Flight Centre, Greenbelt, Md., 50 p.

————, 1972, Shock metamorphic features in the Onaping formation and basement rocks of the Sudbury structure, Ontario: A review: *in* Guy-Bray, J.V., ed., New Developments in Sudbury Geology: Geological Association of Canada Special Paper 10, p. 19-28.

Gibbins, W.A. and McNutt, R.H., 1975a, The age of the Sudbury Nickel Irruptive and the Murray Granite: Canadian Journal of Earth Sciences, v. 12, p. 1970-1989.

————, 1975b, Rubidium-strontium mineral ages and polymetamorphism at Sudbury, Ontario: Canadian Journal of Earth Sciences, v. 12, p. 1990-2003.

Greenman, L., 1971, The grey breccia, host to the ore at the Strathcona mine, Sudbury: Geological Association of Canada, Abstract, Annual Meeting, Sudbury, p. 30.

Guy-Bray, J.V., ed., 1972, New developments in Sudbury geology: Geological Association of Canada, Special Paper 10, 124 p.

Guy-Bray, J.V. and Geological Staff, 1966, Shatter cones at Sudbury: Journal of Geology, v. 72, no. 2, p. 243-245.

Hawley, J.E., 1962, The Sudbury ores: Their mineralogy and origin: Canadian Mineralogist, v. 7, p. 1-207.

————, 1965, Upside-down zoning at Frood, Sudbury, Ontario: Economic Geology and Bulletin of the Society of Economic Geologists, v. 60, p. 529-575.

Hewins, R.H., 1971, The mafic norite and the sub-layer: The petrology of the basal zone of the Sudbury nickel irruptive: Geological Association of Canada, Abstract, Annual Meeting, Sudbury, p. 32-33.

Knight, Cyril W., 1917, Geology of the Sudbury area: Royal Ontario Nickel Report, Toronto, Ontario, p. 105-211.

Krogh, T.E. and Davis, G.L., 1974, The age of the Sudbury Nickel Irruptive: Carnegie

Institute Yearbook, 1973-74, p. 567-569.
Kullerud, G., 1963, The Fe-Ni-S system: Annual Report of the Director, Geophysical Laboratory Yearbook, 62, p. 175-189.
Kullerud, G. and Yoder, H.J., Jr., 1963, Sulphide-silicate relations: Annual Report of the Director, Geophysical Laboratory Yearbook, 62, p. 215-218.
Lochhead, D.R., 1955, The Falconbridge ore deposit, Canada: Economic Geology and Bulletin of the Society of Economic Geologists, v. 50, p. 42-50.
MacLean, W.H., 1969, Liquidus phase relations in the FeS-FeO-Fe_3O_4-SiO_2 system and their application in geology: Economic Geology and Bulletin of the Society of Economic Geologists, v. 64, p. 865-884.
Mitchell, G.P. and Mutch, A.D., 1956, Geology of the Hardy mine, Sudbury, Ontario: Canadian Institute of Mining and Metallurgy, Bulletin 49, no. 526, p. 75-81.
——————, 1957, Hardy mine: *in* Structural Geology of Canadian Ore Deposits, Congress Volume, Canadian Institute of Mining and Metallurgy II, p. 350-363.
Murray, Alexander, 1857, Report for the year 1856: *in* Geological Survey of Canada Reports of Progress for the years 1853-54-55-56, p. 180.
Naldrett, A.J., Bray, J.G., Gasparrini, E.L., Podolsky, T. and Rucklidge, J.C., 1970, Cryptic variation and petrology of the Sudbury Nickel Irruptive: Economic Geology and Bulletin of the Society of Economic Geologists, v. 6, p. 122-155.
——————, Greenman, L. and Hewins, R.W., 1972, The main irruptive and sub-layer at Sudbury, Ontario, 24th International Geological Congress, Proceedings, v. 4, p. 206-214.
Naldrett, A.J. and Kullerud, G., 1967, A study of the Strathcona mine and its bearing on the origin of the nickel-copper ores of the Sudbury district, Ontario: Journal of Petrology, v. 8, no. 3, p. 453-531.
Ontario Department of Mines, 1969, Sudbury mining area; Map No. 2170, Scale 1 inch to 1 mile, coloured geological.
Pattison, E.F. and Phipps. D., 1971: *in* Field Guide to the Sudbury Nickel Irruptive: Joint Meeting, Geological Association of Canada and Mineralogists Association of Canada, Sudbury, p. 12.
Peredery, W.V., 1972, Chemistry of fluidal glasses and melt bodies in the Onaping formation: Geological Association of Canada, Special Paper 10, p. 49-50.
Peredery, W.V. and Naldrett, A.J., 1975, Petrology of the upper Irruptive rocks, Sudbury Ontario: Economic Geology and Bulletin of the Society of Economic Geologists, v. 70, p. 164-175.
Phemister, T.C., 1925, Igneous rocks of Sudbury and their relation to the ore deposits: Annual Report of the Ontario Department of Mines, XXXIV, part 8, p. 1-61.
——————, 1937, A review of the problems of the Sudbury Irruptive: Journal of Geology, XLV, p. 1-47.
——————, 1957, The Copper Cliff rhyolites in McKim township, District of Sudbury: Annual Report of the Ontario Department of Mines LXV, part 3, p. 91-116.
Souch, B.E., Podolsky, T. and Geological Staff, The International Nickel Company of Canada, Ltd., 1969, The sulphide ores of Sudbury: Their particular relationship to a distinctive inclusion-bearing facies of the Nickel Irruptive: Economic Geology Monograph 4, p. 252-261.
Speers, E.C., 1957, The age and origin of common Sudbury breccia: Journal of Geology, v. 65, p. 497-514.
Stevenson, J.S., 1961, Recognition of quartzite breccias in the Whitewater series, Sudbury basin, Ontario: Royal Society of Canada, Transactions, 3rd Series, Section IV, LV, p. 57-66.
——————, 1963, The upper contact phase of the Sudbury micropegmatite: Canadian

Mineralogist, v. 7, p. 413-419.
——————, 1964, Sudbury in terms of upper mantle petrology: American Association for the Advancement of Science, joint meeting with the Geological Society of America, Montreal, Abstract p. 18.
——————, 1972, The Onaping ash-flow sheet, Sudbury, Ontario: *in* Guy-Bray, J.V., ed; New Developments in Sudbury Geology: Geological Association of Canada, Special Paper 10, p. 41-48.
Stevenson, J.S. and Colgrove, G.L., 1968, The Sudbury Irruptive: Some petrogenetic concepts based on recent field work: 23rd International Geological Congress, v. 4, p. 27-35.
Thompson, J.E., 1957, Geology of the Sudbury basin, Ontario: Annual Report of the Ontario Department of Mines, LXV, part 3, p. 1-56.
Thompson, J.E. and Williams, Howell, 1959, The myth of the Sudbury lopolith: Canadian Mining Journal, v. 80, no. 3, p. 57-62.
Wager, L.R., Brown, G.M. and Wadsworth, W.J., 1960: Types of igneous cumulates: Journal of Petrology, v. 1, p. 73-75.
Walker, T.L., 1897, Geological and petrological studies of the Sudbury nickel district, Canada: Geological Society of London, Quarterly Journal, v. 53, p. 40-66.
Wandke, A. and Hoffman, R., 1924, A study of Sudbury ore deposits: Economic Geology and Bulletin of the Society of Economic Geologists, v. 19, p. 169-204.
Williams, Howell, 1957, Glowing avalanche deposits of the Sudbury Basin: Annual Report of the Ontario Department of Mines LXV, part 3, p. 57-89.
Wilson, H.D.B., 1956, Structure of lopoliths: Bulletin of the Geological Society of America, v. 67, p. 289-300.
Yates, A.B., 1938, The Sudbury intrusive: Transactions of the Royal Society of Canada, v. 32, 3rd series, Section IV, p. 151-172.
——————, 1948, Properties of International Nickel Company of Canada, *in* Structural Geology of Canadian Ore Deposits, Jubilee Volume: Canadian Institute of Mining and Metallurgy, p. 596-617.
Zurbrigg, H.F. and Geological Staff, 1957, Frood-Stobie mine: *in* Structural Geology of Canadian Ore Deposits, Congress Volume, Canadian Institute of Mining and Metallurgy, v. II, p. 341-350.

History of Concepts in Precambrian Geology, edited by W.O. Kupsch and W.A.S. Sarjeant, Geological Association of Canada Special Paper 19, 1979

EVOLUTION OF CONCEPTS ON PHYSIO-GEOGRAPHICAL ENVIRONMENTS DURING THE PRECAMBRIAN

Vladimir V. Tikhomirov
Geological Institute, Academy of Sciences, U.S.S.R.
Pyzhevski per. 7, Moscow 17, U.S.S.R.

ABSTRACT

Investigations of the physico-geographical conditions of the early Earth were begun by Johannes Walther in 1909. His studies of the Scottish Torridonian and Norwegian Algonkian deposits led him to deduce that life had a lengthy pre-Palaeozoic history. Walthers' concepts of a global desert were enlarged by A. P. Pavlov, who believed that relief in the Precambrian was very pronounced and that the effects of a violent climate made erosive processes especially effective. Pavlov also believed that there were no Precambrian oceans, only seas that periodically shallowed and dried up.

These concepts were modified by L.S. Berg, who believed that the Algonkian climate was cool and recognized the existence of glaciations in the Archaean and Proterozoic. Berg presented evidence to show that a climatic zonation was established quite early in the Precambrian. He believed also that a land vegetation was in existence by the Lower Archaeozoic.

Subsequent work has filled in this general picture. Poldervaart (1965) suggested that the atmosphere changed character in the Upper Precambrian. Strakhov (1962) linked this with the first appearance of the blue-green algae at the beginning of the Proterozoic, but Ronov (1964) believed photosynthesis to have begun much earlier. Recent research has shown that the earliest organic remains date as far back as 3.5 x 10^9 years.

RESUME

Les investigations sur les conditions physico-géographiques de la Terre primitive débutèrent avec Johannes Walther en 1909. Ses études des dépôts du Torridonien d'Ecosse et de l'Algonkien de Norvège le conduisirent à déduire que la vie avait une longue histoire pré-paléozoïque. Les concepts de Walther sur un désert général furent élargis par A.P. Pavlov, qui croyait en un relief très fort au Précambrien, et aux effets d'un climat très dur qui rendaient les processus d'érosion particuliérement actifs. Pavlov pensait également qu'il n'y avait pas d'océan au Précambrien, mais seulement des mers qui périodiquement perdaient de leur profondeur et s'assèchaient.

Ces concepts furent modifiés par L.S. Berg, qui croyait que le climat de l'Algonkien était froid et reconnaissait l'existence de glaciations à l'Archéen et au Protérozoïque. Berg présenta des évidences pour montrer qu'une zonation climatique s'était établie assez tôt au

Precambrien. Il croyait aussi qu'une végétation terrestre existait à l'Archaeozo que Inférieur.

Les travaux qui suivirent abondèrent des le sens de ce tableau général. Poldervaart (1965) suggéra que l'atmosphère changea de nature au Précambrien Supérieur. Strakhov (1962) relia cela avec la première apparition des algues bleues vertes au début du Protérozoïque, mais Ronov (1964) pensait que la photosynthèse avait commencé bien plus tôt. Les recherches récente sont montré que les premiers restes organiques remontent à aussi loin que 3.5 x 10^9 années.

INTRODUCTION

The composite metamorphism of most Precambrian deposits was the reason for their insufficient study for a lengthy time. Concepts on palaeogeographic environments during the Archaean and Proterozoic were mostly based on logical deductions derived from the Kant-Laplace geogenic hypothesis.

The German geologist Johannes Walther was among the first who tried to investigate in detail the physico-geographical conditions that existed during the early stages in the history of the Earth. Attempting to reconstruct the environments that existed on the surface of the Earth during the Algonkian (i.e., Proterozoic), he analysed the specific features of a sequence exposed in Northern Scotland (the Torridonian rocks). He came to the conclusion that it had been a mountainous land, the sharp peaks of which had not been protected by vegetation and were consequently subjected to erosion. Along the foot of steep slopes, screes of rock fragments accumulated, having been carried down from the mountains by seasonal streams or by landslips, rock slippage or collapse (Walther, 1909). Dried-up lakes existed in which clays were deposited while winds transported sand and formed mobile dunes. He thought that this had been an absolutely lifeless desert, which combined the temperature contrasts of present day deserts with the tropical showers and snow storms characteristic of humid climatic regions today.

Recording that Algonkian deposits in Norway contain a suite of limestones, Walther came to the conclusion that these were an evidence of a lengthy marine transgression that was later succeeded by a continental regime. He believed the limestones had accumulated in a shallow basin that intermittently dried up. Walther thought it possible that, though there was no life on land, there was already life in the seas at the time the limestones were deposited. Admittedly he did not discover any direct signs of living organisms, but he assumed that the carbonate mass could have been formed from the remains of decayed lime-secreting plants and animals, their decomposition determining the dark colour of the rock.

The fact that Cambrian deposits contain remains of numerous and varied organisms of a relatively high organization, which could have originated only as a result of a lengthy prior period of evolution, was correctly regarded by Walther (1908) as evidence for the lengthy pre-Palaeozoic existence of life.

THE CONCEPTS OF PAVLOV (1910)

Walther's contemporary, the Russian geologist A.P. Pavlov, also studied the problem of Precambrian physico-geographic environments. His thoughts on this subject are expressed in greatest detail in his paper "On the oldest deserts of the Earth" (1910). Agreeing on many points with Walther, he assumed that a desert landscape was more typical for the oldest surface of the Earth than were the oceans.

Pavlov wrote that "There certainly has been no World Ocean like that about which Werner was dreaming, whereas the existence of a global protogenic desert is highly probable" (1910, p. 317). In his opinion, the Moon is a wonderful model of the ancient surface of the Earth, its relief having originated under waterless conditions. During this early stage in the history of the Earth ". . . . the unevenly heated land, with numerous volcanoes of a type somewhat different from those of the present and with hot springs, virtually completely dominated the surface of the globe" (Pavlov, 1910, p. 317). Pavlov thought that during the remote Precambrian the relief of the Earth was very pronounced, with mountains even higher and slopes even steeper than on the Moon. In his opinion, during this lifeless period of the history of the Earth is was frequently subjected to strong storms, accompanied by torrential rains. Raging streams brought down large rock fragments from the mountains, depositing them on the plains. The clastic material was poorly rounded and had been subjected to only a very slight sorting, mainly by the fragment size and specific gravity of the rocks. In the most depressed places, short-lived lakes were found in which the finest particles and mica scales were deposited. Among geological factors that transformed the original surface of the planet, Pavlov mentioned not only rains and ephemeral streams but also strong winds and "incandescent clouds" that dispersed hot little crystals of minerals ejected during volcanic eruptions.

During this epoch preceding the appearance of life on the Earth, lava effusions were frequent and the proximity of magmatic foci caused a non-uniform heating of the surface in different regions. Owing to these substantial differences in surface temperature, the gas envelope was not uniformly developed over the Earth. Later the first aqueous basins began to appear and conditions were created in which organic life could originate, becoming extensively developed by the Cambrian period.

In this way, the ideas of Walther and Pavlov, which were in general characteristic for the first two decades of the twentieth century, can be summarized in the following way: a) Precambrian land occupied the greater part of the planet, consisting of a lifeless desert, b) there were no oceans, but there existed seas periodically becoming shallow and drying up, in which primitive organisms appeared rather early and then developed during a lengthy interval of time into the rich and varied flora and fauna of the Cambrian, c) meteorological conditions were very complicated, with ice storms succeeded by tropical showers that formed ephemeral, violent streams and small, rapidly drying-up lakes. Volcanic activity was much more intense than during the present epoch.

THE CONCEPTS OF L.S. BERG (1918-1943)

Owing to subsequent research, new factual material gradually accumulated enabling other geologists to extend and somewhat modify these fundamental concepts. A prominent place among such scientists belongs to Soviet naturalist L.S. Berg, who assigned a substantial space in his papers to consideration of the problems of climates and conditions of soil formation on Precambrian continents. In his first paper along these lines (1918) Berg said that, during the Algonkian, a cool climate was predominant; this was evidenced by the traces of a continental glaciation and the insignificant amount of calcium carbonate in marine sediments. He thought that the land was mountainous, with high elevations covered by glaciers. Three decades later, on the basis of new geological data, Berg (1947) somewhat modified his ideas on the oldest climates. He wrote, among other things, that the glaciation involved certain areas of the planet as early as in the Upper Archaean. In the Proterozoic, glaciation was general and, as Berg correctly stated, it cannot be explained either by a shift of poles or by a drift of continents; instead it was the result of a general colder spell. (This idea is not contradicted by recent data showing that the Upper Proterozoic glaciation was mostly of a mountain type.)

Berg stressed the similarity between Proterozoic climates and the later ones (Quaternary particularly), pointing out that viewpoints then current on the great role of the inner heat of the Earth in bringing about ice ages were erroneous. Referring to modern data showing that the internal heat of the Earth only very slightly effects the heating of its surface (not over 0.1° C), Berg stated that this insignificant effect of the inner heat would have been no greater in the Archaean.

Worthy of stress is L.S. Berg's opinion on the existence of a climatic zonality not only in the Proterozoic, but even earlier. Seasonal changes he believed took place as early as in the Archaean; warm and cold or dry and rainy (Berg, 1948). Berg made this judgement on the basis of a variety of indirect data such as the presence in Precambrian deposits, beginning with the Lower Archaean, of banded ferruginous quartzites (jaspilites) which originated, according to Berg, as a result of the life activity of microorganisms and indicated seasonal changes in the amount of plankton in the marine basins. As another evidence of climatic zonality, Berg mentioned the substantial amount of corundum (Al_2O_3) in Archaean crystalline schists. In his opinion this was produced by a metamorphism of laterites, the formation of which was possible only in the specific conditions of weathering typical for a humid and hot climate.

However, lateritic weathering in Archaean rocks was not universal; Berg noted that in Finland, where there are no signs of lateritization in the Archaean, a temperate to cold climate must have prevailed, with average temperatures not exceeding 10°C.

From the Lower Archaean onwards, solar activity, according to Berg, did not hinder the process of the development of life; he believed that, by the end of the Proterozoic, this activity had already settled into the eleven-year periodicity that can be observed right up to today.

In a lecture delivered in 1943 and published in 1944, Berg expressed his opposition to the generally accepted idea (B.F. Howell, E. Schuchert, A. Dunbar, V.A. Obruchev, A.N. Krishtofovich, N.M. Strakhov and others) that, during the Precambrian, continents were devoid of any vegetation and constituted a lifeless desert and that, even up to the Devonian, vegetation was either entirely absent from the land or was extremely scanty. Berg gave a series of arguments of a physico-geographic and biological nature as to why, from the Lower Archaean onwards, the organic world must have included not only the sea but the land as well. He believed that there would probably have been primitive blue-green algae, bacteria and fungi which could develop on land. The evidence for their existence was considered to be provided by a disseminated graphitic substance, usually highly metamorphosed, found in the oldest sedimentary rocks. Archaean kaolins, he believed, were altered soils. He therefore stated that "beginning with the Lower Archaeozoic, the processes of mechanical and chemical weathering were taking place on the Earth, a land vegetation was in existence, soils were developed, climatic zones were established and there was an alternation of seasons within the year: warm and cold, dry and rainy" (Berg, 1948, p. 294).

DEVELOPMENTS SINCE BERG

During the period of nearly a quarter of a century since the publication of L.S. Berg's paper, only a comparatively small amount of research has dealt seriously with the physico-geographic conditions of the Precambrian.

M. Schwarzbach (1959) recorded the presence among Precambrian deposits of rocks with features of both arid zones and sea basins. He believed that, during the Algonkian, climates did not essentially differ from those of the present; periodically they might have been colder than at present, but he claimed that they could not have been substantially warmer.

A. Poldervaart (1955) assumed that the hydrosphere on a small scale originated at the beginning of the Archaean. During the Upper Precambrian the gas envelope of the Earth changed its composition: instead of an originally reducing nature it became of an oxidizing type. In the opinion of N.M. Strakhov (1962), life originated at the beginning of the Archaeozoic; it was primitive in nature, not differentiated into plants and animals. Only at the boundary of the Proterozoic, when blue-green algae made their appearance, was the process of photosynthesis started. At this time free oxygen began to be supplied to the atmosphere and hydrosphere, which contributed to rapid development in the organic world and to a general biomass increase.

A.B. Ronov (1964) on the whole supported N.M. Strakhov's concepts but thought it necessary to make more remote the epoch when photosynthesis began to be operative. At this point, oxidation became more intense and the amount of CO_2 in the atmosphere and hydrosphere increased. Processes of chemical and physical weathering thus occurred early in the Archaean and gradually became more intense with the progressive formation of the biosphere.

The ideas of a number of other authors of our time are, generally speaking, close to the concepts voiced in the papers just mentioned.

CONCLUSIONS

In this fashion, the study of Precambrian deposits has advanced greatly during the seven decades of our century so far and there has been a steady evolution in ideas on the palaeogeographic features of the early stages in the history of the Earth. The concept of differential heating from the subsurface of individual sections of the surface of our planet, under the effect of shallow magmatic foci, has been abandoned. Instead, the viewpoint has gained ground that, since the beginning of the Archaean, the inner heat of the Earth has had no essential influence upon the temperature of the surface. Contrary to earlier views that the climates of the remote past were markedly dissimilar from those of later times, researchers have come to the conclusion that – probably in the Archaean and certainly in the Proterozoic – there was already a pronounced climatic zonality, there were seasons in the year and, in general, the climates of even the remote Precambrian did not differ much from present day climates.

The approach to the question on the time when life appeared has also changed.

The ideas that life is very ancient has been confirmed by the use of the latest research methods, which have permitted the establishment of the age of the earliest organic remains as 3.5 x 10^9 years. Present data indicate that the Precambrian organic world developed in the seas only and, for the time being, there are no convincing proofs in favour of L.S. Berg's opinion that life existed on the Proterozoic, or even on the Archaean, land.

However, Berg's arguments deserve the most careful attention and emphasize the necessity for organizing special studies of Precambrian sedimentary rocks using the most modern methods (electron microscopy included). There is every reason to expect that, in a few years, the curtain still existing may be lifted and that the pattern of physico-geographic environments which existed on our planet during its earliest stages of geological history may come to be scientifically documented.

REFERENCES

Berg, L.S., 1918, Klimaty geologicheskogo proshlogo [Climates of the geological past]: Priroda, no. 1, p. 3-28.

——————, 1944, Zhizn i pochvoobrazovanie na dokembriiskikh materiakh [Life and soil formation on Precambrian continents]: Priroda, no. 2, p. 28-35.

——————, 1947, Klimaty geologicheskogo proshlogo [Climates of the geological past]: *in* Klimaty i shizn, Moscow: Izdatel, "Nauka", p. 308-324.

——————, 1948, Klimaty v drevneishie geologischeskie vremena [Climates of the oldest geological periods]: Zemlevedenie, New Series, v. 2, p. 81-91.

Pavlov, A.P., 1910, O drevneishikh na Zemle pustyniakh [On the oldest deserts of the Earth]: *in* Dnevnik XII s'ezda russkikh estestvoispytatelei i vrachei: Otdel 1, v. 3, Moscow, p. 302-330.

Poldervaart, A. 1955, Chemistry of the Earth's Crust: Geological Society of America Special Paper 62, p. 119-144.

Ronov, A.B., 1964, Obschie tendentzii evolutzii sostava zemnoi kory, okeana i atmosfery [General tendencies in the composition evolution of the crust, the ocean and the atmosphere]: Geokhimia, v. 8, p. 715-743.

Schwarzbach, M., 1950, Das Klima der Vorzeit: Stuttgart, 211 p.

Strakhov, N.M., 1962, Etapy razvitia vneshnikh geosfer i osadochnogo porodoobrazovania y istorii Zemli [Development stages of outer geospheres and of the sedimentary rock formation in the history of the Earth]: Izvestiya Academy of Sciences of the U.S.S.R., Geological Series, No. 12, p. 3-22.

Walther, J., 1908, Geschichte der Erde und des Lebens: Leipzig, 570 p.

——————, 1909, Ueber algonkische Sedimente: Zeitschrift der Deutschen Geologischen Gesellschaft, Bd. 61, Jg., 1909, pt. 3, p. 283-305.

History of Concepts in Precambrian Geology, edited by W.O. Kupsch and W.A.S. Sarjeant, Geological Association of Canada Special Paper 19, 1979

WILLIAM MACLURE'S CONCEPT OF PRIMITIVE ROCKS ("BASEMENT COMPLEX")

George W. White
Department of Geology, University of Illinois
Urbana, Illinois 61801

ABSTRACT

William Maclure (1763-1840), called by some the "Father of American Geology" was born in Scotland, achieved a fortune in business early and then devoted much time to geological travels in Europe and America. His 1809 description of the geology of the United States, better known in the 1817 revision, was the first written in English.

Maclure used the Wernerian terms "primitive", "secondary", "transition", and "alluvium", not because he subscribed to Werner's doctrines, but because these terms "are generally understood". He suggested that primitive rocks had originally been sedimentary and volcanic rocks that had later been "smelted, roasted, and liquified". He was sure that basalt was of volcanic origin, that trap and greenstone had once been lavas, and he thought that gneiss had probably once been sedimentary rock. He believed that veins were formed by "vapours" ascending into the cracks which lava could not penetrate, from "the seat of volcanic action either in or below the primitive." He proposed an almost unlimited span of geologic time and would have nothing to do with the Deluge as a geologic agent.

RESUME

William Maclure (1763-1840), appelé par certains le "Père de la Géologie Américaine", naquit en Ecosse, fit fortune de bonne heure dans les affaires et consacra ensuite beaucoup de temps à des voyages géologiques en Europe et en Amérique. La description de 1809 de la géologie des Etats-Unis, mieux connue dans sa forme révisée de 1817, fut la première écrite en anglais.

Maclure utilisa les termes wernériens "primitif", "secondaire", "transition" et "alluvion" non parce qu'il souscrivait aux doctrines de Werner, mais parce que ces termes "etaient généralement compris". Il suggéra que les roches primitives avaient été à l'origine des roches sédimentaires et volcaniques qui avaient été ensuite "fondues, roties, et liquéfiées". Il était sûr que le basalte était d'origine volcanique, que trapp et roches vertes avaient été des laves, et pensait que les gneiss avaient été probablement des roches sédimentaires. Il croyait que les veines avaient été formées dans des fissures que les laves ne pouvaient pénétrer, par des "vapeurs" ascendantes à partir "du siège de l'action volcanique située soit dans soit sous le primitif". Il proposa une durée du temps géologique presqu'illimitée et ne voulait rien savoir du Déluge en tant qu'agent géologique.

Figure 1. William Maclure at age 60.

INTRODUCTION

William Maclure (Fig. 1) was born in Ayr, Scotland, October 27, 1763, and died at St. Angelo, Mexico, March 23, 1840 (Morton, 1841). His first visit to America was in 1782, after which he engaged in a mercantile business and soon became wealthy. This enabled him to retire early with a comfortable fortune: thereafter Maclure travelled extensively through most of the countries of Europe, to examine geology and to visit geologists. In 1796 he crossed and recrossed the then United States from New England to Florida and later published his 1809 report and map on American geology. This was reprinted in France in 1811 with a much better colored map. With Alexander Lesueur as his cartographer he examined the geology of many West Indies islands in 1815-16. Accompanied by Lesueur, he re-studied the geology of the United States in 1816 and 1817 and published a revised map and report in 1817, which is the principal source of most people's knowledge of Maclure's geology and ideas.

In 1819 Maclure moved to a large estate in Spain, which he left in 1824 because of the reestablishment of a reactionary administration. He then joined Robert Owen and invested $82,000 in the idealistic venture at New Harmony, Indiana, and lived there for a time, later moving to Mexico where he died in 1840. Before his death he presented a large part of his library to the Academy of Natural Sciences of Philadelphia. The record of these 5232 books and many hundreds of maps (Catalogue of the Academy, 1837) indicates that Maclure's scientific library was probably the most extensive in private hands in America; it consisted of current and classical works on travel, geology, other sciences, and philosophy, as well as scores of complete sets of periodicals and society transactions all major European languages.

In addition to his book on *Geology of the United States,* Maclure published many articles in the *American Journal of Science*, the *Journal of the Academy of Natural Sciences of Philadelphia,* and at the press at New Harmony. Maclure's *Geology of the United States* (1809; 1817) is a description of the rocks and their distribution as outlined in his text and on his now famous map. It has little to say about origin; it is in the much less known *Essay on Formation of Rocks* (1818) and in several later articles that his concepts of origin of the rocks are set forth.

The principal commentators, Merrill (1924, p. 31-37) and Fenton and Fenton (1945, p. 128), appear to have based their discussion of Maclure entirely on the 1817 edition of *Geology of the United States* and make no reference to other publications. Merrill stated (1924, p. 34) "the classifications of the formations adopted by Maclure . . . was naturally largely Wernerian", which implies that Maclure adhered to the theories of Werner. Fenton and Fenton say that Maclure "brought a Wernerian type of geology to the United States (Fenton and Fenton, 1945, p. 128). These authors imply by "Wernerianism" the belief in neptunian (aqueous) origin not only of obviously sedimentary rocks, but also of granite, basalt, and related rocks, and exhibit a very incomplete and in part erroneous understanding of Maclure's ideas. The anonymous review in the *Edinburgh Review* (1818) was ambiguous, regarding Maclure as "a disciple of Werner" because of his terminology, but recognizing that "his geological creed is not that of Werner" and giving quotations to support that belief.

PRIMITIVE ROCKS

Maclure's concept of the form, composition and distribution of primitive rocks is shown in his 1809 *Geology of the United States* and is only slightly changed in the 1817 revision. Rather than devise a new system of nomenclature – he was sharply critical of the proliferation of names – he decided to use the Wernerian classification of rocks as "Primitive, Transition, Floetz or Secondary, and Alluvial," because he felt this usage was understood. He stated that he would have preferred the term "chrystalline" to "primitive", and used the former term rather than the latter in some places. He apparently derived the terminology not from the earliest classification of Werner (1786, p. 274-276; 1971, p. 45), but from an exact reproduction of the later amplified one which Werner's student Jameson published in 1808 (p. 99). Maclure says (1817, p. 28-29): "Without entering into any investigation of the origin, or first formation of the various substances, the following nomenclature will be used:

CLASS I. — Primitive Rocks.

1. Granite,
2. Gneiss,
3. Mica Slate,
4. Clay Slate,
5. Primitive Limestone,
6. Primitive Trap,
7. Serpentine,
8. Porphyry,
9. Sienite,
10. Topaz-rock,
11. Quartz-rock,
12. Primary Flinty-slate,
13. Primitive Gypsum,
14. White-stone."

In his ensuing descriptions he dealt at length not only with the distribution in the United States of these rocks, as illustrated in his accompanying map, but also in northern North America, Scandinavia, Russia, and other parts of Europe. He did not give an organized description of the various primitive rocks; he seemed to assume that his readers were familiar with their composition. He obviously agreed with the most detailed American description of rocks according to Werner's system, that by Cleaveland (1816, p. 597-617).

Granite

Maclure assumed that this rock always contained quartz, feldspar and mica. Maclure stated: "The first primitive rock is granite, which is a granular aggregate of chrystals, decomposing into a gravelly mass: this rock proceeding slowly through the other stages of decomposition . . ." (1817, p. 70-71). He found two kinds of granite based on texture – the coarse-grained granite and a "middling grained Granite, often with much quartz in it [which] has more resemblance to some of the feldspatic lavas, than it has to any rock known to be of Neptunian origin. It likewise approaches the volcanic . . . without any regular stratification" (1818, p. 340). He was impressed by the deep weathering of granite to depths of 40 feet in the southern states (1817, p. 45) and the release of "quartz in small chrystalline particles [to form] siliceous sand" (1817, p. 72).

Gneiss

Maclure regarded gneiss as a banded, coarse-grained rock, with "extensive and regular stratification" (1818, p. 337), usually having a granite-like assemblage of minerals, but with more mica; "gneiss, from its fissile structure and additional quantity of mica, is of easier decomposition and forms a soil a little more argillaceous [than granite]" (1817, p. 71). Of the gneiss in the primitive rocks of the United States he says (1817, p. 45-46):

> "Gneiss extends perhaps over half of this formation, and includes in a great many places beds from three to three hundred feet thick, of a very large grained granite, which run in the same direction, and dip as the gneiss does; it is in those beds generally where the emerald, phosphat of lime, tourmaline, garnet, cymophane, octahedral iron ore, graphic granite, etc. etc. are found. These beds are mixed, and alternate occasionally in the same gneiss, with the primitive limestone, the beds of hornblende and hornblende slate, serpentine, magnetic iron ore, and feldspar rocks. In some places this gneiss contains so much mica, as to run into mica slate; in others, large nodules of quartz or feldspar; in others, hornblende takes the place of the mica; in short, I scarcely know any of the primitive rocks that may not occasionally be found included in the gneiss formation."

Mica Slate and Clay Slate

Maclure regarded mica slate as a "species of gneiss" (1818, p. 337), more finely banded and finer grained than gneiss and containing much more mica. Clay slate is described in one place as a rock made of very fine-grained minerals in thin laminae.

Primitive Limestone

Maclure regarded primitive limestone as crystalline calcareous or dolomitic rock, at places somewhat like the finer grained, fossiliferous secondary (horizontal) limestone, but without any fossils. He occasionally used the term "marble" for this rock and said that it "perhaps only differs [from undoubted Neptunian limestone] in total absence of the remains of organic matter" (1818, p. 338).

Primitive Trap

Maclure's definition and opinion on primitive trap is contained in a letter to Benjamin Silliman (1819) in which he states:

> ". . . primitive trap is one instance [of a name from Werner] which I do not use because I never met with [it]. I do not use trap as substantive, except in describing that kind of trap which Werner calls the newest flaetz [sic] trap, the nearest of which is your trap (ridges of Greenstone near New-Haven), which covers the oldest red sandstone.
>
> What Werner calls primitive trap may perhaps be compact hornblende, or perhaps the newest flaetz trap, when it happens to cover the primitive; for, this species of trap, like the currents of lava, covers indescriminately all classes of rocks, and is one reason why I consider it as the remains of ancient lava.
>
> Transition trap is a rock that I have not met with, and may perhaps be a part of the flaetz trap that happened to cover the transition, without any immediate connexion, but like a current of lava, overlying all the classes of rocks it meets with. This misapplication of names naturally arises from the system of neptunian origin, on which the nomenclature of Werner is founded."

Other Primitive Rocks

Maclure had little to say about the other primitive rocks in his list. He did mention serpentine and porphyry as interbedded with some of the banded primitive rocks such as gneiss and mica slate. Maclure realized that "porphyry" was used at the time not only for granite and other crystalline rocks with noticeable phenocrysts but also for amygaloidal basalt, which he recognized had been formed by filling of vesicles. He specifically stated that he had not found primitive gypsum in the United States and was uncertain about topaz rock. He said nothing about "White-Stone", which Werner used for a rock composed of "compact feldspar, with very little mica" (Jameson, 1808, p. 145), which seems to describe a rock now called aplite.

PRIMITIVE ROCKS — ORIGIN

Maclure again and again argued vehemently against constructing theories on the basis of insufficient observations and asserted that it was not yet known how certain rocks were formed. In his great *Essay on the Formation of Rocks* he said (1818, p. 268): We can only "proceed from the known, which we see daily forming, towards the unknown, through a chain of reasoning strictly analogous, [which] is perhaps all our present Knowledge will permit us to do." He proceeded from the known to the unknown in a closely reasoned treatment of both aqueous and volcanic deposits: 1) formed by nature under our observation: 2) resembling these and by analogy of the same origin; and 3) rocks ("formations") partly separated by erosion, but allowing interpretation similar to 2 above. He then proceeded to the much more uncertain area of the primitive rocks.

Maclure had seen lava flows,. fragmental material, ash layers (tuff), scoria, mudflows ("mud lava"), and volcanic breccia which had come from volcanic vents. He related this knowledge of active volcanoes to the extinct volcanoes and their deposits which he had seen in the Auvergne, Spain, Italy, the Rhine Valley and Hungary. Passing to other regions with similar rocks, but with volcanic craters no longer visible, he interpreted these similar materials as also volcanic and not Neptunian (except as for the influence of water in producing breccia and mudflows).

Maclure then traced sedimentary ("Neptunian") rocks from undoubted sediments – clastic, organic, and chemical (1818, p. 261-276, 286-307) – through sandstone, limestone, coal, and other rocks clearly derived from these sediments, to rocks of less certain, but of probable, sedimentary origin – "gray wacke", silica cemented sandstone, anthracite, some slates, and "siliceous shist."

Proceeding farther into conjecture – and clearly stating the uncertainty – he said (1818, p. 337) that "those rocks which conjecture might be disposed to place in the rocks of Neptunian origin" are gneiss, because of its stratification; mica slate – a "species of gneiss"; primitive limestone [marble]; clay slate; and very doubtfully, serpentine.

Those rocks "which analogy might be disposed to place in the Volcanic Origin [are] Hornblende rocks, both greenstones and sienites [which] resemble some species of lava, porphyry resembles much some lavas" (1818, p. 339). He very hesitatingly included granite as volcanic, but thought that fine-grained granite "resembles some of the feldspatic lavas." It seems that Maclure did not consider consolidation and crystallization of granite and other crystalline rock from intrusive magma, but only from crystallization from lava flows at the surface. And yet at another time he came close to magmatic intrusion when writing about origin of veins from "vapours" (1831, p. 406-408).

Maclure's interest and knowledge of primitive ("chrystalline") rocks was much deeper than that of any other geologist of the time. In a letter to Silliman, Maclure (1829b, p. 162) urged that more attention be given to the "primitive formation . . . granite, gneiss, sienite (including protogene gneiss or talcous gneiss) and clay slate." In this connection he insisted that primitive rock was characterized by crystallization and that "one effect also, and not the least . . . had been the uplifting of strata, of which none ought to doubt, who favors the igneous origin of the rocks of the primitive class."

Whereas almost all Americans, including Benjamin Silliman, Amos Eaton, and (probably) Parker Cleaveland, assumed a short time – 6,000 years or only a little longer – for the age of the earth, Maclure assumed a very long – essentially an unlimited – time. He was sure that primary rocks were very old and that their original composition and structure could only be conjectured. Furthermore, he postulated that the original form of the primitive rocks was not that seen today (1829c, p. 351), but that they were probably derived from preexisting sedimentary and volcanic rocks which had been "smelted, roasted and liquified" (1829a, p. 385). He recognized that some basalts had been modified by later solutions forming amygdules and probably causing recrystallization, and that greenstones were altered basalt and a species of trap. He believed that veins were formed by ascending hot solutions and he may have thought these also affected the wall rock. He was willing to entertain other methods of changing rocks (1817, p. vi):

"Until lately we have restricted nature to two modes of acting; by fire, and by water: now, it is found, that she can change and metallize rocks in the dry way, without any solution or fluidity; and the galvanic pile may be formed in the stratifications of a mountain, as well as in a chemist's laboratory. These are two other modes wherein we must now allow her to change and modify the surface of this earth; and who can say how many more means yet unknown, she may possess? each of which, when found out by accurate and impartial observation, must make a change in former theories."

Maclure's concept of veins must be included in any discussion of his concept of origin of primitive rocks. Whereas Parker Cleaveland (1816, p. 594-496) had discussed veins, but apparently not from personal knowledge, Maclure's opinions were derived from his experience with Mexican mines and minerals and had also seen them in mines in several European countries. He pointed out the connection between metal ores and volcanic regions and suggested that "precious metals converted into vapour would penetrate through chinks that would not permit lava to pass" (1831). By implication he included gangue minerals in the same process. He thought "the seat of volcanic action is either in or below the primitive" (1829a, p. 385).

As influencing consciously or unconsciously Maclure's thinking about the origin of primitive rocks, it is important to note that he did not believe in the "onion skin" concept of universal formations, but insisted that secondary rocks varied markedly in different parts of the world and were not continuous even within a single continent (1819, p. 213):

". . . the secondary is deposited in basins alongside of one another and that each basin has a different order of supraposition according to the nature of the agents employed in the deposition, that it is a partial, and by no means a general deposition . . . I am therefore inclined to think that in geology, the best mode for the greatest part of the secondary would be to give the relative position of the strata in each valley or basin and I am rather of opinion that they would all differ from one another."

It was then but one short step to explain the great variation in composition and arrangement of primitive rocks, and to assert that he could not see the invariable age sequence of Werner and his followers (1817, p. 16).

CONCLUSIONS

Maclure had derived his opinions from long and careful observations in Europe and North America. He had seen more primitive ("chrystalline") rocks than anyone else of his time in either continent, and was particularly concerned with their distribution, description and origin.

He believed that rocks had formed in separate basins and therefore that there was no expectation of finding either primitive or later rocks in world-wide "onion-skin" continuity and sequence.

He asserted that ancient deposits, similar to present day ones might have been modified by various means, some known and some as yet unknown, to form the primitive rocks we see today.

Maclure used the term "primitive" in the sense of "basement complex" and was sure that the transition, secondary, and volcanic rocks had been deposited upon it; he believed they must be later because they contained fragments of earlier rocks. He had seen the unconformable contact in valleys near Woldinaria (Russia) in the Hudson Valley, and elsewhere. He was sure of the volcanic origin of basalt or trap of any position or age and in this connection says (1822, p. 198): "The geology of the United States . . . is a strong argument against the Wernerian system – all these theories have had their day, and are fast going out of fashion."

Gneiss was once sedimentary, but although he thought granite might be a part of an original molten crust, he regarded the "granite problem" as beyond certain solution. It is curious that he did not consider Hutton's and Playfair's ideas of intrusive origin of granite. However, Maclure gave good chemical reasoning for believing that granite was not an aqueous deposit. He considered that greenstones had once been lavas. He recognized the difference between dikes and veins, and believed that veins were formed by hot "vapours" ascending into cracks into which lava could not penetrate.

Maclure was uncertain about the length of geologic time, but was so little influenced by the then current general belief in a 6,000 year age of the earth and in the Deluge, that he rarely mentiond them, except in a disparaging way (1824, p. 261; 1838, p. 175). He stated (1824, p. 263) "that we have no right to limit [time] nor to estimate with our lilliputian senses the operation of nature." He was not a catastrophist but an actualist, saying that (1817, p. iv-v):

> In all speculations on the origin, or agents that have produced the changes on this globe, it is probable that we ought to keep within the boundaries of the probable effects resulting from the regular operations of the great laws of nature which our experience and observation has brought within the sphere of our knowledge. When we overleap those limits, and suppose a total change in nature's laws, we embark on the sea of uncertainty, where one conjecture is perhaps as probable as another; for none of them can have any support, or derive any authority from the practical facts wherewith our experience has brought us acquainted.

REFERENCES

Anonymous, 1818, (Review of) Observations on the geology of the United States of America, by William Maclure, Philadelphia, 1817, 8 vol., 127 p.: Edinburgh Review, v. 30, p. 374-382.

Catalogue of the Library of the Academy of Natural Science of Philadelphia: 1837, Philadelphia, J. Dobson, 300 p.

Cleaveland, Parker, 1816, An elementary treatise on mineralogy and geology, being an introduction to the study of these sciences . . . : Boston, Cummings and Hilliard, 668 p.

Fenton, C.L., and Fenton, M.A., 1945, The story of the great geologists: Doubleday, Doran and Co., Inc., Garden City N.Y., xvi + 301 p.; revised second ed. [entitled Giants of geology], 1952, xvi + 333 p.

Jameson, Robert, 1808, Elements of geognosy, being vol. III, and part II of the system of mineralogy: Edinburgh, William Blackwood, 368 p.

Merrill, George P., 1924, The first one hundred years of American geology: New Haven Yale Univ. Press; (reprinted 1964, Hafner Publishing Company, New York and London), xxi + 773 p.

Maclure, William, 1809a, Observations on the geology of the United States, explanatory of a geology map: Transactions of the American Philosophical Society, v. 6, p. 411-428.

————, 1809b, Extract d'une lettre de M. William Maclure, à J.C. Delametherie, sur la géologie des Etats-Unis. Observations sur la géologie des Etats-Unis, servant à expliquer une carte géologique: Journal de Physique, de Chimie et Histoire Naturelle, v. 69, p. 201-215.

————, 1811, Suite des observations sur la géologie des Etats-Unis servant a l'explication de la carte ci-jointe: Journal de Physique, de Chimie et Histoire Naturelle, v. 72, p. 137-165.

————, 1817, Observations on the geology of the United States of America; with some remarks on the effect produced on the nature and fertility of soils by the decomposition of the different classes of rocks; . . . in reference to the accompanying geological map: Philadelphia, Abraham Small, 127 p., also in Transactions of the American Philosophical Society, New Series, v. 1, p. 1-91.

————, 1818, Essay on the formation of rocks, or an inquiry into the probable origin of their present form and structure: Journal of the Academy of Natural Sciences of Philadelphia, v. 1, p. 261-276, 285-310, 327-345. Reprinted, Philadelphia, 1818, 60 p. Revised and reprinted, New Harmony, 1832, 53 p.

————, 1819, Hints on some outlines of geological arrangement, with particular reference to the system of Werner, in a letter to the Editor . . . dated Paris, 22 August, 1818: American Journal of Science, v. 1, p. 209-213.

————, 1822, Extracts of a letter from Wm. Maclure, Esq. to the Editor, dated, Madrid, Dec. 4, 1821: American Journal of Science, v. 5, p. 197-198.

————, 1824, Miscellaneous remarks on the systematic arrangement of rocks, and on their probable origin, especially of the secondary . . . in a letter dated Alicante in Spain, April 29, 1923, addressed to the Editor: American Journal of Science, v. 7, p. 261-264.

————, 1829a, Remarks on the theory of a central heat in the earth, and on other geological theories; . . . : American Journal of Science, v. 15, p. 384-386.

————, 1829b, Extract of a letter to the Editor, from an American resident in Mexico, dated Halcota, Near Temascaltepec, July 13, 1828: American Journal of Science, v. 16, p. 159-163.

————, 1829c, Remarks on the igneous theory of the earth, in a letter to the Editor . . . , dated Jalapa, Mexico, February 8, 1829: American Journal of Science, v. 16, p. 351-352.

————, 1831, Geological remarks relating to Mexico &c. in a letter dated Mexico, May 30, 1830: American Journal of Science, v. 20, p. 406-408.

————, 1838, Geology of the earth – Geological observations: Opinions on various subjects dedicated to the industrious producers, 3 volumes, New Harmony, v. 3, p. 175-178. Reprinted, with commentary, Journal of Geological Education, v. 18, p. 127-128, 1970.

Morton, S.G., 1841, A memoir to William Maclure, Esq.: Philadelphia, T.K. and P.G. Collins, 37 p.; also in American Journal of Science, v. 47, 1844, p. 1-17.

Werner, A.G., 1786, Kurze Klassification und Beschreibung der verschiedenen Gebirgsarten: Abhandlungen der Böhmischen Gesselschaft der Wissenchaften, v. II, p. 272-297.

————, 1971, Short classification and description of the various rocks. Translated (with an introduction and notes) by Alexander M. Ospovat: New York, Hafner Publishing Company, 194. p.

History of Concepts in Precambrian Geology, edited by W.O. Kupsch and W.A.S. Sarjeant, Geological Association of Canada Special Paper 19, 1979

CHARLES D. WALCOTT — AMERICA'S PIONEER IN PRECAMBRIAN PALEONTOLOGY AND STRATIGRAPHY

E.L. Yochelson
U.S. Geological Survey, E-501 Museum of Natural History,
Washington, D.C. 20560, U.S.A.

ABSTRACT

Walcott's first major contact with Precambrian strata was late in 1882 on the north side of the Grand Canyon; his earlier geologic work had been almost exclusively in highly fossiliferous lower Palaeozoic strata. Although he first assigned the rocks of the inner gorge to the Cambrian, he reconsidered this assignment, possibly on evidence from regional stratigraphy. Presumed animal remains were subsequently discounted and may not have influenced this revision in age.

Beginning in the late 1890s, Walcott worked for a number of seasons in the Belt Supergroup of Montana and in adjacent areas. He subdivided the beds and found algal stromatolites and a few presumed animal remains. His stratigraphic framework has remained satisfactory, but the "metazoans" have been discounted by later workers.

Walcott described Precambrian algae in several early papers. His 1914 paper laid the groundwork for further studies, but he did not return to this subject. This paper pointed out the importance of stromatolites within the Precambrian and was one of the first to demonstrate the diversity of this group.

Because of the characteristic absence of animal fossils, Walcott presumed great thicknesses of Precambrian strata to be of nonmarine origin. He compared the stromatolites with fresh water algae in recent lakes and thermal springs and was one of the first to suggest bacterial action as important in limestone deposition. Physiological explanations are now invoked to explain the absence of metazoans throughout most of the Precambrian, but at the time it was proposed Walcott's hypothesis of nonmarine deposition was logical.

RESUME

Le premier contact important de Walcott avec les strates précambriennes fut vers la fin de 1882 sur le flanc nord du grand Canyon; son précédant travail géologique concerna presqu'exclusivement les strates très fossilifères du Paléozoique Inférieur. Bien qu'il assigna tout d'abord au Cambrien les roches de la gorge intérieure, il reconsidéra cette détermination peut-être à partir d'évidence sur la stratigraphie régionale. De présumés restes animaux furent ultérieurement éliminés et ne peuvent avoir influencé cette révision d'âge.

Commençant à la fin des années 1890, Walcott travailla de nombreuses saisons dans le Supergroupe du Belt du Montana et des régions adjacentes. Il subdivisa les lits et trouva des stromatolites alguaires et quelques présumés restes animaux. Sa charpente stratigraphique est restée satisfaisante, mais les "métazoaires" ont été écartés par les chercheurs ultérieurs.

Walcott décrivit des algues précambriennes dans plusieurs de ses premiers articles. Son article de 1914 laissa les bases pour des études ultérieures mais il ne revint pas sur ce sujet. Cet article souligna l'importance des stromatolites au Précambrien et fut l'un des premiers à montrer la diversité de ce groupe. A cause de l'absence caractéristique de fossiles animaux, Walcott présuma l'origine non marine de grandes épaisseurs de strates précambriennes. Il compara les stromatolites avec les algues d'eau douce de lacs récents et de sources thermales et fut l'un des premiers à suggérer l'importance de l'activité bactérienne dans le dépôt des carbonates. Les explications physiologiques sont invoquées maintenant pour expliquer l'absence de métazoaires dans la majorité du Précambrien, mais à l'époque où elle fut proposée, l'hypothèse de Walcott sur le dépôt non marin était logique.

INTRODUCTION

Charles D. Walcott is best known for his investigations of the Cambrian of North America. His studies of the Precambrian form a small percentage of his published articles, in sheer volume perhaps second only to those of James Hall in the United States; but the areas of Precambrian outcrop Walcott investigated were enormous and his efforts to find evidence of early life met with at least limited success.

Although one might assume that Walcott's interest in the Precambrian was a natural and logical outgrowth of interest in the Cambrian faunas, there is no indication of this in any of his earlier work. Rather, the evidence suggests that his efforts in the Precambrian were a direct result of a U.S. Geological Survey field assignment ordered by John Wesley Powell.

THE START OF A PROFESSIONAL CAREER

Walcott's life and career have been discussed elsewhere (Darton, 1928; Yochelson, 1967) and need only be outlined in geologic terms to support the suggestion presented above. He was born March 31, 1850, in New York Mills, New York, on the western outskirts of Utica. The Hudson River Valley in this area is cut in Middle Ordovician shales. During his early years he saw many fossils, partly from the spoil banks of the Erie Canal; collecting fossils was a local schoolboy tradition. At the start of the Civil War in the early 1860s, he spent several summers at Trenton Falls, the type area of the richly fossiliferous Trenton Limestone, on the southwest margin of the Adirondack dome. The underlying Lower Ordovician-Upper Cambrian rocks are virtually unfossiliferous. The strongly metamorphosed and intruded Precambrian rocks of the Adirondacks are not the sort to gladden the heart of a paleontologist.

At the age of 21, Walcott moved to the Rust farm at Trenton Falls, where he divided his time between farming and fossil collecting; the latter turned out to be a reasonably profitable business, both financially and scientifically. Biological work on trilobites also was started during his stay at Rust Farm. While at Trenton Falls, some fossil exchanges were made with other collectors, including G.F. Matthews; Walcott spent a few days visiting him and collecting from the Cambrian at St. John, New Brunswick. Still, his acquaintance with any pre-Trenton fossils was very slight.

In later years, Walcott (1916a, p. 254) recounted an anecdote connected with the chance breaking of a rock by a wagon he was driving to Trenton Falls when he was 17. Although he was already well acquainted with the Trenton fauna, this isolated block contained unfamiliar older trilobites which sparked his interest.

In November 1877, Walcott moved to Albany as assistant to James Hall. Most of his fieldwork for Hall was in the Devonian of New York and the Silurian of Ontario, but he made a brief reconnaissance to Saratoga, New York. He later spent two weeks of "vacation" collecting Late Cambrian fossils at Saratoga, New York, and had the opportunity to see the *Cryptozoon* reefs. The collections at Albany obviously were more extensive than at Rust Farm and geologic duties at the State Museum were more varied than he had undertaken previously. In spite of this broadening of horizons, there is no indication that his primary interests were other than in fossils of the Trenton. In terms then current, his concern was with Early Silurian fossils rather than Cambrian ones; certainly there was nothing in his background to suggest any interest in the Primordial.

Figure 1. Charles Doolittle Walcott; a formal photograph, probably taken in the late 1880's (Photo by courtesy U.S. Geological Survey).

BEGINNING WORK WITH THE GEOLOGICAL SURVEY

On July 21, 1879, Walcott was appointed to the newly formed U.S. Geological Survey and immediately headed west for fieldwork. His first season was in the upper Palaeozoic and younger strata of the area near Kanab, Utah; he spent the winter in Washington. In the summer of 1880 he was attached to Arnold Hague's party, working in Nevada on the geology of the Eureka mining district. This provided his first opportunity to come in contact with a thick, richly fossiliferous Cambrian section.

Writing descriptions of the Cambrian fauna and those of the overlying Palaeozoic units occupied about a year; Walcott worked at the American Museum of Natural History in New York City, again an area and institution not conducive to pre-Palaeozoic inquiry.:

In July, 1882, he returned to the Eureka district for additional fieldwork and after six weeks transferred back to the Kanab region. In mid-November in the House Rock Valley, he met Major John Wesley Powell, who directed him to undertake work in the Grand Canyon. On November 26, 1882, Walcott started for Nankoweap Valley, having been instructed by Powell to study the pre-Carboniferous strata as far south as possible. Walcott finally returned to House Rock Valley on February 6, 1883; by the time a few earlier observations were rechecked and collections packed, he had spent a full eight months in the field.

THE GRAND CANYON 1882-1883

The rationale behind Walcott's Grand Canyon assignment is given in an 1883 letter from Clarence Dutton to Archibald Geikie (Craig and Rabbitt, 1969).

> ". . . G. K. Gibert, basing his inference on the discovery of a single fossil, has assigned the lower part of the conformable series to the Cambrian. If Gilbert were right, then an apparently conformable series of beds was Carboniferous above and Cambrian below . . . and the very fresh-looking rocks below the unconformity were pre-Cambrian without any known equivalent in that part of the country . . . Walcott had gone into the Canyon along a new trail and found an abundant fauna 'most decisively and unquestionably Cambrian' for 500 ft. above the unconformity. After two months' search he found only a single *Stromatopora* in the beds below."

A summary report of the fieldwork is available (Walcott, 1884b). By almost any standards it was arduous and dangerous work; some quotations from the account are appropriate:

> "The strata below the Tonto Group was first studied in Nunko-weap Valley, then in Kwa-gunt, and on to Chuar Valley. Trails were built over the high ridges separating these valleys, as it was found impractical to follow along the shore of the Colorado River." "Mr. Haskell is an excellent collector, and I regretted his going at that time, as I was left with but one man for two weeks, and was obliged to undertake many dangerous climbs on the canyon walls alone." "To reach the Archean, south of Vishnu's Temple, the Tonto terrace (2,200 feet above camp) was scaled and followed along for several miles to a narrow cleft which give a passage down to the Archean, 1,700 feet below." "In going out one mule was killed outright and two badly injured."

The question of the age of the rocks was discussed in a series of reports considered later in this paper. However, the Grand Canyon trip was responsible for two other papers which enhanced Walcott's reputation as a field geologist. The first is a small but interesting work on structural geology (Walcott, 1899). The suggestion that certain structural features along the Butte fault within the inner gorge are related to the thickness of overlying sedimentary rocks has a modern ring to it.

The second paper is his principal contribution to igneous geology (Walcott, 1894). In point of fact this work on Precambrian lava flows belies its title; it is concerned primarily with the sedimentary rocks and is the definitive Walcott summary of the Grand Canyon Precambrian. Walcott seldom gave a talk which was not later published in some form, and preliminary organization for this paper could have come from preparation for a lecture (Anonymous, 1891).

THE AGE AND NOMENCLATURE OF THE GRAND CANYON SERIES

In his paper on the stratigraphy of the area, Walcott (1883, p. 440) divided the pre-Tonto, post-Archaean Grand Canyon group of Powell into an overlying Chuar group and an underlying Grand Canyon group.

> "But for the discovery of a small Discinoid shell, a couple of specimens of a Pteropod allied to Hyolithes triangularis and an obscure Stromatopora-like group of forms, the two and ½ month search for fossils in these groups would have been without result. They serve, however, to retain the groups within the Cambrian and also point to a fauna which must be searched for elsewhere, as they alone could not have been the only representatives of the life in the sea at that time."

Walcott then compared the Grand Canyon rocks to the Kewenawan Series and suggested that they too were Cambrian! In a lecture on the Cambrian (Walcott, 1884a, p. 99), he noted;

> ". . . the presence of a few fossils that enable us to refer it [Grand Canyon Series] to the Cambrian but not to refine its stratigraphic horizon. That is done on the evidence of the position it occupies with reference to the Tonto."

There may be a discrepancy between Walcott's writings in 1883 and the tantalizing fragment of Dutton's correspondence previously cited. Dutton in quoting Gilbert was apparently convinced that his own earlier opinion on the age of the older Grand Canyon rocks was wrong; Walcott's conclusions supported Gilbert. Still, other interpretations are possible; perhaps this point may be resolved when more of the Dutton correspondence is available.

In 1884, Walcott was detailed to look at a reported coal deposit in Arizona. During his return trip to Washington he made a hasty reconnaissance visit to the central mineral region of Texas (Walcott, 1884c, p. 431). He named the Llano series and noted:

> "At all localities where the base of the Potsdam was observed, it rests, *unconformably*, on a great formation that is stratigraphically the equivalent of Powell's Grand Cañon series (Grand Canon and Chuar groups)."

He correlated the Llano rocks with those of the Grand Canyon because of the unconformity and the similarity of the overlying Cambrian. The Llano had previously been considered "granitic, metamorphic and igneous" Archaean. No fossils were found, but Walcott still considered the older rocks to be Cambrian. He reinforced his original interpretation, for in a footnote (Walcott, 1884c, p. 432) he mentioned that, after his earlier paper had been written, a trilobite fragment was found in a bit of the Chuar shale.

Walcott was impressed with the apparently fresh physical condition of the Grand Canyon Series. In several papers he compared the fresh-appearing sediments and alternating lithologies of the upper 6,000 feet of the Chuar with the Trenton limestone and Utica shales.

In 1886 (p. 43), he changed his concept of the age of the Grand Canyon Series and wrote:

> "Ripple marks and mud cracks abound in many of the layers, but not a trace of a fossil or the trail of a mollusk or annelid was seen.
>
> Midway of the lower portion of the shales and limestones of the overlying Chuar strata the presence of a fauna is shown by a minute Discinoid or Patelloid shell, a small Lingula-like shell, a species of Hyolithes, and a fragment of what appears to have been the pleural lobe of the segment of a trilobite belonging to a genus allied to the genera Olenellus, Olenoides, or Paradoxides. There is also an obscure Stromatopora-like form that may or may not be organic. The fauna, as given above, is very unsatisfactory, but it shows the presence of a fauna that is Cambrian in character as far as we know, although it may be a trace of a fauna preceding that of the Lower Cambrian of the Atlantic border; and as the stratigraphic evidence favors this view I do not think we can consider it of Cambrian age."

Thus, the first hazy outlines of present-day North American Precambrian stratigraphy are given in the brief summary:

> "The Grand Cañon, Llano and Kennenaw series are probably of pre-Cambrian age, and are omitted from the Lower Cambrian although I have heretofore made a provisional reference of them to the Cambrian." (Walcott, 1886, p. 64).

One of the characteristics of this era was a tendency of authors toward repetition, and Walcott was no exception. Some lines and even paragraphs of earlier papers were repeated in Walcott's papers, as illustrated above; however, each paper did amplify or modify the preceding one. When the full details of the stratigraphy were presented with a colored geologic map (Walcott, 1894) years after the fieldwork, they were also presented in a more generalized account (Walcott, 1895).

The term Algonkian was employed by Walcott in 1890 (p. 50) for the Grand Canyon rocks; he included in this the discordantly underlying, strongly metamorphosed Vishnu Schist and followed that usage for many years. Walcott (1894, p. 506; 1895, p. 316), introduced the term Unkar Group as a substitute for the term Grand Canyon Group he had employed a dozen years earlier. The Unkar Group, the Nankoweap Limestone and the overlying Chuar Group now form the Grand Canyon Supergroup (Elson and Scott, 1976). Stratigraphic revision of the Chuar Group now fittingly includes a Walcott Member within the Kwagunt Formation (Ford and Breed, 1972a, p. 8). This most recent geologic map of the Grand Canyon (Grand Canyon Natural History Association, 1976) shows a "Walcott graben" off the Vishnu fault, extending south of Nankoweap Valley in the eastern part of the park.

SPECULATIONS ON A CHANGE OF MIND

Between 1884 (when he completed his report on the Llano area) and 1886, Walcott revised his ideas on the age of the pre-Tonto rocks. This interval may have been even shorter, for an illustration in his 1886 paper (p. 44) refers these rocks to the Cambrian – a point Walcott noted. Presumably, figures were prepared in advance of the final text; perhaps they were prepared to conform with an earlier manuscript draft.

During the 1885 field season, Walcott worked in the Highland Range of Nevada and the Wasatch Mountains of Utah. The Precambrian is particularly impressive in the Wasatch, where thousands of feet of sedimentary rocks occur, including a major tillite. He observed a tremendous thickness of nonfossiliferous rocks which were overlain in turn by fossiliferous Cambrian rocks containing faunas older than that of the Tonto in the Grand Canyon sequence. After fieldwork, Walcott spent that winter reviewing Cambrian stratigraphy for his major 1886 work. One may presume that either further acquaintance with the older Cambrian of western stratigraphy or the anomalously great stratigraphic thickness of the Cambrian in the Grand Canyon area when compared with other areas of undoubted Cambrian age, or both, were the factors responsible for the change in his interpretation.

THE GRAND CANYON FOSSILS

Although pre-Tonto fossils were reported in 1883, description and illustration of them were not immediately forthcoming. In later years Walcott seldom delayed so long in describing a biota. From this time lag, it is evident that restudy of the fossils in the laboratory could not have influenced his change in age of the series, for comments on the fossils made in 1883 remained unchanged in 1886.

The most clear-cut of these early fossils was the "Stromotopora-like" form. Walcott sent this to Sir William Dawson, who studied and returned it in 1883 (Walcott, 1899, p. 233). Dawson (1897a, p. 208) finally described it formally as *Cryptozoon? occidentale*. Walcott (1899) gave the text of Dawson's original 1883 note, provided additional illustrations, and compared the species with the Late Cambrian *Cryptozoon proliferum*. Later he referred it to *Collenia* (Walcott, 1914a, p. 111) and added several more illustrations of specimens from the Grand Canyon. Stromatolites of the Grand Canyon were considered briefly (White, 1928a) and then ignored for many years; several beds of stromatolites occur in the region.

Chuaria circularis was named and illustrated at this time (Walcott, 1899, pl. 27, figs. 12, 13). He suggested that the specimens "appear to be the remains of a compressed conical shell." This species has not been subjected to restudy until recently (Ford *et al.*, 1969; Ford and Breed, 1972b). The specimens are not marine invertebrates but appear to be small spherical bodies which have been flattened. They are larger than known spores and the largest are about half the size of floats seen on modern sargasso weed. However, having examined the type material, I can well understand how they may have been confused with a shell of subcircular outline; the best preserved of Walcott's specimens do show a concentric pattern of wrinkles which simulate ornament. It is only by examining additional specimens that the irregularity caused by compression becomes evident.

The presumed hyolithid was figured by Walcott in the same paper. He was careful to point out the impressive resemblance to undoubted hyolithids but concluded that the specimens were of mechanical origin. I have examined this material and concur with Walcott that they are inorganic. The trilobite fragment reported in 1884 (1884c, p. 432) was also discounted because of ". . . the experience gained by a study of the pre-Cambrian fossils of the Belt . . ." Once again I concur that the item in question is not organic.

The "Lingula-like" shell mentioned in 1886 cannot be traced, though this may be the "obscure remains of a shell which, if seen in the Cambrian, would be referred to the genus Acrothele" (Walcott, 1899, p. 235). Walcott in 1893 (p. 660) alluded to a limestone in the Chuar showing "abundant evidence of the presence of spiculae of sponges, and what appear to be worn fragments of some small fossils." Apparently he later discounted the spicules for, in spite of numerous secondary references in textbooks, they are not mentioned again by Walcott. Microfossils have only recently been found in Precambrian cherts of the Grand Canyon (Schopf *et al.*, 1973) and in the shales, a lithology hitherto little explored for organic remains (Bloesser *et al.*, 1977).

THE NORTHEAST

After completion of his "Paleontology of the Eureka district" (Walcott, 1884d), Walcott plunged into the Taconic question. To make sense of this problem, of necessity he extended his interest in Cambrian fossils and the problems of Cambrian correlation. The outcrops of strongly metamorphosed rocks in the Green Mountains and Adirondacks did not help with his search for earlier fossils. However, the noted *Eozoon*, which some geologists considered a giant primordial foraminifer, was reported from both Massachusetts and Adirondack outcrops. Possibly, Walcott saw some of these exposures. Possibly he fell under the spell of the charming Sir William Dawson, who later in life admired Walcott's Early Cambrian work as well as his Grand Canyon animal fossils and even described the stromatolite Walcott collected there. "The *Eozoon* question", as it is so delicately labelled by Merrill (1906, p. 635-646; 1924, p. 564-578), became an even more heated controversy during the mid-1880s. For reasons which are obscure but which might have been personal rather than professional, at the 1888 International Congress of Geology when Walcott was formally queried on the question, "In your opinion is *Eozoon Canadense* of organic origin?" he voted in the affirmative (Hofmann, 1971, p. 13). Only two others of the fourteen geologists voting concurred with him in this error.

New England experience does not loom large in Walcott's Precambrian work but, in his first significant paper on its fossils, he illustrated a bed of graphite mined in the Adirondack Mountains near Hague, New York. He indicated that the unit had the appearance ". . . of a fossil coal-bed, the alteration having changed the coal to graphite . . ." (Walcott, 1899, p. 227). Although this is a curious remark in the perspective of today, it was quite in line with what was then a generally prevalent opinion. Thus, Van Hise (1896, p. 583, 760) accepted Precambrian graphite as being of organic origin. For Sir William Dawson, such graphite came from ". . . a vast vegetable growth, either on land or sea, or both." (Dawson, 1897b, p. 92). Since Dawson was a staunch supporter of creation theory against the supporters of evolution theory, he saw in the ". . . graphite indirect evidence of creation on the third day . . ." (O'Brien, 1971, p. 52-53), it being formed originally from the Biblical covering of the land. However, except for the brief remarks noted above, Walcott tended to avoid the scientific controversy regarding iron ore, limestone, and graphite and other indirect evidence of possible organic deposition in Precambrian rocks and he steered clear of any religious interpretation of early life.

The U.S. Geological Survey, as an enlightened scientific organization, did not confine Walcott's geological investigations to the United States. He wrote:

> "In the summer of 1888 I made a rapid trip from Saint John Harbour [Newfoundland] to Topsail Head, Conception Bay, which gave me a clear impression of the details" (Walcott, 1899, p. 218).

This did give Walcott the opportunity to consider the Avalon succession and to reexamine the presumed Precambrian fossils described by Billings (see Hofmann, 1971, p. 14-17). Though he did not specifically commit himself, Walcott (1899, p. 229-232) gave the distinct impression that he did not believe these "fossils" of Maritime Canada to be organic.

Walcott was more definite about this pseudofossil after spending part of the 1899 field season at Random Island and Smith South, Newfoundland:

> "The Random terrane is considered to be the upper member of the Avalon series. Animal life existed during the deposition of a portion of it, as is evidenced by clearly marked annelid trails. A collection of the form known as *Aspidella terranovica* was made from the Momable terrane of the Avalon series. It proved the supposed fossil to be a spherulitic concretion and this removes it from among the possible pre-Cambrian forms of life." (Walcott, 1900, p. 5).

A field note with several slabs he collected is instructive:

> "*Aspidella* *June 28.*
>
> Examination made of occurrence at Ferryland, Newf. Extend thro' fifty feet or more of siliceous and arenaceous slaty shale.
>
> *Aspidella terranovica* occurs on many of the layers varying in size from 1/16 in. to 3 in. & in form from circular to oval. The impression made is that the *Aspidella* is of inorganic origin. Many of them recall the spherulitic concretions or *Palaeotrochis* – others suggest incipient 'cone-in cone' and some a patelloid shell. The latter variety was named and illustrated by Billings."

"Aspidella" is now generally considered to be the result of compaction or other diagenetic effects in shale (Cloud, 1968, p. 67; Hofmann, 1971, p. 16). The annelid trails are discussed by Hofmann (1971, p. 41-42).

For the next half dozen years, Walcott was increasingly occupied with administrative work, succeeding Powell as Director of the U.S. Geological Survey. Fieldwork was limited, and when he undertook extensive field investigations it was in a new area, the northwestern United States.

THE NORTHWEST

A field season in Montana or Idaho may not have been quite as hazardous or difficult as that in the Grand Canyon, but even here the rocks were not easy of access. Walcott summarized a great deal of arduous work through rocks measured in thousands of feet and over an area measured in thousands of square miles in a few sentences:

"During September, 1895 I made a rapid trip along the line of the Great Northern Railroad from Belton to the vicinity of Essex and thence north to the head of Nyack Creek" (Walcott, 1906, p. 12). "The observations thus made led me to visit the region again in 1898 and make a general study of the pre-Cambrian rocks of the Big and Little Belt mountains and of the exposures of the Prickly Pear Valley northwest and southeast of Helena. A visit was also made to the section on the Gallatin River, and special attention was given to the study of the relations of the Flathead Cambrian sandstones to the subajacent Belt formations" (Walcott, 1899, p. 203). "The results of my investigations were the discovery of a great straigraphic unconformity between the Cambrian and the Belt formations; that the Belt terrane was divisible into several formations, and that fossils occurred in the Greyson Shales nearly 7,000 feet beneath the highest beds of the Belt terrane" (Walcott, 1899, p. 204).

The Belt Series (now Belt Supergroup) was known from earlier investigations and had been subdivided. However, Walcott was the first to demonstrate the major unconformity; his stratigraphic units and their regional correlation have stood up fairly well.

"In 1900, I crossed the Belt mountains and endeavored to trace the connection, north of Helena, of the Belt terrane and the pre-Cambrian rocks of the Rocky Mountain 'front' at Lewis and Clark pass and north on the south fork of the Dearborn River" (Walcott, 1906, p. 2).

During this decade, U.S. Geological Survey parties examined the Precambrian in the Coeur d'Alene, Idaho, and extended reconnaissance eastward into western Montana. Walcott was able to integrate this work for he

". . . also had the opportunity of studying the pre-Cambrian rocks of northwestern Montana in 1905, and of measuring the section from northwest of Scapegoat mountain on the Continental divide, westward, crossing the Swan and Mission ranges, to the canyon of Jacko Creek above Ravalli" (Walcott, 1906, p. 2).

Some years later, in 1914, Walcott (1915a, p. 2) ". . . proceeded to White Sulphur Springs, Montana, for the purpose of studying the ancient sedimentary Pre-Paleozoic rocks of the Big Belt Mountains." During this trip he observed algal structures and looked for more specimens of *Beltina*, a form discussed below.

"It was found that the algal remains were deposited very much in the same manner as those that are now being deposited in many fresh-water lakes and that many of the forms had a surprising similarity to those being deposited in the thermal springs and pools of the Yellowstone National Park" (Walcott, 1915a, p. 9).

The following year in 1915 Walcott (1916b) tested these ideas by detailed observations at the hot springs of Yellowstone. Later he camped in Deep Creek Canyon in the Big Belt Mountains. "About two tons of pre-Cambrian specimens were collected in this vicinity, before the storms of late September closed the season's field work" (Walcott, 1916b, p. 27). This season essentially closed his field investigations on the Belt Series in Montana. Unfortunately, the large collection of stromatolites was never fully investigated by Walcott; it would have reinforced his systematic

Figure 2. "Slaty shale in which the Prepaleozoic crustacean fossils were found near mouth of Deep Creek Canyon, Big Belt Mountains, 16 miles east of Townsend, Montana." The man in the picture has not been identified. (Photo by courtesy U.S. Geological Survey. (A similar photograph, in which the figure is seated, appeared in Smithsonian Misc. Coll. vol. 65 no. 6, fig. 4 and was credited to Walcott).

studies of this group.

Walcott faced one major challenge to his Precambrian work. Rothpletz (1915) indicated that Walcott had confused strata near Helena, Montana and that his supposed Precambrian was actually Cambrian. In reply, Walcott (1916c) dealt with the geology of the area in a masterful summary of evidence to support his interpretation. He also discussed the Precambrian more generally and again referred to the Grand

Canyon work; he supplied some magnificent photographs of the unconformity, possibly taken in 1901 (Darton, 1928, p. 89). If there is a classic example of an older professional defending his early work against a later worker, it is this paper; the rejoinder to Rothpletz was systematic, gentlemanly, and devastating.

The Belt Supergroup is a thick, widespread and complex series of rocks (Ross, 1970). These rocks and their equivalents continue to attract the attention of large numbers of geologists in western North America (Idaho Bureau of Mines, 1973); Walcott's early efforts, even after more than half a century of additional work, have stood the test of time.

THE BELT FOSSILS

In marked contrast to Walcott's work on the Grand Canyon, which was published in a series of papers throughout more than a decade, his work on the Belt Series was published more promptly and in a more authoritative manner. This may have been a consequence of Walcott's position as Director, but more than likely it was the result of his considerable experience.

Pre-Cambrian fossiliferous formations (Walcott, 1899) is *the* definitive work on this subject of the 19th Century. Granted that this is a limited subject area, a variety of papers had been written because of the high interest in this problem. Walcott made this paper a summary of both the American and foreign literature, as well as of his own work. Still, the principal part of this study was a description of the Belt stratigraphy and the apparent fossils he had found in 1898.

For Walcott, the most exciting organic remains were those of the presumed merostome crustacean *Beltina danai*, ". . . a crustacean of a much higher type than most paleontologists would have predicted for this horizon" (Walcott, 1899, p.236). Walcott was impressed with the similarity of fragments from the Belt rocks to fragments of eurypterids he had collected in the lower Paleozoic of New York. He thought that several forms were present but that this material was too inadequate to differentiate more than one generalized type. "Only a few hours were available for collecting material in the field" (Walcott, 1899, p. 238). Additional material was sent to him from southern Canada (Walcott, 1911, p. 21). Not only did Walcott maintain the arthropod nature of *Beltina*, he continued to suggest relationship to the eurypterids and even republished an earlier plate of illustrations (Walcott, 1914a, pl. 22).

Subsequent workers have not dealt kindly with *Beltina*. Fenton and Fenton (1936, p. 616; 1937, p. 1949), among others, judged at least some of the material to be noncalcareous algae; Cloud (1968, p. 53) considered *Beltina* as either nonvital or algal. The specimens that I have reexamined are fragmental and nondescript. Nevertheless, in many respects, they really do look like Palaeozoic eurypterid fragments! However, on one unfigured specimen in the museum collections, the shape is so irregular and with so many sharp angles that it is inconceivable that this piece was a broken fragment of a stiff carapace.

Two quotes effectively end the *Beltina* story:

> "A very few fragments, selected from thousands, do remotely resemble parts of eurypterids. This may be said of four of the thirty-one specimens figured by

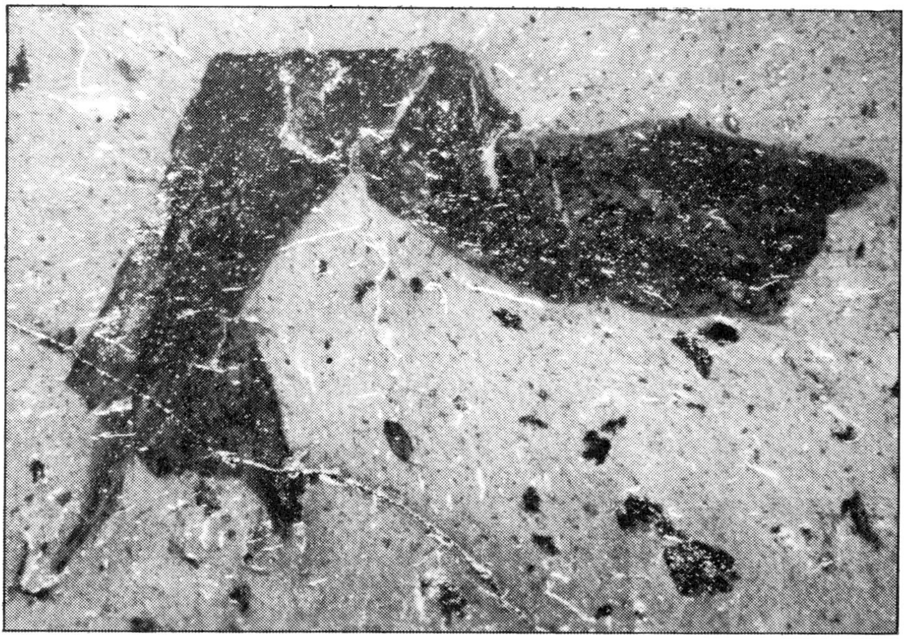

Figure 3. "Beltina danai" Walcott. One of Walcott's original specimens (U.S.N.M. no. 245129). Near the centre of the photograph is a right-angle bend where the piece has been flexed and then compressed. Many fragments occur on this slab. (Photo by courtesy U.S. Geological Survey).

Walcott . . . There can be little doubt, however, that the fragments are of organic nature, for they are widespread in Montana and British Columbia. It is not at all improbable that they are of algal origin, perhaps remains of brown algae, though nothing definite can be ascertained from their structure" (Raymond, 1935, p. 382).

"Later students of the North American Pre-Cambrian have dismissed its interpretation as arthropod remains with rare unanimity" (Glaessner, 1962, p. 478).

Beltina clearly seems to be a case of hope exceeding data.

Walcott (1899, p. 236) also described presumed five animal markings from Belt strata and, as with *Beltina*, republished the illustrations of them later without further comment (Walcott, 1914a, pl. 21). Two were assigned to *Planolites*, one confidently to *Helminthoidichnites* and two questionably to that genus. Glaessner (1962, p. 480) noted these five and other Precambrian trails in a review of Precambrian life but has never had the opportunity to examine the museum specimens (Glaessner, written commun. 1971).

Walcott interpreted each of the *Planolites* as a cast of a burrowing worm in the

Figure 4. "Looking after *Paradoxides*". Charles D. and Helena B. Walcott on their honeymoon in 1888, collecting at Manuel's River, Conception Bay, Newfoundland. Mrs. Walcott later assisted in the collecting of "Beltina". (Photo by courtesy U.S. Geological Survey).

brief text and as a burrow in the figure descriptions. Cloud (1968, p. 55) considered both species of *Planolites* as algal(?). In my opinion, neither of the *Planolites* is convincingly interpreted as a cast, trail, burrow, or simliar animal structure. They are far more likely to be primary sedimentary structures or a consequence of diagenetic or structural factors than organic in origin.

Planolites superbus Walcott is represented by three pieces nearly a foot in total length; were it either a worm casting or a body cast – one cannot determine for certain which one Walcott meant – it would be most atypical of either. The only specimen of *P. corrugatus* is preserved in three dimensions, but its corrugations are only on one side. Both species are unlikely to be burrows.

The three taxa assigned to *Helminthoidichnites* are all different, and they all could be organic; Cloud (1968, p. 55) listed them as "probably algal." Walcott considered *H.? neithartensis* as a trail possibly made by a slender worm, minute mollusc or crustacean, though on the plate legend it is given under annelid trails. Some specimens could easily be dismissed as inorganic staining. For two illustrated specimens, Walcott suggested an animal coming and going along parallel routes, quite unlike worm feeding patterns. One of these specimens is the only example showing even a trace of relief, it being a barely depressed trench. On another illustrated example, an apparently closely-spaced meandering pattern is not clearly traceable. Several unillustrated smaller slabs show elongated narrow marks, differentiated only by color rather than relief. They are not trails. The various small spots on the slabs appear physically similar to the presumed trails. No obvious carbon film is left but it is easier to interpret this form as the residue of "algal fronds" than it is to assign it to the animal kingdom.

H.? spiralis Walcott, cited as an annelid trail, is the most convincing of the three as being organic; it does not look like an obvious soft algal remnant but, likewise, it does not resemble any animal trace with which I am familiar. There is no relief and the width, though narrow, is far wider than that of a few cells in a growing filament. The spiral form varies widely among specimens in the degree of curvature and number of whorls present. This shape and its variation might be the result of a strand or elongate fragment of some sort, coiling as it dropped to the bottom, such as occurs on a much larger scale with long crinoid stems. The latest interpretation is that these two species are algal and constitute distinct genera (Walter *et al.*, 1975).

The only one of the three which shows anything more than the slightest relief is *H. meeki*, and even this is extremely shallow and flat, unlike younger Precambrian trails (Glaessner, 1969). It seems much too smooth and uniform to be interpreted as a "worm" trail. The only slab has either two examples crossing or some elaborate sort of figure 8 folding. The examples of *H. meeki* are about three times as wide as most specimens of *H.? spiralis*. If there is a relationship between the width of the "stem-like" remains and their radius of spiral when dropping into soft bottom sediments, the apparent differences between these two forms are not so striking. More specimens of all the Belt "trails" are needed for study, but especially of this last form.

Walcott's later work (1906) was primarily a stratigraphic study of the Belt rocks. Rightly or wrongly, he used the occurrence of *Beltina dania* as one basis for establishing stratigraphic relationships and correlations; in this he was followed by other workers in Canada (Daly, 1907, p. 94). He also noted the presence of *Cryptozoon*, both to identify beds as Algonkian in age and, apparently to correlate from one area to another. In a rather slight piece of paleontologic work in this paper he introduced the name *Cryptozoon frequens*, basing this new species entirely on a photograph without any accompanying description.

His most comprehensive work on the Belt algal structures appeared eight years after Walcott's principal stratigraphic paper (Walcott, 1914a). In this study he named eight new genera, six of which were monotypic; six other new species were also described. Fenton and Fenton (1931, p. 670) expressed surprise that Walcott did not record several other prominent stromatolite beds in the Belt besides those which yielded his specimens. It was noted that, "A few people have maintained a somewhat skeptical attitude toward these 'plants' " (Raymond, 1935, p. 377). This negative attitude toward the Belt organic remains did not persist long (Fenton and Fenton, 1936, 1937).

Walcott did not study the famous stromatolites of the Medicine Bow Mountains in Wyoming, but he was aware of them (see Fenton and Fenton, 1939, p. 92); he also examined specimens from Globe, Arizona. He wrote nothing about Canadian Algonkian specimens, though obviously familiar with them in the field. Additional material was available, and though he never returned to a further investigation of this group, Walcott (1914a, p. 79) was quite emphatic that he had merely written, ". . . a preliminary paper on a fossil algal flora . . ." To subject it to a critical species-by-species analysis does a certain injustice to this early effort. "Stromatolites are notoriously variable" (Glaessner, 1968, p. 587) and this cursory treatment of algal structures relative to Walcott's other efforts is entirely a reflection of my limited knowledge; stromatolites are complex and there is much which may be learned from their study (Hofmann, 1973).

Today, it is generally accepted that algal structures occur worldwide in the Precambrian and might even have some biostratigraphic utility (Cloud and Semikhatov, 1969; Walter and Preiss, 1972). With this acceptance of large stromatolites as truly organic has come a growing awareness that they were formed in a marine

Figure 5. Stromatolites in the Belt Supergroup. a) "Lenticular mass", approximately 61.25 cm (24.5") vertically by 90 cm. (36") horizontally, grading laterally into undurlatory stromatolitic material. Wallace Formation just east of Clark Fork, Bonner Co., Idaho.

environment (Fenton and Fenton, 1933). The less obvious forms also can be subjected to environmental interpretation (O'Connor, 1972, p. 592). The literature is large, and I have made no attempt to follow the current systematic disposition of any of Walcott's species and genera; Rezak (1957) has published a comprehensive study based in part on his material. Walcott's generic names, especially *Cryptozoon* and *Collenia,* tend to be misused. Some taxonomists who deal largely with columnar branching stromatolites, rather than with the simpler forms described by Walcott, do not employ them.

Although some of Walcott's stromatolites are diagenetic phenomena or concretions (Fenton and Fenton, 1936, p. 613-615), the majority are unquestionably of organic origin. Even if several supposed species were inadequately differentiated,

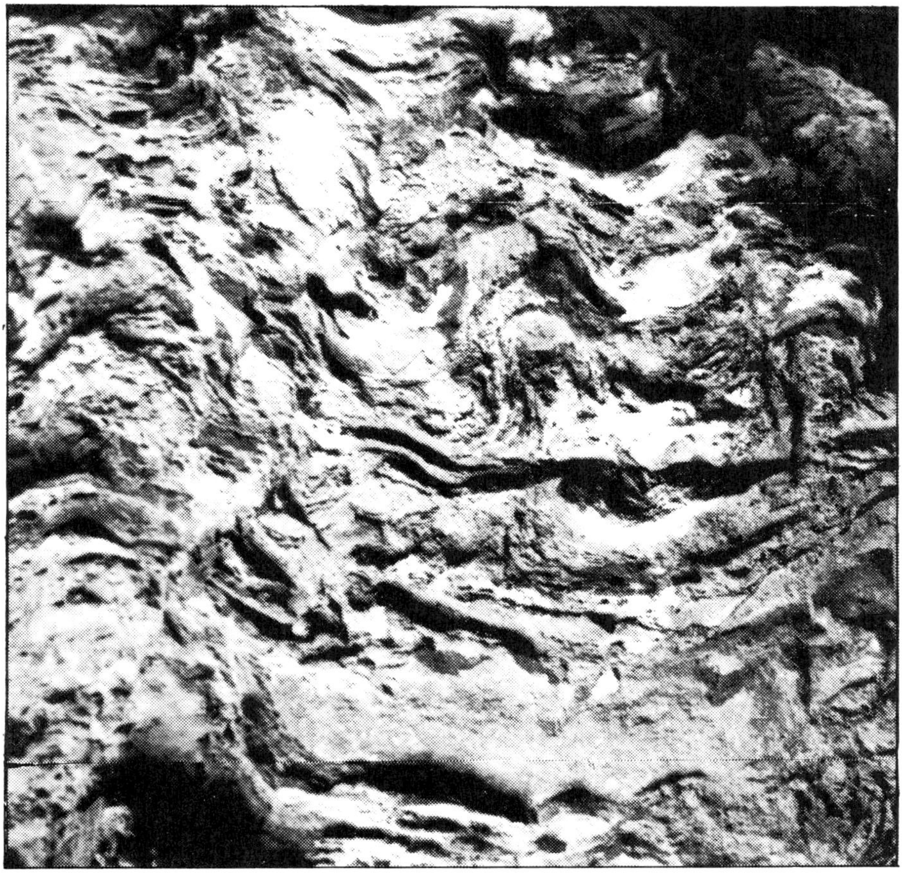

b) Reef structures of stromatolites: the exposed face measures 2.7 m (9′) vertically by 1.5 m (5′+) horizontally. Wallace Formation, John Creek, central Benewah Co., Idaho. (Photos by courtesy Idaho Bureau of Mines and Geology).

Walcott's monograph was a key factor in leading to this general acceptance of the existence of Proterozoic algae. One can argue that the "Pre-Cambrian Algonkian Algal Flora" marked the start of a new era in investigations of Precambrian life.

MISCELLANEOUS INVESTIGATIONS

In 1911 Walcott briefly visited Steeprock, west of Port Arthur, Ontario, Canada, an area mapped by A.C. Lawson. Fieldwork was from a canoe with two companions and turned out to be hazardous. After a boating accident, one geologist drowned; Walcott was pulled unconscious from the water but recovered to do another 15 years of field study.

His trip to the Port Arthur area was to observe and collect a presumed fossil. The genus *Atikokania* was described with two included species; Walcott (1912a; 1914a, p. 98) considered it related to the sponges or archaeocyathids. He also illustrated a *Cryptozoon??* from the same outcrop, later named *C. walcotti* (Rothpletz, 1916, p. 80; Hofmann, 1971, p. 58-59). Almost immediately his interpretation of *Atikokania* as a fossil was refuted (Abbott, 1914). Walcott (1914b), in reply, mentioned that after having seen more Precambian algal remains, he was not inclined to assign it to either animal group he had originally suggested. Apparently he did not refer to this form again. The inorganic nature of *Atikokania* has been generally accepted for four decades (Raymond, 1935, p. 381; Glaessner, 1962, p. 472; Hofmann, 1971, p. 24-27). Walcott and Abbott exchanged letters for a number of years regarding similar appearing inorganic structures in the Permian limestone of England, but no publication to elaborate upon the organic interpretation of this pseudofossil ever resulted.

In 1915, Walcott (1915b) published a scant two-page paper in the inaugural volume of the National Academy of Science Proceedings, recording the occurrence of Algonkian bacteria within an algal specimen from the Gallatin River in Montana; he had reported it three days earlier to the Botanical Society of Washington. Walcott was excited about this discovery for it supported his ideas on the formation of many Precambrian limestones. The paper was published one week after its submission. Walcott's find was confirmed in the early 1920s by other discoveries of presumed early bacteria, but these in turn were dismissed, the underlying concepts simply not being accepted by critics.

> "He makes no argument in favor of the identification, and leaves it on faith that one organism without hard parts, and less than 0.01 millimeter in diameter, would be preserved in identifiable condition from pre-Cambrian times to the present" (Raymond, 1935, p. 378).

So far as I can determine, Walcott's original material has not been reinvestigated, even though his slides are still available at the U.S. National Museum. A body of literature now supports the fact that Precambrian microorganisms *are* preserved (see Licari and Cloud, 1968; Cloud and Licari, 1968; and others). The press of activity during World War I certainly prevented Walcott from proceeding with further work in this area, but in any case the lack of sophisticated optical equipment would have inhibited investigations. Still, had this find been taken more seriously and had some of his speculations about the role of bacteria in deposition of limestone

(Walcott, 1914b) – written before his find – been pursued, the pace of investigations of Precambrian life might have been speeded up. In this point, Walcott was far ahead of his time:

"... the fundamental association of cherts, stromatolites, and Precambrian microfossils suggested by the studies of Walcott and Gruner has been fully confirmed and with minor modifications has formed the basis of the productive investigations of recent years" (Schopf, 1970, p. 32a).

The 1915 season in Yellowstone at the thermal springs certainly deserves to be mentioned in this heading. Although a lavishly illustrated short account of the work appeared (1916b), nothing scientific was produced. Again, it is likely that World War I activities prevented further study.

Walcott had a clear notion of the direction of this facet of his reesarch, for he wrote to W. Howchin, then lecturer in geology and paleontology at the University of Adelaide:

"During the summer of 1915 we made large collections of both siliceous and calcareous deposits in the hot springs and geysers of the Yellowstone National Park. The field study led me to conclude that much of the deposit was made through algal growths. In some instances, however, the algal growth clearly controlled the form. This taken in connection with the so called 'lake balls' [see Walcott, 1916b, p. 4, figs. 1 and 2] leads me to still conclude that Cryptozoon and a number of other Paleozoic and Pre-Paleozoic forms owe their shape to algal growth" (letter November 24, 1916).

Study of the deposits confirms both the importance of bacteria and algae in deposition (Walter *et al.*, 1972) and the validity of the hot springs deposits as a model for some Precambrian algal deposits (Walter, 1972).

Even though Walcott never published on the hot-springs specimens or extensive algal collections made in 1915, part of his material was for many years on public display at the U.S. National Museum.

WALCOTT'S LATER YEARS

In 1907, Walcott left the Geological Survey to become fourth secretary of the Smithsonian Institution, following S.P. Langley. This change of position allowed him more time for research and a greater latitude in fieldwork. Most of his effort was transferred to the Canadian Far West. Except for several weeks of the 1909 field season (Walcott, 1910a), in the Bow River valley, little of this work was directly related to the Precambrian; it is sketched here for the sake of completeness. After several summers investigating primarily Cambrian strata, Walcott found the incredible Burgess Shale (Middle Cambrian) in 1909 (Walcott, 1912b). The 1910, 1911, and 1913 seasons were mainly devoted to exploiting this find. If anything, it served to reinforce his notion of the paucity of Precambrian life.

After Walcott's work in Yellowstone Park and the Montana Belt outcrops, noted previously, he returned to the Canadian Rockies. He spent the 1917 season back at the Burgess Shale quarry. Walcott continued a full schedule of research in spite of monumental efforts during World War I; according to one account he was an active participant on 33 committees.

Figure 6. Charles Doolittle Walcott when Director of the U.S. Geological Survey, 1894-1907. Portrait in oils by Charles H. Ruffins, hanging in the U.S. Geological Survey, National Center, Reston, Virginia. (Photo by courtesy U.S. Geological Survey).

His final work on the Precambrian of western Canada was peripheral to broad reconnaissance of the lower Palaeozoic strata. In general terms, Walcott extended the stratigraphy of the Belt equivalents (Walcott, 1927, 1928), in part synthesizing the studies of other field geologists, but nothing new or startling in regard to the Precambrian resulted from this work. He did speculate briefly whether the Proterozoic geosyncline was divided into separate troughs of deposition, in an Ulrichian sense, but this was not profound.

WALCOTT'S PRECAMBRIAN CONCEPTS

A classic dilemma near the turn of the last century was the short time-span for the age of the earth insisted upon by the physicists and the far longer span deduced by the geologists. Walcott summarized various estimates of time, made some of his own calculations based on rate of sedimentation, and concluded that the Palaeozoic and the Algonkian were each of about 17,500,000 years duration; the total age of the earth was perhaps 60,000,000 years old (Walcott, 1893, p. 675). He was not fixed in this position, for later (Walcott, 1916a, p. 235) he remarked that Joly's work pointed to a minimum age of 90,000,000 years for the earth. Still this is a far more restricted conceptual time frame than is currently accepted.

The environment of deposition may never have been far from Walcott's mind. In 1910 (Walcott, 1910b, p. 2), he indicated that, after 18 years of search:

> ". . . the conclusion has gradually been forced upon me that on the North American Continent we have no known pre-Cambrian *marine* deposits containing traces of organic remains, and that the abrupt appearance of the Cambrian fauna results from geological and not from biotic conditions."

He suggested that the Algonkian rocks

> ". . . were deposited in fresh or brackish water seas to which the marine life of the extracontinental seas very rarely had access."

The concept of fresh water origin for the Algonkian beds was elaborated upon in connection with his algal studies (Walcott, 1914a). He compared the Precambrian structures favorably with those of recent fresh water, so-called lake balls. He also briefly summarized the possible role of bacteria in deposition of limestone and commented on them further in a more popular paper (Walcott, 1916a, p. 240-242).

These references are readily available and need not be paraphrased. The best summary of his thoughts on the environment of deposition is given in a report of a 1915 lecture:

> "Mr. Walcott described the stratigraphic position of the great Prepaleozoic Beltian series of central Montana, which he considered to be of fresh or brackish water origin. They were deposited in a large inland lake or lakes covering approximately 6,000 square miles of area, also on river flood plains as sand and gravel, or as fine dust carried by the wind. The formations now consist of sandstone, siliceous shales, calcareous shales and beds of limestone, the last varying in thickness from a few inches to several thousand feet. The same type of deposits also occurs in the Grand Canyon region of Arizona, and they extend northward along the main ranges of the Rocky Mountains far into Alberta and British Columbia.
>
> At a horizon approximately 9,000 feet below the base of the Cambrian numerous reefs of algal deposits occur at several horizons in the Newland limestone

formation of the Beltian in Montana, and isolated concretionary-like forms occur scattered at various levels in the overlying Spokane shales of the Belt Mountains. The algal remains occur in many forms, some of which are surprisingly similar to those of the fresh-water lake and stream, blue-green algal deposits of Pennsylvania, New York, Michigan, etc. Others are similar in appearance to the blue-green and green algal deposits of the thermal waters of Yellowstone National Park" (Spaulding, 1915).

In an historical inqury one may search for an event which begins to channel the thought on a subject in a particular mold. Walcott's work on the abrupt appearance of the Cambrian fauna (Walcott, 1910) was specifically written for a symposium at the International Geological Congress in Sweden, and perhaps that invitation stimulated his inquiry. Alternatively, the American Association for the Advancement of Science meeting in Baltimore at the close of 1908, might have provided earlier impetus. Other scientific societies, including the Geological Society of America, gathered, and at the start of the new year, scientific ferment continued as Johns Hopkins University noted the centenary of Darwin's birth and 50 years of "The origin of the Species". The program for AAAS, Section E (Geology) consisted of sixteen papers on the subject of correlation in historical geology. Walcott produced one of his few short summary papers for this occasion. He wrote:

"The traces of pre-Cambrian life, though very meager, are sufficient to indicate that the development of life was well advanced long before Cambrian time began. The characteristic fossil of the known pre-Cambrian fauna is *Beltina danai,* a crustacean probably more highly organized than the trilobites. The associated annelid trails indicate that this phase of the fauna was also strongly developed. Stratigraphically, this fragment of what must have been a large fauna occurs over 9,000 feet beneath an unconformity at the base of the upper portion of the Lower Cambrian in northern Montana [(footnote in text) C.D. Walcott, *Observations of 1908*.] This fact indicates that it is practically hopeless to search for the first forms of life – those that could leave a trace of their existance – in strata now referred to the Cambrian or early Paleozoic. With this thought in mind we shall consider what is known of the life of the early Lower Cambrian . . ." (Walcott, 1909, p. 196)

Walcott then went on to emphasize the geographic limitations of Lower Cambrian outcrops relative to the broader expanse of Middle and Upper Cambrian areas. He noted the occurrence of Early Cambrian coral-like organisms and the large number of individual fossils of various kinds also present in the Early Cambrian and suggested that both points indicated uniform and warm climates. He indicated the rapid and accentuated development of Middle Cambrian fossils and attributed this to the increase of shoreline and shallow water areas as seas moved over the continent, to the shifting of shorelines, and to the breakdown of restrictive barriers. He also emphasized the cosmopolitan nature of Early Paleozoic faunas as a consequence of more or less world-wide movement of currents. "In other words it is evident that the evolution of the early Paleozoic faunas was profoundly influenced by their environment" (Walcott, 1909, p. 202). This paper is not antique in its thinking and probably would be accepted for publication today. Had Walcott not had a *fauna* in the Belt rocks, his concept of Precambrian lakes need not have been developed.

1908 was also the year when W.K. Brooks, the eminent biologist from Johns Hopkins University, died. Walcott acknowledged his debt to Brooks by quoting his work on the importance of microscopic forms, not known directly in the early Paleozoic fossil record. The fossils which were in the Cambrian ". . . were mainly carnivorous, and were adapted either to straining minute organisms from the water, or the gathering them up from the bottom" (Walcott, 1909, p. 202). Brooks was the first biologist, in America at least, to seriously develop the concept of marine pelagic organisms as precursors of benthonic life. In Walcott's copy of Brooks' seminal paper (1903), a few paragraphs are marked in pencil and seem to indicate the thoughts which most impressed him; in turn it might be mentioned that Brooks was impressed by Walcott's (1891) demonstration of the encroachment of Cambrian seas onto the Precambrian rocks.

The key paragraphs in Brooks' paper begin: "On the land the mineral elements of plant-food are slowly supplied as the rains dissolve them . . ." (p. 138); "The oldest petropods or lamellibranchs or crustacean or echinoderm or vertebrate which we know from fossils exhibits its own type of structure with perfect distinctness . . ." (p. 157); "Embryology teaches that each line had its own pelagic representatives . . ." (p. 157); "The belts are elevated areas where far enough from shore to be free from sediments . . . (p. 160); and finally, "Colonies which started in shallow water were exposed to accidents . . ." (p. 161).

The size of North America was a particularly significant point in Walcott's fully developed hypothesis (Walcott 1910b). In several papers, Walcott reiterated the notion that that continent was larger in the Precambrian than in later times. Walcott had accepted from Brooks the view that life was originally oceanic rather than beginning in fresh water. Any early fossiliferous remains would have been deposited in marine sediments, far from the margins of present-day North America and would not then be found by paleontologists. When the Algonkan continent was reduced in size, shallow water fossiliferous marine sediments were finally deposited as the Cambrian ocean steadily encroached upon the continent.

The term "Lipalian is proposed for the era of unknown marine sedimentation between the adjustment of pelagic life to litteral conditions and the appearance of the Lower Cambrian fauna. It represents the period between the formation of the Algonkian continents and the earliest encroachment of the Lower Cambrian seas" (Walcott, 1910b, p. 14). This is an ingenious suggestion to allow for development of a diverse fauna without its leaving any record. Walcott mentioned the Lipalian in several of his later papers; he only elaborated on this original description to emphasize the stratigraphic and time break below the Cambrian (Walcott, 1915c, p. 171). The rare presence of the presumed crustacean *Beltina*, the presumed annelid trails, and abundant protozoans (the assignment of *Cryptozoon* at that time) in Algonkian rocks of the Belt Super-group was considered evidence of the occasional encroachment onto the continent of marine tongues and, therefore, was further indirect support for the notion of fresh water origin of most of the sedimentary units.

There is one final enigmatic footnote to Walcott's notion that

". . . Belt sediments accumulated in lacustrine basins. Mrs. Walcott (pers. commun.) reports that he abandoned the hypothesis because of the great thicknesses of limestone and dolomite in northern Montana, and because of abundant salt crystal casts in clastic beds" (Fenton and Fenton, 1937, p. 1934; the stratigraphic work by Daly and Schofield in the Canadian Belt equivalent which showed crystal casts is summarized by Fenton and Fenton, 1937, p. 1924).

This at least suggests that Walcott could have considered alternative ideas but there is no basis for further speculation.

The notion of fresh water deposits in the Precambrian, followed by the Lipalian interval representing a long time hiatus in sedimentation, is an unusual combination and warrants further consideration. It is generally viewed as illogical and currently dismissed out of hand; one may consider it a form of special pleading. Nevertheless, the hypothesis is not constructed of whole cloth. Walcott had investigated fresh water limestones in some detail. There is also new evidence that Walcott was inadvertently assisted in this misdirection toward fresh water deposition as a reason for the absence of invertebrate fossils.

After having worked in the Canadian equivalents of the Belt rocks, Daly (1907) made his contribution to the question of the marked difference in fossil content between Cambrian and Precambrian strata in writing *The Limeless Ocean of Pre-Cambrian Time*. In essence Daly suggested a modification and clarification of the classic hypothesis of W.K. Brooks, of a long time for evolution of life in a pelagic state because,

". . . for *practical physiological purposes* lime salts were nonexistant in the sea-water for most of the pre-Cambrian life-period" (Daly, 1907, p. 95).

Because no calcium carbonate was available to form hard parts, the conspicuous absence of invertebrate fossils could be explained.

Walcott generally did not write in his books and papers – a potential source of information on development of ideas – but rather uncharacteristically he did leave two slips of paper in a reprint of Daly's work. The first is a note: "carbonate of lime not deposited in cold water," followed by two references. The second is more extensive and deserves to be quoted in full:

"Limeless Sea. Daly August 5/07

A very well worked out hypothesis. The Pre-Cambrian fossils of the *Belternan* series. *Cryptozoon?* that occur in such great quantities in Montana is strongly against the Daly hypothesis.

Have fossils analized to ascertain if they are dolomitic? calcareous?

Also test of Lower Cambrian trilobites, etc. from Inyo Co. Calif.

Ask Prof. F.W. Clarke of Geol. Sur. [Chief Chemist at that time] to study the chemical portion of Daly's paper.

CDW"

This note suggests that Walcott must have carefully considered a physiological approach to explain the absence of Precambrian invertebrate remains and he ultimately rejected it. Having rejected physiological considerations, he then had to fall back on a physical reason, namely deposition of the sediments in fresh water, to account for the lack of animal remains. It was a curious situation, for the field geologist (Daly) appealed to a biological answer to this enigma, but his conclusion influenced the biologist (Walcott) to resort to a field approach!

With the benefit of more than half a century of study of the calcification process in animals, but especially with the advent of radioactive dating, one can reconsider the absence of obvious small animal remains in the Precambrian of North America. There is now much data to confirm that the Belt Supergroup is vastly older than was realized fifty years ago. Walcott and his contemporaries were willing to think in millions of years or even in tens of millions, but that these fresh-appearing sediments could have been deposited hundreds of millions of years ago (Obradovich and Peterman, 1968) was an alien concept. Apparently the Belt sediments do not contain shelled invertebrates simply because shelled invertebrates had not developed at the time these sediments were deposited. Indeed, it is likely that the Precambrian strata which have yielded soft-bodied fossils (Glaessner, 1972; Sokolov, 1972) are younger than the rocks of the Belt Supergroup and probably those of the Grand Canyon.

The question of precisely why animals develop hard parts is still an open one, subject to new ideas (Towe, 1970; Rhodes and Morse, 1972); and even the simpler question of how the process works today is not fully understood. The entire climate of opinion concerning the origin and evolution of the physical and biological world has changed in the last few decades (Tappan and Loeblich, Jr., 1970; Cloud, 1972, 1976; and others). Profound additions to basic knowledge have taken place in the fields of biology, chemistry, and geology.

Viewed in the context of those times, I believe Walcott had considered the various possibilities when, after reviewing in a more popular context the ideas he published in 1910, he wrote:

> "I fully realize that the conclusions above outlined are based primarily on the absence of a marine fauna from the Algonkian rocks, but until such is discovered I know of no more probable explanation of the abrupt appearance of the Cambrian fauna" (Walcott, 1916a, p. 238).

SUMMARY

Charles Doolittle Walcott can properly be called North America's pioneer in the study of Precambrian life. His work followed shortly after the ill-founded *Eozoon* controversy (O'Brien, 1970; Mitchell, 1971) and placed investigations of the subject on a rational, nonemotional basis, providing a series of increasingly comprehensive papers for three decades.

> "Pre-Cambrian fossils so interested the late Dr. Charles D. Walcott that he spent many months of his long, busy life in the discovery and interpretation of their remains" (Bassler, 1941, p. 519).

Unfortunately, he did not establish a "school" and his investigations were not directly followed by others. In a sense his accomplishments, major as they were, occurred in a vacuum.

Walcott's work on Precambrian animal remains has now been largely discredited. In contrast, his research on stromatolites gave an excellent start for further studies and formed the basis for one of the most active areas of Precambrian paleontology. His limited work on microfossils was highly premature, possibly in part because of the lack of sophisticated techniques and equipment. A realistic summary of Walcott's impact on the field of Precambrian paleontology is given by Schopf (1971, p. 1013):

"Walcott was an acknowledged leader in the early search for Precambrian fossils and was one of the first to stress the probable algal origin of Precambrian laminated stromatolites."

Even a reviewer (Raymond, 1935, p. 388), not noted for being an enthusiastic supporter, wrote,

"The man who probably expended more time and energy than any other individual in trying to find pre-Cambrian fossils was Walcott."

EPILOGUE

Walcott's influence on Precambrian investigations did not die with him. It is appropriate to reproduce in its entirety the public notice for the Walcott medal (White, 1928b).

"Under the terms of a trust recently made by Mrs. Mary Vaux Walcott, in memory of her late husband, Dr. Charles Doolittle Walcott, a fund to be known as the Dr. Charles Doolittle Walcott Fund for the promotion of knowledge of pre-Cambrian life was in April, established under the auspices of the National Academy of Sciences. Notable features of the terms of the gift are:

"The fund is to be known as the CHARLES DOOLITTLE WALCOTT FUND, and is to encourage and reward individual achievement in advancing our knowledge of pre-Cambrian life and its history in any part of the world.

"The income of the fund, except the small amounts that may be allowed for administrative expenses, shall be used for the award of medals and honoraria to persons between the ages of 21 and 48 years, the results of whose published researches, explorations, and discoveries in pre-Cambrian life and history shall be judged by the Trust Fund Board to be most meritorious.

The awards shall be made without respect to nation, race, sex, or academic degree.

"From the income there shall be provided a medal, to be known as the Charles Doolittle Walcott Medal 'for the promotion of the knowledge of pre-Cambrian life.' which medal shall be awarded every five years, beginning with 1932. Should there fail at the end of any five-year period work worthy of the award, in the judgment of the Board, the funds shall accumulate until an award is deemed justified, and a new series of 5-year periods shall commence with such delayed award.

The medal shall be cast in bronze or some other expensive metal, and the accrued income, after paying the expenses of coinage and minor expenses, shall, on the same occasion, be awarded, in the round sum of the highest even hundreds of dollars, as an honorarium to accompany the medal.

The award shall not be made to the same recipient on two sucessive periods.

The funds, amounting to $5,000, shall be held in trust by the National Academy of Sciences in the same manner as similar funds now in its keeping.

All awards of medals or honoraria from this fund shall be made on the recommendation of a board of five members, to be known as the CHARLES DOOLITTLE WALCOTT TRUST FUND BOARD, which is composed as follows:

a) The Secretary of the Smithsonian Institution (ex officio).

b) Two members eminent in the paleontology of early Paleozoic formations, one to be appointed by the Institut de France, and one by the Royal Society of London.

c) Two members distinguished in Paleozoic paleontology to be appointed by the President and Council of the National Academy of Sciences.

"The Board shall report to the Council of the National Academy of Sciences, to which its recommendations covering the awards of medals and honoraria shall be submitted for approval, not later than February 1 of the year in which the award is to be made."

This foundation should have marked effect in the stimulation of interest on the part of paleontologists in the search for and systematic study of traces of pre-Cambrian life, a field to which Dr. Walcott contributed more than any other paleontologist. Mrs. Walcott faithfully reflects Dr. Walcott's expressed views in providing that by far the greater part of the income, which in the course of five years should exceed $1,000, shall go as an honorarium, instead of for precious metal in the medal; and, further, that the awards are to aid and encourage pre-Cambrian studies by young paleontologists."

Subsequently, the terms of the Walcott bequest were modified to remove the age requirement (National Academy, 1935, p. 6) and broadened to include Cambrian paleontology (National Academy, 1935, p. 7). In a sense, one may consider Walcott's career, once he became established, as a search for ever older life forms, and, in that regard, perhaps his wishes were not followed to the letter.

In another sense, Walcott appreciated thoughtful work in paleontology regardless of the age of the fossils. The Walcott medalists have been: David White (1934); A.H. Westergaard (1939); Alexander G. Vologdin (1947); Franco Rasetti (1952); Pierre Hupé (1957); Armin A. Opik (1962); Allison R. Palmer (1967); Elso S. Barghoorn (1972); and Preston Cloud (1977). One expects that he would have been satisfied with the selections made in his name.

ACKNOWLEDGEMENTS

A number of colleagues, knowing of my interest in C.D. Walcott, have directed my attention to papers which bear upon subsequent work in areas he studied. I am particularly indebted to four people for reading this paper and improving it by their comments and references: P.E. Cloud, U.S. Geological Survey; Martin F. Glaessner, University of Adelaide; Hans Hofmann, University of Montreal, and Malcolm R. Walter, Australian Bureau of Mines, Geology and Geophysics.

REFERENCES

Anonymous, 1891, [Report on meeting] March 9, 1891: Rochester Academy of Sciences, v. 1, p. 151-153.

Abbott, George, 1914, Is *"Atikokania lawsoni"* a concretion?: Nature, v. 94, p. 477-478.

Bassler, R.S. 1941, A supposed jellyfish from the Pre-Cambrian of the Grand Canyon: Proceedings of the United States National Museum, v. 89, p. 519-522.

Bloesser, B., Schopf, J.W., Horodyski, R.J. and Breed, W.J., 1977, Chitinozoans from the Late Precambrian Chuar Group of the Grand Canyon, Arizona: Science, v. 195, p. 676-679.

Brooks, W.K., 1893, Salpa in its relations to the evolution of life: John Hopkins University, Studies from the Biological Laboratory, v. 5, no. 3, p. 129-211.

Cloud, P.E., 1968, Pre-metazoan evolution and the origins of the Metazoa: *in* Evolution and environment: New Haven, Connecticut, Yale University Press, 72 p.

——————, 1972, A working model of the primitive earth: American Journal of Science, v. 272, p. 537-548.

——————, 1976, Beginnings of biospheric evolution and their biochemical consequences: Paleobiology, v. 2, p. 351-387.

Cloud, P.E. and Licari, A.R., 1968, Microbiotas of the banded iron formations: National Academy of Sciences of the U.S.A. Proceedings, v. 61, p. 779-786.

Cloud, P.E. and Semikhatov, M.A., 1969, Proterozoic stromatolite zonation: American Journal of Science, v. 267, p. 1017-1061.

Craig, G.Y. and Rabbitt, M.C., 1969, Sir Archibald Geikie – Letters from American geologists: Geotimes, v. 14, p. 21.

Daly, R.W., 1907, The limeless ocean of Pre-Cambrian time: American Journal of Science, v. 4, no. 23, p. 93-115.

Darton, N.H., 1928, Memorial of Charles Doolittle Walcott: Geological Society of America Bulletin, v. 39, p. 80-116.

Dawson, J.W., 1897a, Note on *Cryptozoon* and other ancient fossils: Canadian Record of Science, v. 7, p. 203-219.

——————, 1897b, Relics of primeval life: beginning of life in the dawn of geological time: New York, Fleming H. Rewell Co., 336 p.

Elston, D.P. and Scott, R.G., 1976, Unconformity at the Cardenas-Nankoweap (Precambrian) Grand Canyon Supergroup, northern Arizona: Geological Society of America Bulletin, v. 87, p. 1763-1772.

Fenton, C.L. and Fenton, M.A., 1931, Algae and algal beds in the Belt Series of Glacier National Park: Journal of Geology, v. 39, p. 670-686.

——————, 1933, Algal reefs or bioherms in the Belt Series of Montana: Geological Society of America Bulletin, v. 44, p. 1135-1142.

——————, 1936, Walcott's "Pre-Cambrian Algonkian algal flora" and associated animals: Geological Society of America Bulletin, v. 47, p. 609-620.

——————, 1937, Belt series of the north; stratigraphy, sedimentation, paleontology: Geological Society of America Bulletin, v. 48, p. 1873-1970.

——————, 1939, Pre-Cambrian and Paleozoic algae: Geological Society of America Bulletin, v. 50, p. 89-126.

Ford, T.D., Breed, W.J. and Downie, Charles, 1969, Preliminary geologic report on the Chuar Group, Grand Canyon, Arizona: *in* Four Corners Geological Society. Geology and natural history of the Grand Canyon egion, Fifth Field Conference, Powell Centennial River Expedition, 1969 (Durango, Colo.): Four Corners Geological Society, p. 114-122.

Ford, T.D. and Breed, W.J., 1972a, The Chuar Group of the Proterozoic, Grand Canyon, Arizona: 24th International Geological Congress, Montreal, Section 1, p. 3-10.

——————, 1972b, The problematical Precambrian fossil *Chuaria*: 24th International Geological Congress, Montreal, Section 1, p. 11-18.

Glaessner, M.F., 1962, Pre-Cambrian fossils: Cambridge Philosophical Society Biological Reviews, v. 37, p. 467-494.

——————, 1968, Biological events and Precambrian time scale: Canadian Journal of Earth Sciences v. 5, p. 585-590.

——————, 1969, Trace fossils from the Precambrian and basal Cambrian: Lethaia, v. 2, p. 369-394.

————, 1972, Precambrian fossils – a progress report: 1968 Proceedings of the International Paleontological Union, 23rd International Geological Congress, p. 377-384.

Grand Canyon Natural History Association, 1976, Geologic map of the Grand Canyon National Park, Arizona, scale 1:62,500.

Hofmann, H.J., 1971, Precambrian fossils, pseudofossils, and problematica in Canada: Geological Survey of Canada Bulletin 189, p. 1-146.

————1973, Stromatolites; characteristics and utility: Earth-Science Reviews, v. 9, p. 339-373.

Idaho Bureau of Mines, 1973, Belt symposium. 2 vols: Special Publication Idaho Bureau of Mines, v. 1, 322 p.; v. 2, 138 p.

Licari, G.R. and Cloud, P.E., 1968, Reproductive structures and taxonomic affinities of some nannofossils from the Gunflint Iron Formation: Proceedings of the National Academy of Sciences of the U.S.A., v. 59, p. 1053-1060.

Merrill, G.P., 1906, Contributions to the history of American geology: United States National Museum Report for 1904, p. 189-734.

————, 1924, The first one hundred years of American geology: Yale University Press, New Haven, Connecticut 733 p.

Mitchell, M.E., 1971, On *Eozoon canadense*: Isis, v. 62, p. 380-382.

National Academy of Sciences, 1943, Report of National Academy of Sciences fiscal year 1932-1933, 172 p.

National Academy of Sciences, 1935, Report of National Academy of Sciences fiscal year 1933-1934, 196 p.

Obradovich, J.V. and Peterman, Z.E., 1968, Geochronology of the Belt Series, Montana: Canadian Journal of Earth Sciences v. 5, p. 737-747.

O'Brien, C.F., 1970, *Eozoon canadense,* the dawn animal of Canada: Isis, v. 61, p. 206-233.

————, 1971, Sir William Dawson: a life in science and religion: Memoirs of the American Philosophical Society, v. 84, p. 1-207.

O'Connor, M.P., 1972, Classification and environmental interpretation of the cryptalgal organosedimentary 'molartooth" structure from the late Precambrian Belt-Purcell Supergroup: Journal of Geology, v. 80, p. 592-610.

Raymond, P.E., 1935, Pre-Cambrian life: Geological Society of America Bulletin, v. 46, p. 375-392.

Rezak, Richard, 1957, Stromatolites of the Belt Series in Glacier National Park and vicinity: United States Geological Survey Professional Paper 294-D, p. 1-154.

Rhodes, D.C. and Morse, J.W., 1972, Evolutionary and ecologic significance of oxygen-deficient marine basins: Lethaia, v. 4, p. 413-428.

Rothpletz, August, 1915, Ueber die systematische Deutung und die stratigraphische Stellung der ältesten Versteinerungen Europas und Nordamerikas mit besonderer Berucksichtigung der Cryptozoen und Oolithe. I. Teil, Die Fauna der Belt-formation bei Helena in Montana: Abhandlungen der Mathematisch-Physikalischen Classe der Königlich Bayerischen Akademie der Wissenschaften, v. 28 (1), 468 p.

————, 1916, Ueber die systematische Deutung und die stratigraphische Stellung der ältesten Versteinerungen Europas und Nordamerikas mit besonderer Berücksichtgung der Cryptozoen and Oolithe. II. Teil, Ueber *Cryptozoon, Eozoon* and *Atikokanja*: Abhandlungen der Mathematisch-Physikalischen Classe der Königlich Bayerischen Akademie der Wissenschaften, v. 28, no. 4, 92 p.

Ross, C.P., 1970. The Precambrian of the United States of America - northwestern United States, the Belt States. *In* The geologic systems - The Precambrian, v. 4, Kalervo Rankama, editor. Interscience - John Wiley & Sons, New York, London, p. 145-251.

Schopf, J.W., 1970, Precambrian micro-organisms and evolutionary events prior to the origin of vascular plants: Biological Reviews of the Cambridge Philosophical Society, v. 45, p. 319-352.

———, 1971, Organically preserved Precambrian microfossils: North American Paleontological Convention, Chicago, 1969, Proceedings H, p. 1013-1058.

Schopf, J.W., Ford, T.D. and Breed, W.J., 1973, Microorganisms from the Late Precambrian of the Grand Canyon, Arizona: Science, v. 179, p. 1319-1321.

Sokolov, B.S., 1972, The Vendian Stage in early history: 24th International Geological Congress, Montreal, Section 1, p. 78-84.

Spaulding, Paul, 1915, Botanical Society of Washington [Report of 109th meeting]: Science, v. 41, p. 879.

Tappan, Helen and Loeblich, A.R., Jr., 1970, Geobiologic implications of fossil phytoplankton evolution and time-space distribution: *in* Symposium on Palynology of the Late Cretaceous and Early Tertiary: Geological Society of America, Special Paper 127, p. 243-340.

Towe, K.M., 1970, Oxygen-collagen priority and the early Metazoan fossil record: National Academy of Sciences of the U.S.A. Proceedings, v. 64, p. 781-788.

Van Hise, C.R., 1896, Principles of North American Pre-Cambrian Geology with an appendix on flow and fracture of rocks as related to structure by Leander Miller Hoskins: United States Geological Survey 16th Annual Report (1), p. 571-874.

Walcott, C.D., 1883, Pre-Carboniferous strata in the Grand Canyon of the Colorado, Arizona: American Journal of Science, v. 3, no. 26, p. 437-442, 484 (fig).

———, 1884a, The Cambrian system in the United States and Canada [Abst.]: Bulletin of the Philosophical Society of Washington, v. 6, p. 98-102.

———, 1884b, Report of Mr. Charles D. Walcott: *in* Administrative reports of chiefs of divisions: United States Geological Survey, 4th Annual Report, 1882-1883, p. 44-48.

———, 1884c, Note on Paleozoic rocks of central Texas: American Journal of Science, v. 3, no. 28, p. 431-433.

———, 1884d, Paleontology of the Eureka District: Monographs of the United States Geological Survey, 8, 298 p.

———, 1886, Second contribution to the studies on the Cambrian faunas of North America: United States Geological Survey Bulletin, 30, 369 p.

———, 1890, Study of a line of displacement in the Grand Canyon of the Colorado, in northern Arizona: Geological Society of America Bulletin, v. 1, p. 49-64.

———, 1893, Geologic time, as indicated by the sedimentary rocks of North America: Journal of Geology, v. 1, p. 639-676.

———, 1894, Pre-Cambrian igneous rocks of the Unkar terrane, Grand Canyon of the Colorado, Arizona; with notes on the petrographic character of the lavas, by J.P. Iddings: United States Geological Survey, 14th Annual Report, Part 2, p. 497-524.

———, 1895, Algonkian rocks of the Grand Canyon of the Colorado: Journal of Geology, v. 3, p. 312-330.

———, 1899, Pre-Cambrian fossiliferous formations: Geological Society of America Bulletin, v. 10, p. 199-244.

———, 1900, Random, a pre-Cambrian upper Algonkian terrane: Geological Society of America Bulletin, v. 11, p. 3-5.

———, 1906, Algonkian formations of northwestern Montana: Geological Society of America Bulletin, v. 17, p. 1-28.

———, 1909, Evolution of early Paleozoic faunas in relation to their environment: Journal of Geology, v. 17, p. 193-202.

————, 1910a, Cambrian geology and paleontology I. No. 7, Pre-Cambrian rocks of the Bow River valley, Alberta, Canada: Smithsonian Miscellaneous Collection, v. 53, p. 423-431.

————, 1910b, Cambrian geology and paleontology, II, No. 1, Abrupt appearance of the Cambrian fauna on the North American continent: Smithsonian Miscellaneous Collection, v. 57, p. 1-16.

————, 1911, Cambrian geology and paleontology, II. No. 2, Middle Cambrian Merostomata: Smithsonian Miscellaneous Collection, v. 57, p. 17-40.

————, 1912a, Notes on fossils from limestones of Steeprock Lake, Ontario, Canada: Geological Survey Branch Memoirs, Department of Mines, Canada, 28, p. 16-22.

————, 1912b, Studies in Cambrian geology and paleontology in the Canadian Rockies: *in* Expeditions organized or participated in by the Smithsonian Institution in 1910 and 1911: Smithsonian Miscellaneous Collection, v. 58, p. 39-45.

————, 1914a, Cambrian geology and paleontology, III. No. 2, Pre-Cambrian Algonkian algal flora: Smithsonian Miscellaneous Collection, v. 64, p. 77-156.

————, 1914b. *Atikokania lawsoni* Walcott: [Letter in] Nature, v. 94, p. 478.

————, 1915a, Geological explorations in the Rocky Mountains: *in* Explorations and field work of the Smithsonian Institution in 1914: Smithsonian Miscellaneous Collection, v. 95, p. 1-10.

————, 1915b, Discovery of Algonkian bacteria: Proceedings of the National Academy of Sciences of the U.S.A., v. 1, p. 256-257.

————, 1915c, The Cambrian and its problems in the Cordilleran region: *in* Problems of American Geology: A series of lectures dealing with some of the Canadian Shield and of the Cordilleras: New Haven, Yale University Press, p. 162-233.

————, 1916a, Evidences of primitive life: Smithsonian Institution Annual Report, 1915, p. 235-255.

————, 1916b, Geological explorations in the Rocky Mountains: *in* Explorations and field work of the Smithsonian Institution in 1915: Smithsonian Miscellaneous Collection, v. 66, p. 1-27.

————, 1916c, Cambrian geology and paleontology, III. No. 4, Relations between the Cambrian and pre-Cambrian formations in the vicinity of Helena, Montana: Smithsonian Miscellaneous Collection, v. 64, p. 259-301.

————, 1927, Cambrian geology and paleontology, V. No. 4, Pre-Devonian Paleozoic sedimentation in southern Canadian Rocky Mountains: Smithsonian Miscellaneous Collection, v. 75, p. 147-173.

————, 1928, Cambrian geology and paleontology, V. No. 5, Pre-Devonian Paleozoic formations of the Cordilleran provinces of Canada: Smithsonian Miscellaneous Collection, v. 75, p. 175-368.

Walter, M.R., 1972, A hot spring analog for the depositional environment of Precambrian iron formations of the Lake Superior region: Economic Geology and Bulletin of the Society of Economic Geologists, v. 67, p. 965-980.

Walter, M.R. and Preiss, W.V., 1972, Distribution of stromatolites in the Precambrian and Cambrian of Australia: 24th International Geological Congress, Montreal, Section 1, p. 85-93.

Walter, M.R., Bauld, John and Brock, T.R., 1972, Siliceous algal and bacterial stromatolites in hot spring and geyser effluents of Yellowstone National Park: Science v. 178, p. 402-404.

Walter, Malcolm, Oehler, J.H. and Oehler, D.Z., 1975, Megascopic algae 1300 million years old from the Belt Supergroup, Montana: a reinterpretation of Walcott's *Helminthoidichnites:* Journal of Paleontology, v. 50, p. 872-891.

White, David, 1928a, Algal deposits of Unkar Proterozoic age in the Grand Canyon, Arizona: Proceedings of the National Academy of Sciences of the U.S.A., v. 14, p. 597-600.

———, 1928b, Encouragement of Pre-Cambrian paleontology: Journal of Geology, v. 36, p. 743-744.

Yochelson, E.L., 1967, Charles Doolittle Walcott, March 31, 1850 - February 9, 1927: Biographical Memoirs, National Academy of Sciences of the U.S.A., 39, p. 471-540.

The Geological Association of Canada Special Papers

8. **A Comparative Atlas of Textures of Archean and Younger Volcanic Rocks**
   W. W. Moorhouse — Edited by J. J. Fawcett, 1970
9. **Geoscience Studies in Manitoba**
   Edited by A. C. Turnock, 1972
10. **New Developments in Sudbury Geology**
    Edited by J. V. Guy-Bray, 1972
11. **Variations in Tectonic Styles in Canada**
    Edited by R. A. Price and R. J. W. Douglas, 1972
12. **Huronian Stratigraphy and Sedimentation**
    Edited by G. M. Young, 1973
13. **The Cretaceous System in the Western Interior of North America**
    Edited by W. G. E. Caldwell, 1975
14. **Metallogeny and Plate Tectonics**
    Edited by D. F. Strong, 1976
15. **Conodont Paleoecology**
    Edited by C. R. Barnes, 1976
16. **Volcanic Regimes in Canada**
    Edited by W. R. A. Baragar, L. C. Coleman, J. M. Hall, 1977
17. **Late Silurian and Early Devonian Graptolite, Brachiopod and Coral Faunas from Northwestern and Arctic Canada**
    By D. E. Jackson, A. C. Lenz and A. E. H. Pedder, 1978
18. **Western and Arctic Canadian Biostratigraphy: Percival Sydney Warren Memorial Volume**
    Edited by C. R. Stelck and B. D. E. Chatterton, 1978
19. **History of Concepts in Precambian Geology**
    Edited by W. O. Kupsch and W. A. S. Sarjeant, 1979

Orders and requests for information should be sent to:
Geological Association of Canada Publications,
Business and Economic Service, Ltd.,
111 Peter Street, Suite 509,
Toronto, Ontario M5V 2H1.